AGENT ORANGE

History, Science, and the Politics of Uncertainty

EDWIN A. MARTINI

University of Massachusetts Press
Amherst & Boston

Copyright © 2012 by University of Massachusetts Press
All rights reserved
Printed in the United States of America

LC 2012028716
ISBN 978-1-55849-975-1 (paper); 974-4 (library cloth)

Designed by Sally Nichols
Set in Minion Pro
Printed and bound by Thomson-Shore, Inc.
Library of Congress Cataloging-in-Publication Data

Martini, Edwin A., 1975–
  Agent Orange : history, science, and the politics of uncertainty / Edwin A. Martini.
    p. cm. — (Culture, politics, and the cold war)
  Includes bibliographical references and index.
  ISBN 978-1-55849-975-1 (pbk. : alk. paper) — ISBN 978-1-55849-974-4 (library cloth : alk. paper) 1. Vietnam War, 1961–1975—Chemical warfare. 2. Agent Orange—War use. 3. Agent Orange—Toxicology. 4. Agent Orange—Environmental aspects. 5. Veterans—Diseases—United States. 6. Veterans—Diseases—Australia. I. Title.
  DS559.8.C5M37 2012
  959.704'38—dc23
                    2012028716

British Library Cataloguing in Publication data are available.

For Gracen Taylor and Kyan Jude

Contents

Acknowledgments ix

List of Abbreviations and Acronyms xv

Introduction: Approaching Agent Orange 1

1. Only You Can Prevent Forests: The Chemical War and the Illusion of Control 17
2. Hearts, Minds, and Herbicides: The Politics of the Chemical War 53
3. Incinerating Agent Orange: Dioxin, Disposal, and the Environmental Imaginary 97
4. The Politics of Uncertainty: Science, Policy, and the State 146
5. "All Those Others So Unfortunate": Vietnam and the Global Legacies of the Chemical War 197

Conclusion: Agent Orange and the Limits of Science and History 238

Notes 249

Index 291

Acknowledgments

I wake up every day very thankful that I have a job, one that I love, at a great university and in an amazing department. I wake up every day even more grateful for an incredible network of family and friends who not only support me in everything I do but also help keep me very grounded. There is not enough space here to thank them all sufficiently, but I would like at least to make an honest effort, since this book truly could not have been completed without them.

Several friends, colleagues, and even strangers helped make possible the travel and research necessary for this project. I knew early on that I would be returning to Vietnam for this book and was extremely lucky to be led once again on that journey by Tran Dinh Song, a multilingual veteran of the Army of the Republic of Vietnam (ARVN) who was both fearless and shameless in guiding me and translating for me during my two trips to that amazing country. Susan Hammond, Merle Ratner, and the entire staff at the headquarters of the Vietnam Association of Victims of Agent Orange/Dioxin (VAVA) in Hanoi were helpful in arranging various meetings throughout the country. The members of the Agent Orange Working Group listserv (not to be confused with the White House Agent Orange Working Group [AOWG] described in chapter 4), however much we have disagreed over the years, helped steer me toward a number of important resources and forced me constantly to rethink my assumptions. My cousin Kim housed and fed me during part of a long research trip that took me to Texas, Colorado, and Wyoming. Arnold Schecter and Al Young opened up their lives, research files, and even their own homes to me on that same trip, for which I am especially grateful.

I could not have completed this book without the help of a number of outstanding and dedicated archivists. Special mention goes to Rich Boylan at the National Archives in College Park, Maryland, who helped me track down a number of hard-to-find documents. The entire staff of the Alvin

Young Collection at the National Agricultural Library made me feel as though I was still in their good graces even when their patience with me was clearly exhausted. Everyone at Archives New Zealand in Wellington was warm and welcoming, even as their country suffered through the devastating earthquake in Christchurch in the spring of 2011. The staff at the Vietnam Archive at Texas Tech University and at the Kennedy, Ford, Carter, and Reagan Presidential libraries, as well as at a number of smaller archives, put up graciously with a number of fishing expeditions, many of which returned no reward for me, let alone for them.

I am exceedingly grateful to the many talented graduate research assistants who helped me at various stages on this project. Chris Breyer, Amy Wilson, and especially Katherine Ellison and Bill Watson helped me transcribe, scan, copy, edit, and contextualize countless pages of documents and articles. Over the past several years I have been able to work through a number of nagging questions surrounding this book with my students, both undergraduate and graduate, at Western Michigan. In my courses on the Vietnam War and other topics in recent U.S. history, countless undergrads have asked probing questions and often shared with me their family stories about Vietnam and Agent Orange. My beloved cohort of graduate students has been supportive of me throughout this project, going far beyond what one would expect of one's students. I am incredibly lucky to have had Sara Bijani, Francis Bonenfant, Skylar Bre'z, Katherine Ellison, Jill Gibson, Rex Hafer, Chris Jannings, Tyler Miller, Bill Watson, and David Zwart, among others, to teach me about their own work and to ask questions about mine.

This book has been entirely conceived, researched, and written during my tenure at Western Michigan University in Kalamazoo, where I was extremely lucky to land a job several years ago. I am eternally grateful to my colleagues at Western, who inspire, challenge, and support me in myriad ways. Outside of the History Department, the AAUP young Turks collective of Sarah Hill, Lisa Minnick, and Jon Adams has made academic life (particularly meetings) a great deal more fun than it really should be. Steve Bertman put up with incessant sophomoric questions about dioxins and Ah receptors, while Joan Hawxhurst introduced me to many of the valuable resources of greater Kalamazoo, including many local Vietnam veterans. Jill and Jess Hermann-Wilmarth are treasured friends who helped see our family through a number of difficult times over the past few years.

Inside the department, while everyone has been encouraging and helpful, a few require special mention: Buddy Gray and Joe Brandão have been incredibly supportive of me from the very start of my career, and both were crucial in providing me with the flexibility and resources required to finish this book. Bob Berkhofer, Andrea Berto, Sally Hadden, Lynne Heasley, Catherine Julien, Mitch Kachun, Cheryl Lyon-Jenness, Jim Palmitessa, Eli Rubin, Bill Warren, and Takashi Yoshida have all been wonderful friends as well as wonderful colleagues. Candy List and Brenda Brewer helped with all manner of logistical and financial challenges. All of us at Western lost a dear colleague when Nora Faires died of cancer in early 2011. Nora was my professional mentor and one of my closest friends. She always pushed, inspired, and supported me, and she would have loved to see this book published.

The bulk of the research I conducted for this book was funded by grants from Western Michigan University. An international research award from the Faculty Research and Creative Activities Support Fund allowed me to travel to Vietnam in 2008 and to take several archival research trips. The Burnham-Macmillan endowment in the Department of History granted several generous awards that allowed me to present drafts of various chapters at conferences, travel to Europe and New Zealand, and offset the cost of publishing the final product. Without such assistance, this book would still be years away from completion.

Portions of this book were presented at dozens of conferences around the country and around the world, far too many to mention here. I am especially grateful, however, to the Center for Southeast Asian Studies at the University of Michigan for inviting me to give an early version of the essay that became the core of chapter 2. A later version of that paper was published in *Diplomatic History* 36, no. 5 (November 2012). Parts of chapters 2 and 3 were published as "More Dangerous than Bombs or Bullets: Agent Orange, Dioxin, and the Environmental Imaginary," in *New World Coming: The Sixties and the Shaping of Global Consciousness*, ed. Karen Dubinsky et al. (New York: Palgrave-Macmillan, 2009). As this book went to press, sections of chapter 1 and chapter 3 were forthcoming as articles in *War and Society* and the *Journal of Military History*, respectively. My thanks go to all of the reviewers and editors for their insightful and thoughtful comments on those essays and for granting me the right to reprint them.

I continue to be exceedingly fortunate to have my first two books published by the University of Massachusetts Press. The UMass Press is a wonderful operation that turns out great and beautiful books. Among the many people who have helped shepherd my projects, Bruce Wilcox and Carol Betsch have been especially helpful. Sally Nichols created a beautiful design that complements the book wonderfully. Lawrence Kenney provided insightful, thorough, and timely copy editing, which greatly improved the narrative. No author working on projects related to the Vietnam War could ask for better editors than Chris Appy and Clark Dougan. As the author of what I still consider to be two of the best books on the Vietnam War, Chris is an expert second to none, and he contributed valuable feedback on key sections of this book. Clark, an accomplished historian and Vietnam expert in his own right, has become a trusted friend and adviser as well as a fabulous editor, and for that I am especially grateful. Clark and Chris are the best in the business.

A number of other colleagues and friends read the entire manuscript at various points. I am indebted to Jeremi Suri, David Zierler, David Kieran, Steve Bertman, Bill Zeller, and Mike Carr for their thoughtful comments. The final version is immeasurably better as a result. I will never publish nearly enough books to offer sufficient thanks to H. Bruce Franklin for being the generous friend, colleague, mentor, and adviser that he is. And what can I say about Scott Laderman? It's really not fair that someone as brilliant and prolific a writer as Scott should also be a dedicated, talented teacher. But what makes it worse is that he is simply the nicest person one could ever meet. I am eternally grateful to Scott for reading early drafts of several chapters and the entire final manuscript, for offering many helpful suggestions along the way, for continuing to conspire with me on other projects, and for being a wonderful friend.

As always, I could not even begin to imagine undertaking, let alone completing, a project like this without the tireless support of my close friends and family. The Morisettes—Jeff, Elizabeth, Clementine Rose, and Twyla Lynn—have always been there for my family and me and have themselves become family in the process. Here in Michigan we are equally favored to have the Colosky family at our side. Allison lent her considerable copy-editing skills to an early draft. Jason offered more indirect but equally important aid by always having a guitar and an Oberon at the ready. Linda Pritchard and Mary Osborne have become treasured friends, whose support of me and my family has been especially generous. The entire Smallmouth crew has also provided lots of love and good times. Pete Haugh, Steve Shebuski, Brian Hayslip, Brian Tucker, and Matt Lindberg continue to be the best friends anyone could ask for. Even though we are now scattered around the country, they still encourage and inspire me.

My mother and father, Shari and "Big" Ed, continue, as they always have, to support me, my work, and my family in every way possible. Shari made a number of trips to Kalamazoo to stay with my family while I was away doing research, and Big Ed (aka RA no. 2) actually accompanied me on a few research trips, including the one to New Zealand in 2011. I continue to be grateful for all their love. My sister Kris, her wife, Jen, and my beautiful niece Alex also helped sustain us throughout this project. I have always said that anyone is lucky to have one set of understanding and accommodating parents. I am undeservedly graced to have two. Bill and Gennis Zeller helped make this book possible in countless ways, one of which was having Gennis stay with my family while Bill (aka RA no .1) accompanied me to Vietnam in 2008, to Europe in 2009, and to New Zealand in 2011. Bill's companionship, fearlessness, and curiosity have been invaluable on these trips, and his thorough, thoughtful comments on the manuscript greatly improved the book.

Finally, I can barely summon the words to thank the three most important people in my life: my wife, Genanne, and my two sons, Gracen and Kyan. Gracen came into our lives during my last project, and Kyan appeared right in the middle of this one. They both put up patiently with a dad who was alternately away doing research, locked in his office writing, and, when home, too often distracted with herbicidal thoughts. While they think it is very cool that this book is dedicated to them, I hope someday they will both realize fully how proud I am of them, how much I love them, and how they make each day worth living.

Nothing would be possible without Genanne Zeller. We have been through so much during the course of writing this book, all of which has made me more appreciative than ever that she continues to be my best friend and my partner in all things. I am so proud of her for all that she does for our family, our friends, and everyone she meets. Of all the things I wake up grateful for each day, she is and will always be first and foremost.

Abbreviations and Acronyms

AAAS	American Association for the Advancement of Science
AOWG	White House Agent Orange Working Group
ARPA	Advanced Research Project Agency
ARVN	Army of the Republic of Vietnam
CDC	Centers for Disease Control
CICV	Combined Intelligence Center, Vietnam
CORDS	Civil Operations and Revolutionary Development Support
CWS	Chemical Warfare Service
DMZ	Demilitarized Zone
DOD	Department of Defense
DOS	Department of State
DVA	Department of Veterans Affairs (Australia)
EIS	Environmental Impact Statement
EOI	Exposure Opportunity Index
EPA	Environmental Protection Agency
ESG	Army Environmental Support Group
GAO	Government Accountability Office
GVN	Government of Vietnam (often used interchangeably with RVN)
IOM	National Institute of Medicine
IWD	Ivon Watkins Dow, Ltd.
JCS	Joint Chiefs of Staff
MACV	Military Assistance Command, Vietnam
NAS	National Academy of Sciences

NCBC	Naval Construction Battalion Center
NLF	National Liberation Front for the Liberation of South Vietnam
OTA	Office of Technology Assessment
PAVN	People's Army of Vietnam
ppm/b/t	parts per million/billion/trillion
RAD	Residents Against Dioxin (New Zealand)
RVN	Republic of Vietnam
TCDD	2,3,7,8-tetrachlorodibenzo-p-dioxin
USAF	United States Air Force
USAID	United States Agency for International Development
USFS	United States Forest Service
VA	United States Veterans Administration/Department of Veterans Affairs
VAVA	Vietnam Association of Victims of Agent Orange/Dioxin
VVA	Vietnam Veterans of America
VVAA	Vietnam Veterans Association of Australia
VVANZ	Vietnam Veterans Association of New Zealand

AGENT ORANGE

INTRODUCTION
APPROACHING AGENT ORANGE

In the spring of 1966 a United States Navy electrician was on his way home. Although he had not seen any action in Vietnam, spending less than half a day in Saigon simply to transfer flights, he was anxious to get stateside. Arriving at the airport in Saigon, he "bought a pack of cigarettes and snapped a few photos" before leaving on the flight that would take him back to the States. Decades later this American veteran developed Type 2 diabetes. Under regulations adopted by the Veterans Administration (VA) in 2010, he was now eligible for a service-related disability claim because he was a Vietnam-era veteran who had developed a condition now recognized by the VA as having a "suggestive" link to exposure to Agent Orange. Most important, he had physically set foot in the Republic of Vietnam (RVN), if only for a few hours. Had he been routed back through Guam, Manila, or Bangkok, he would not have been eligible for any benefits. In 2011 this veteran was collecting several thousand dollars a year from an Agent Orange disability claim despite almost certainly having never been exposed to the herbicide or its dioxin contaminant while in Vietnam.[1]

At the Phu My Orphanage for Agent Orange Victims, just a few miles from where the navy electrician spent his one afternoon in Vietnam, dozens of mentally and physically handicapped children believed to be third- and fourth-generation victims of Agent Orange listen to stories, work on craft projects, and gaze with curiosity at the American visitors. Visiting in 2008, I asked the director of the orphanage how she knew that the children's conditions were caused by dioxin exposure from Agent Orange. She replied that they "will never know for sure." There are hardly any records to indicate where the majority of the children at the orphanage came from; most of them were simply left there, some literally on the doorstep, without any information from their families. None of the children have been tested to see if they have elevated dioxin levels. The test costs around a thousand dollars, and at the time of my visit there was not yet a facility in Vietnam that could perform the lab work. For a thousand dollars, the director explained, the orphanage can provide food, clothing, and physical assistance for one child for about six months. That is more important, she relates through a translator, "than telling us what we already know."[2] Although these children are more likely to have been exposed to dioxin in Vietnam than the navy electrician, no one knows for sure. One thing is clear: the children are not and never will be receiving several thousand dollars per year from the U.S. government.

These scenes from southern Vietnam, separated by more than forty years, illuminate the ongoing battles over the global legacies of the chemical war waged by the United States in Southeast Asia. From 1961 to 1971 the United States and its South Vietnamese allies sprayed nearly seventy-three million liters (over nineteen million gallons) of chemical agents over two and a half million acres of southern and central Vietnam to defoliate the landscape and limit the access of the National Liberation Front (NLF, commonly known in the United States as the Viet Cong) to local food supplies. Of those seventy-three million liters, about 62 percent—over forty-five million liters—of the chemicals deployed consisted of Agent Orange, a 1:1 mixture of the herbicides 2,4-D and 2,4,5-T that by the late 1960s was known to contain often dangerous levels of dioxin, specifically 2,3,7,8-tetrachlorodibenzo-p-dioxin (TCDD), one of the most deadly toxins ever created.[3] Since the late 1960s, when the world became aware of Agent Orange and the other so-called rainbow herbicides used by the United States and its South Vietnamese allies during the war, veterans and civilians around the world have sought to understand the implications of the herbicide and its associated dioxin. But Agent Orange has consistently provided far more questions than answers. Despite decades of study, questions of exposure, causality, compensation, and justice remain at the forefront of scientific, legal, political, and diplomatic debates over the legacies of the chemical war.

To this day, both the short- and long-term effects of these chemicals remain a great source of controversy in many nations, communities, and academic fields. Soldiers and civilians who claim they were exposed to Agent Orange blame the U.S. government and the chemical companies that produced it for a variety of medical conditions they and their families have experienced. Many in Vietnam claim that the dioxin found in Agent Orange is responsible for birth defects and other conditions currently being found in the third and even fourth generation of exposed populations, such as the children at Phu My. At the same time, several ongoing scientific and epidemiological studies dispute many of these claims.

Questions about the effects of Agent Orange on human and environmental health have, with the realization of its global nature, become more profound and more complicated over time. While the situation in Vietnam remains, rightly, at the center of contemporary concerns about Agent Orange, the scenes from Saigon and the tensions they represent have been repeated around the world over the past twenty years as the full scale of the global apparatus for producing, testing, shipping, and storing Agent Orange has become clearer. From New Zealand to Canada, from Missouri to Korea, veterans and local citizens, including those born decades after the end of the Vietnam War, continue to grapple with the mysteries and frustrations and the very uneven resolution of the legacies of Agent Orange.

I began searching for answers to these questions in the summer of 2001, while traveling in Vietnam and doing research for my first book. There I witnessed the scars of war that remain part of everyday life for millions of Vietnamese. Many of those scars—landmines, unexploded ordnance, lakes formed by bomb craters—made their way into that work. But the ongoing devastation caused by Agent Orange and other chemical agents used by the United States during the war resisted easy answers and simple explanations. It was too big and complex an issue to shoehorn into that project. As I began to follow the trail of Agent Orange, reading everything I could on the subject, I realized that historians had almost completely avoided the topic.

Histories of the Vietnam War devote scant, if any, attention to the use of herbicides and chemical agents by the United States in Vietnam. Some of the most popular textbook accounts contain only a few sentences or paragraphs devoted to the topic.[4] Much of the rest of the historical scholarship is polarized by, on the one hand, military history approaches that too often take at face value the documentation proffered through official channels, and, on the other, works of political advocacy that seek to reclaim the voices of the victims of Agent Orange, but often at the expense of historical context.[5] Now a new generation of scholars has begun focusing on Agent Orange, resulting in the publication of a number of works that are interdisciplinary in method and international in scope and that shed additional light on the legacies of the chemical war.[6] Yet aside from David Zierler's book *The Invention of*

Ecocide few have succeeded in illuminating the politics of Agent Orange without themselves falling prey to them. Zierler's work, which masterfully traces the history of herbicides and the idea of ecocide all the way back to Charles Darwin, places Agent Orange against the backdrop of Cold War diplomacy and environmental politics, through the end of the Vietnam War in 1975.[7] Still missing from the scholarship is a book that fills in the details of both the history and the legacies of herbicidal warfare and does so on a global scale. As a result, the use of herbicides and other chemical agents by the United States in Vietnam—what I call the chemical war in Southeast Asia—has become one of the most recognizable, yet least understood, aspects of the Vietnam War. This book seeks to fill that gap.

My own assumptions about dioxin, Agent Orange, and the chemical war have been challenged, often turned completely on their head, and in many cases disproven altogether by my research. I assumed that Agent Orange was responsible for all of the horrific birth defects shown in the pictures that seem always to accompany news stories about its use in Vietnam. I discovered that the science on the relationship between dioxin and birth defects, as well as a host of other conditions, is very much divided and inconclusive. I assumed that the military and perhaps even the policymakers knew about the dangers of Agent Orange in the early 1960s but ignored them in choosing to weaponize the herbicides, only to find out that no historical evidence supports such a claim.

On these and other issues, I have treated Agent Orange very much like a mystery or a puzzle to be solved. My approach to the topic is driven not by a political agenda but by an interest in following the trails of evidence in a variety of places and in reconstructing the history of this controversial substance by placing that evidence in a broader context. By tracing these stories across temporal and geographic boundaries I have sought at the most basic level to understand how actors and communities around the world shaped and were shaped by Agent Orange.

Such an undertaking has required me not only to sharpen my skills as a historian but also to draw on interdisciplinary approaches, utilizing the insights of scholars in other fields. My archival research has taken me to a number of collections in the United States, most notably the National Archives in College Park, Maryland, the National Agricultural Library in Beltsville, Maryland, and the Vietnam Archive in Lubbock, Texas.[8] It has taken me back to Vietnam as well as to New Zealand, whose National Archives in Wellington house a number of sources related to Agent Orange. I make use as well of sources from Australia and Canada, although the majority of those were gathered through library loans and online research and via their availability in other archival collections. To get the fullest possible story, however, I have also spent a great deal of time researching non-archival sources, including the work of scholars in such fields as toxicology, epidemiology, and public health history, to learn how those disciplines have contributed to public discussions, policy, and contested meanings of Agent Orange. I have done much research in newspapers, examining media coverage of Agent Orange over time as well as in contemporary chatrooms and on listservs, where thousands of people affected by Agent Orange continue to shape and contest its many meanings. Finally, I have conducted a series of interviews, not with American veterans, whose voices on this subject are both important and well documented, but with scientists and others working on the ground in Vietnam and elsewhere, searching for answers to the most vexing questions about Agent Orange. In 2008 I traveled to Vietnam and met with veterans, victims of Agent Orange, caregivers, and representatives of VAVA at the national, provincial, and district levels. The voices of veterans and other alleged victims are important, to be sure, as are those of policymakers, scientists, and citizens. All of these are represented here, but they are consistently balanced against what a mountain of scientific evidence has shown about the historical realities of dioxin contamination caused by exposure to Agent Orange. Far from being static, uniform, and universal, the meaning of Agent Orange has always been contingent upon where, when, how, and by whom it has been encountered. In part, then, my book seeks to recover both the history of Agent Orange as a material artifact—the actual herbicide used by the United States in Southeast Asia, which had very real and very serious effects—and the cultural phenomenon of "Agent Orange," the meaning of which has been steadily made and remade by people around the world.

To get at these contested meanings I have divided the book into five chapters that are organized roughly chronologically but also thematically. The multiple histories generated by Agent Orange have led me to depart from a strictly chronological sequence. While some readers may find this a bit jarring, I feel it offers the best path for following the complexities of the subject over time. The first chapter, "Only You Can Prevent Forests," explores the decision of the administration of John F. Kennedy to weaponize herbicides in Southeast Asia. By placing this decision in its specific context, I show how a variety of historical forces shaped the thinking of policymakers and military leaders at the time. The use of herbicides in Vietnam, I argue, was yet another manifestation of the mindset embodied by the "best and the brightest" occupying the White House, the Department of State (DOS), and the Pentagon at the time. The architects of the American War in Vietnam continually offered technological and military solutions to fundamentally political problems, attempting to impose stringent forms of control over a situation that was at best ill-suited for such mechanisms. Herbicidal warfare was simply one more failed attempt among many to impose control over a nation, a people, and a landscape—indeed, over nature itself—all of which refused to accept the dictates of American power just as stubbornly as American policymakers refused to accept the limitations of that power.

That refusal would have a number of long-term consequences for both the United States and Vietnam. One of the most significant results of the use of herbicides in Vietnam would be the health problems experienced by those who were heavily exposed to them over time. While many recent works have used contemporary oral histories and ethnography to reconstruct Vietnamese encounters with Agent Orange, no study to date has used military records from the time of the war itself to explore how Vietnamese actors responded to the chemical war as it was going on around them. In chapter 2, "Hearts, Minds, and Herbicides," I employ overlooked archival records to reconstruct how soldiers and civilians understood the use of herbicides at the time, as well as what those reactions meant for the larger political and military efforts of the United States in Southeast Asia. As the historical record shows, the military continued to cling to its belief that the tactical value of herbicides far outweighed any potential political consequences of their use, despite the fact that all available evidence points to the exact opposite conclusion: the chemical war was proving to be demonstrably counterproductive in the battle for "hearts and minds," while yielding, at best, limited evidence of military benefits.

Most studies of Agent Orange have focused on either the wartime use of herbicides or the long-term impact of that use after the war was over. Lost in that narrative gap is the story of what happened to the 2.4 million gallons of Agent Orange still in possession of the U.S. military after the administration of Richard Nixon banned its use in 1970. In addressing this surplus, the United States Air Force (USAF) was forced to navigate a host of new challenges that had sprung up in the decade since the start of Operation Ranch Hand, the operation responsible for the vast majority of herbicide missions in Southeast Asia. The predicaments ranged from the rapid growth of the environmental movement to new government bureaucracies devoted to environmental protection. In chapter 3, "Incinerating Agent Orange," I trace the military's efforts to negotiate these challenges alongside parallel stories of how communities like Times Beach, Missouri, and New Plymouth, New Zealand, dealt with the prospect of dioxin contamination resulting from the production of Agent Orange and its constituent components. In each of these cases various groups and actors dealt with the growing paradox at the heart of the legacy of Agent Orange: the more scientists learned about dioxin and the more precisely they could measure and detect it in discrete amounts, the more they were able to identify its *potential* threat to human and environmental health. Better detection, however, did not lead to greater ability to predict with accuracy what the effects of that exposure would be, particularly in human health. The citizens of Times Beach and New Plymouth, like the veterans of the war, were asking questions for which scientists and policymakers had few answers. In the absence of scientific proof, they increasingly relied on personal experience to make their case.

Much has been written about how American veterans of the Vietnam War adjusted to everyday life after the conflict, a story in which Agent Orange plays a role.[9] Few studies have compared U.S. veterans' struggle to seek justice for what they believed to be the consequences of herbicide exposure with those of their counterparts in Australia and New Zealand. In chapter 4, "The Politics of Uncertainty," I show how competing discourses of knowledge about exposure, risk, human health, and scientific uncertainty shaped the legal and political battles of veterans in multiple countries. Drawing on the important work of public health historians and environmental studies scholars, I historicize the concept of scientific uncertainty itself, showing it to be the product of modern scientific thought and disciplinary practice. The concept of uncertainty was embedded in existing relations of power through which chemical corporations and states were often rewarded by legal and regulatory decisions while veterans and others who were potentially exposed were forced to show, in ways nearly impossible to demonstrate, clear lines of causality between chemicals, exposure, and particular health conditions. As veterans and activists around the world developed, proposed, and advocated competing epistemological, legal, and regulatory frameworks that placed the burden of proof on chemical companies and others who denied the risks of chemical exposure, they found themselves hemmed in by the politics of uncertainty; rather than proposing more holistic, ecological approaches, they tried to show the link between specific exposure scenarios and specific health conditions. In all of these cases, I argue, the ultimate decisions about benefits and exposure have been based on political rather than scientific grounds.

In the final chapter, "All Those So Unfortunate," I explore some transnational legacies of the chemical war. Beginning and ending with the effects—demonstrated and alleged—of Agent Orange on Vietnam, I build on the earlier discussion of uncertainty to show how struggles similar to those experienced by American veterans are playing out worldwide. As knowledge about the impact of Agent Orange and the global scope of its production, distribution, and use became more widespread in the late twentieth and early twenty-first centuries, people who had never considered themselves to be part of the history of herbicidal warfare became participants in its construction and negotiation. As veterans, family members, and local activists learned that military herbicides had been used at former military sites in places such as Canada and Korea, among others, they joined a growing list of those who believed, rightly or wrongly, that they too were victims of Agent Orange. Their cases were no easier to prove than earlier ones, but the uneven resolution of their claims has further laid bare the politics of uncertainty. While military veterans in the United States, Australia, and New Zealand have fought for and achieved benefits and compensation regimes that continue to grow in size and scope, Vietnamese civilians find themselves being told by the governments compensating those veterans that no scientific link has yet been found between Agent Orange exposure and a variety of health conditions prevalent in areas around former military installations. Questions of scientific uncertainty have become transnational in scope, but the politics of uncertainty continues to be shaped by power relations within and between nation-states.

I have attempted to write the most complete history of Agent Orange to date, but it is far from comprehensive. No single volume could tell the whole story; there are simply too many voices, too many communities, and too many complex issues involved. I have explored the multiple meanings of Agent Orange before, during, and after the years in which it was most widely deployed as a weapon in Southeast Asia and in a diverse group of communities around the world, using an equally diverse group of sources. In following these contested meanings across temporal and geographic boundaries, I develop four primary themes.

1. The Story of Agent Orange Is Global in Scope

Most of the work written about Agent Orange focuses on the two countries most directly involved with it, the United States and Vietnam. But this ignores a large part of the story. While some scholarship has expanded the scope to discuss non–state actors in those two countries, my book moves beyond a binational approach to put the story into a more global perspective. I include the voices of actors in Canada, Australia, New Zealand, and elsewhere to highlight both the similarities in all communities affected by dioxin and the differences in how other nations and communities have responded to the threat and legacies of Agent Orange. Americans and Vietnamese take the main roles in most of the stories recounted here, but their experiences are better understood in this larger context.

Similarly, the history of Agent Orange relies on various networks that were and continue to be global in nature. In chapter 1, for instance, I trace the global network of testing, production, and distribution that made Operation Ranch Hand possible. Following an average barrel of Agent Orange from its point of production to its final destination in Vietnam gives one a glimpse of how many nations and localities were involved in constructing Agent Orange, literally and figuratively, from the 1950s to the present. In the later chapters I show how the key transnational networks continue to revolve around production and distribution, but production and distribution of knowledge, not of herbicides. Scientists from around the world shared data, findings, and hypotheses, but also served as consultants for states, corporations, and local governments dealing with issues related to Agent Orange. These scientists were instrumental in drawing attention to the dangers of herbicides in the 1960s and 1970s. For example, after a major dioxin disaster in Seveso, Italy, in 1976, American scientists were brought in as consultants partly on the basis of their experiences and understanding of dioxin exposure stemming from the Vietnam War; scientists from this network, which now included Italians from Seveso, consulted with the governments of Australia, America, and New Zealand during state investigations and legal wrangling in the 1980s. In Vietnam the limits of the transnational channels were tested as Western scientists called into question the rigor and evaluation of Vietnamese studies. In a parallel development, groups of activists worldwide, making use of new media, have promoted and shared knowledge about the lingering effects of Agent Orange. Veterans from the United States, Australia, and New Zealand have advanced the cause of victims of Agent Orange in Vietnam, and nongovernmental organizations like the Ford Foundation have been at the forefront of raising money around the world to pursue environmental remediation at dioxin hot spots. The ongoing tensions between the United States and Vietnam are reminders that nation-states and their relative power within the international system still matter a great deal. But any history of Agent Orange that omits discussion of transnational elements will miss a major portion of the story.

2. The Story of Agent Orange Shaped and Was Shaped by the Rise of Environmentalist Thinking

The active years of herbicidal warfare in Vietnam coincided with the rise of the environmental movement in the United States and elsewhere. In January 1962 Operation Ranch Hand began its formal defoliation operations. Only months later Rachel Carson's *Silent Spring* appeared, the book often credited with launching the modern, middle-class, suburban environmental movement in the United States. At the time the book was published hardly anyone had heard of dioxin. Eight years later a great deal had changed. On April 15, 1970, the White House announced, in the face of growing concerns about the dangers of Agent Orange, that the U.S. government was suspending registration of the herbicide 2,4,5-T, effectively making it illegal to sell or transport products containing the compound in most of the country. Exactly one week later the first Earth Day celebration was held, and by the end of the year Congress had authorized the creation of the Environmental Protection Agency (EPA). By the time Operation Ranch Hand ended in 1971, a discernible, vocal, and organized environmental movement in the United States was heightening the public's awareness of the landscape, the human body, the relationship between these and a variety of risks and hazards, and the role of the state in regulating all of these factors. In communities from Times Beach to New Plymouth, veterans, citizens, and activists would develop meanings and politics around Agent Orange that were framed by such thinking.

3. The Story of Agent Orange Shaped and Was Shaped by Challenges to Scientific and State-Based Authority

Closely related to the rise of environmentalist thinking during the lifespan of Agent Orange is the growing contestation of a modern mindset grounded in scientific rationalism. The Enlightenment ideas that are the philosophical foundation of modernity presumed that the natural world is ruled by universal laws that can be known with certainty through rational thought and scientific inquiry. Over the course of the twentieth century such authority was impugned on several fronts, including science itself, as relativity theory and quantum mechanics affirmed the role of contingency and uncertainty in nature. During the second half of the century postmodern thinkers expanded on the implications of this shift, affirming the role of uncertainty, the importance of experience, and the value of alternative epistemologies in the search for truth. The ascendancy of postmodernism lay in the growing resistance to experts who based their professions of authority largely on their privileged access to various forms of sanctioned scientific and historical knowledge. This repudiation often came from everyday citizens, who offered their own experiences, including their bodily trauma, to problematize experts' assumptions, findings, and recommendations. In the wake of a series of epidemiological investigations that found no clear link between exposure to Agent Orange and types of cancer and birth defects, for instance, citizens in communities affected by the agent have practiced a popular or populist epidemiology that combines anecdotal evidence with existing scientific studies and information gathered from official, state-based sources.[10]

The politics of scientific uncertainty, discussed at length in chapters 4 and 5, can also be seen in part as a reaction to the challenging of scientific authority. While scientists themselves would likely be the first to acknowledge that uncertainty is a fundamental part of the scientific process, the uses to which such authority and uncertainty were put in areas like nuclear weapons, vaccines, cancer, and climate change became more and more politicized.[11] Many critics attempted to discredit the scientific process altogether. In 1996 the physicist Alan Sokal, an active participant in the so-called science wars of the 1990s, offered a rejoinder to such critics: be careful not to "conflate epistemic and ethical issues" by confusing the rational, scientific search for knowledge with the uses to which that knowledge is often put by policymakers and politicians.[12] In fact, most of the veterans, activists, and citizens who disputed the science of Agent Orange and practiced a popular epidemiology were seeking alternative sources of data they felt were being ignored or dismissed. Indeed, as the history of Agent Orange shows, citizens and activists around the world turned to science and scientists for resolution on dioxin as often as they indicted science for failing to supply the answers they were searching for. Such continued reliance on scientific and state-based authority is a reminder that insurgent forms of community-based knowledge did not easily tear down the structures of modern states and disciplines; but neither did they go unnoticed. Rather than seeking to support or discredit science itself, this book uses Agent Orange to chart the politics surrounding scientific authority from the 1970s to the early twenty-first century. Far from being directed consistently by corporate and state interests, the politics of uncertainty have been fluid and contested, benefiting chemical companies at some points, the U.S. government at others, and veterans at still others;

they have shifted before, and they will most certainly shift again in the years ahead.

4. The Story of Agent Orange Is Almost Always about Much More than Agent Orange

While most of this book is devoted to reconstructing the story of Agent Orange as a material artifact and how people came to understand it, the history cannot be understood apart from the way in which the *idea* and the *symbol* of Agent Orange operated as a screen onto which diverse actors projected their feelings about the war in Vietnam, the veterans who fought that war, and the civilians who bore the brunt of it; about the United States and its use of chemical agents during the war, its failures and shortcomings in the eyes of its citizens, and its failure to accept responsibility for the long-term effects of Agent Orange in Vietnam and elsewhere; and about the role of chemicals in the modern world, the corporations that produced them, the states that sanctioned them, and the uncertainty that surrounds them. The media played a major role in this process. Often news stories struggled to relate the admittedly complex realities of dioxin exposure; at other times they grossly overstated or misrepresented the facts at hand. More commonly, however, the use of Agent Orange as a screen was simply the product of common meaning-making activities through which everyday people tried to make sense of complex, controversial, and often painful, traumatic events. Given the complexities in the history and legacies of Agent Orange, it has proven to be especially prone to obfuscation, projection, and reinterpretation.

Indeed, such complexity is what makes the story of Agent Orange so fascinating. It is by its very nature an interdisciplinary, international problem. It cannot be understood through scientific analysis alone: the help of historical insights must be enlisted; or only through contemporary recollections: they must be measured against evidence from the historical record. But history alone cannot solve these issues. As I make clear, the archival paper trail is incomplete at best, often riddled with major gaps that have limited the ability of contemporary scientists to determine, for instance, the precise levels of herbicides sprayed by aircraft during the war. All the anecdotal evidence and oral histories in the world that might supplement that record cannot measure the current residual dioxin content in Vietnamese soil. Agent Orange is an interdisciplinary problem, and it requires solutions and stories that embrace interdisciplinarity. In addition to the study of how it has been used and how a variety of groups have made meaning out of it, Agent Orange is a lens through which to examine the history and legacies of the Vietnam War, the environmental movement in the United States and elsewhere, and the politics of scientific knowledge in the twentieth and twenty-first centuries.

Although I have sought to offer context, rather than closure, for the issues surrounding Agent Orange, I believe this book can help supply tentative answers to at least three pressing questions: First, how could the United States have done such a thing? Critics of Ranch Hand, of U.S. policy toward veterans since the war, and of U.S. foreign policy in general often decry the evils of Ranch Hand as though American war planners set out, in the full knowledge of what these herbicides were capable of, to cause long-term environmental damage as well as birth defects in innocent children. Policymakers and military commanders, as I explain, made more than their share of deadly mistakes, and Ranch Hand is certainly one of the biggest and most momentous. But that does not mean they were evil men out to do evil deeds. I have tried to comprehend the policymakers' decisions in terms of the worlds in which they lived and the options available to them. They sought to inflict devastating damage on Vietnam and were willing to sacrifice countless Vietnamese civilians and tens of thousands of American soldiers in the process. In the face of such wanton disregard for human life, it is understandable to attribute nefarious motives to those in the administrations of Kennedy and Lyndon B. Johnson. Yet for all their errors in judgment and their arrogance and shortsightedness, there is simply no evidence that at the outset of the war U.S. policymakers knew about the risks of herbicides to human health.

Critics in some of these same camps regularly refer to Ranch Hand as an example of chemical warfare, comparing it to the use of mustard gas in the First World War and describing it as a war crime tantamount to genocide. These are serious (albeit hyperbolic) charges, and neither chemical warfare nor war crimes are terms that should be tossed around lightly. They have long, complex histories as well as contemporary legal implications. By way of terminology, I employ *chemical war* to describe not simply the defoliation and crop destruction missions of Operation Ranch Hand, but also the larger effort by the United States to apply a variety of chemical agents, including but not limited to herbicides, to impose its will and sense of order on the people and landscape of Southeast Asia. This does not mean, however, that I subscribe to the view that the use of Agent Orange constitutes a historical example of chemical warfare. Legal definitions of chemical warfare require that the weapon in question be directed against humans, specifically against "the person of the enemy." As I note in chapters 1 and 2, little evidence points to humans as being the primary targets of defoliation missions. At the same time, a great deal of data show that human beings were sprayed directly, with the knowledge of the military, and were affected as a direct result of the spray. Whether or not this was the intention of military commanders and their civilian supervisors does not necessarily absolve them of the charges of conducting chemical warfare, but the fact is that defoliation was not targeted primarily at humans, and under the language of international law charges of chemical warfare are thereby very unlikely to hold up in the arenas where they matter most, law and diplomacy.

I distinguish between defoliation missions, which aimed to limit the cover provided by the natural landscape, and crop destruction missions, which sought to deny food to enemy forces. The ability to differentiate between civilian crops and those controlled by revolutionaries was premised on a number of faulty assumptions, the result being the imposition of unnecessary additional hardship on the civilian population of central and southern Vietnam. Although the lasting consequences of defoliation with regard to human and environmental health draw the most attention when charges of chemical warfare and war crimes are leveled at Operation Ranch Hand, it is, in my view, the short-term impact of crop destruction for which there is the clearest evidence for making either case. Targeting the forest with defoliants believed to be safe for humans, in other words, is not chemical warfare when viewed historically. Intentionally destroying food supplies in the knowledge that civilians were directly affected is a different matter entirely. That should be considered among the most unconscionable acts of the war, not because it was accomplished through chemicals but because it was done in the full recognition that civilian food supplies were clearly disrupted. In the final analysis, the language of chemical warfare and war crimes is a matter of international law. Given that they do little to advance one's understanding of either the history or legacies of the chemical war, such labels are not useful in historical discussions of Agent Orange.[13]

Finally, there is the issue of what should be done for victims of Agent Orange. I argue that the total number of Agent Orange victims in Vietnam, the United States, and throughout the world is almost certainly overstated.[14] I also hold that decisions about exposure to Agent Orange and the resulting determinations of compensation have consistently been made on political rather than scientific grounds. Given the difficulties in proving exposure, the nearly impossible task of showing causality, and the limited resources available for both, there is little hard evidence to support most claims regarding Agent Orange, claims that rely on suggestive links between exposure likelihood scenarios and a steadily expanding list of conditions believed to be associated with the agent. In the case of American veterans, the VA now grants presumptive disability benefits for a range of conditions, some, including heart disease and prostate cancer, with thinly suggestive connections to Agent Orange exposure. As matters currently stand, any U.S. veteran who spent even a few hours in Vietnam and now has heart disease, diabetes, or a variety of other illnesses is presumed, by the dictates of VA policy, to be an Agent Orange victim. Similarly, the Vietnamese government has regularly increased the official number of Agent Orange victims to the current level of nearly five million with no accounting of how they derived such numbers. Orphanages throughout the country are filled with children, many of them horribly disabled, who are presumed, most without a shred of proof, to be victims of the chemical war.

As in the case of science, however, one should be careful not to conflate the search for historical truth with the political and ethical choices made by policymakers based on that knowledge. To shed light on the historical process whereby the burden of proof in exposure cases was taken off of American service men and women does not deny that many veterans were indeed exposed to Agent Orange and other herbicides; likewise, calling on VAVA to produce better information about the sources of their data does not equate with denying the lasting damage done to people in Vietnam. Most important, the attempt to historicize these issues in no way absolves the United States of its responsibilities to its veterans and to the Vietnamese. Regardless both of what its intentions were and of the state of knowledge of the enduring effects of herbicides, the U.S. government is responsible for putting millions of people into the potentially deadly path of Agent Orange with neither their knowledge nor their consent. This responsibility is not contingent upon whether there are thousands, hundreds of thousands, or millions who have been affected by exposure to Agent Orange.

At the end of the day, neither scientific nor historical analysis can offer the full range of answers that victims of Agent Orange, real or imagined, desire. The best that historians can do is try to contextualize and understand historical problems based on the evidence at hand. I am sure the stories, evidence, and conclusions offered here will not please everyone. I hope they illuminate more than they obscure and contribute to the understanding of one of the last unresolved chapters in one of longest, costliest, and most divisive wars of the last century.

Sites in Vietnam Related to Agent Orange. (Courtesy Western Michigan University Mapping Center, Jason Glatz Designer)

CHAPTER ONE

ONLY YOU CAN PREVENT FORESTS
The Chemical War and the Illusion of Control

The planes came to Cat Son a little after six o'clock in the morning. The reconnaissance plane came first, followed by two fighter jets that strafed the village. Then came the big cargo planes, three of them, flying in formation, parallel and low to the ground, and spraying a fine mist that looked to the people below like white smoke. The planes sprayed the trees and fields surrounding Cat Son, including the villagers' fruit trees and rice paddies. Upon surveying the damage three days after this mission in October 1965, one eyewitness to the attacks, a member of the main force unit of the NLF in Binh Dinh province and a former farmer himself, described the effects of what Vietnamese soldiers and civilians came to call the spray: "All the fields along the two banks of a small river were utterly destroyed. Even the people's vegetables and fruit-tree gardens near the fields were ruined. According to the people, after the spraying, the tree leaves were wet as if soaked in oil. The water had a film on the surface, which looked like fat skim. A little while later, the leaves became dry and fell on the ground. Rice stalks turned dry, banana trees sank, potato and manioc became soft and rotten, the pineapple was tainted, the coconuts split, and the jet fruits fell on the ground."[1] No eyewitnesses described any immediate health problems among the villagers after the attack, but locals refused to drink the water for several days. The rice seedlings, still fairly young, were destroyed. Most of the village's fruit supply was rendered inedible, and the trees would have to be replanted.[2]

Scenes like this one were repeated thousands of times during the Vietnam War, as the modified C-123 aircraft of Operation Ranch Hand carried out herbicide missions across central and southern Vietnam in an attempt to defoliate the jungle and deny easy access to food to the revolutionary forces of Vietnam. Caught in between the various sides of this long, bloody conflict, millions of Vietnamese civilians watched helplessly as the spray destroyed their crops, trees, and much of the surrounding landscape. Looking back on these events with the hindsight of history and knowing what we now know about the scale of destruction of the Vietnam War, one might regard these chemical attacks as less than shocking. Images of the widespread bombing of Vietnam, Laos, and Cambodia, of the massacre at My Lai, and of children burned by napalm running in the street remain familiar components of the cultural memory of the war. But the absence of televised images and immediate casualties from herbicide missions should not obscure the unusual, even radical, nature of these attacks. How and why did the United States come to carry out chemical attacks on the villages and surrounding areas of millions of peasant farmers half a world way? Why were chemical herbicides produced in places like Midland, Michigan, being sprayed over places like Cat Son village? Why did American planes destroy the crops of those they were ostensibly trying to protect and for whose hearts and minds U.S. forces were ostensibly battling?

Learning how the chemical war came to be requires understanding three unique, interrelated historical contingencies of the early 1960s. First, Operation Ranch Hand cannot be comprehended apart from the political and military circumstances in Southeast Asia and the world that the Kennedy administration faced when it took office in 1961. Both the Cold War and the decolonization of former European empires were at their height when Kennedy was elected. The war in Vietnam quickly took its place alongside Cuba and Berlin as the primary hot spots of the struggle between the United States, the Soviet Union, and their respective allies. Kennedy's decision to deploy herbicides and other chemicals in Vietnam has to be grasped within the context of its interrelated efforts at nation building and counterinsurgency in Southeast Asia, efforts which policymakers at the time saw as part of a global struggle against communism. Just as important, however, as many within the Kennedy and Johnson administrations were acutely aware, the potential fallout from this decision could not easily be separated from the Cold War. No sooner had Operation Ranch Hand begun than the propaganda battle ignited. From its inception the chemical war was intricately entwined not only with the battle for hearts and minds but also with the larger questions about American actions and intentions in the Cold War.³

The political and military components, however, cannot alone explain the material conditions through which the chemical war was prosecuted. Operation Ranch Hand was supported by a global infrastructure of chemical development, production, testing, and distribution that embodied what President Dwight Eisenhower had described as the military–industrial complex. Herbicides were produced, tested, and delivered to Southeast Asia via a network of industrial, corporate, and military connections that literally spanned the globe. The scope of this network underscores the common, widespread application of herbicides and other chemicals during the period in which the United States escalated the war in Vietnam and the ways in which a new environmental consciousness would coincide with that war. As the potential dangers of Agent Orange became known across equally global networks, individuals and communities in these disparate locations would come to have common concerns about the effects of exposure to these chemical agents.

Finally, the weaponization of herbicides in general and the use of Agent Orange in particular cannot be understood apart from the mindset Kennedy and his advisers brought to the White House. The best and the brightest, as they have been infamously labeled, practiced a technocratic management style paired with an air of idealism and superiority about their mission; they came to believe that a technologically advanced society like the United States could manage and control an environment like southern Vietnam and could, through the analysis and manipulation of data and the proper application of modern tools, including herbicides, impose its will on that environment and its inhabitants.⁴ This will to control was palpable not only in the White House, but also in the Pentagon under Secretary of Defense Robert S. McNamara. Alongside endeavors to control the movement of civilians and combatants, the U.S. military tried to manipulate and control the very environment through which its forces moved, most of which was densely vegetated jungle and forest. Throughout the war in Vietnam, the United States and the RVN used many chemical compounds as they strove to control insects, combatants, civilians, and even the jungle itself. American forces first sought to defoliate the forest in order to locate enemy forces more easily; when this proved insufficient, they tried to ignite large-scale forest fires to remove the forest altogether.

Herbicides were designed to kill vegetation, but the United States did not set out to destroy huge swaths of landscape in Southeast Asia any more than it willfully set out to kill more than two million Vietnamese. Herbicidal warfare and the use of chemical agents more broadly were initially conceived as a technological substitute for direct military intervention by American troops. Yet by attempting to impose their will on the people and landscape of the region, American war planners consistently and stubbornly refused to accept the limits of their power—political, economic, military, and technological. This intransigence resulted not only in an ultimately futile prolonging of the war effort, but also in efforts such as Operation Ranch Hand that would have a severe, longtime impact on the landscape and people of Southeast Asia and on all those who served there in the war.

The Origins of the Chemical War

By the early 1960s the Third World had emerged as perhaps the most important site in the Cold War. Although Western Europe would never be far from the minds of policymakers, the emergence of the Chinese as a major power in Asia and the rapid decolonization of former European colonies made places like Southeast Asia critical battlegrounds in the global struggle between capitalism and communism. President Kennedy had no intention of making Vietnam the primary battleground of the Cold War at the start of his presidency, but by the time he was assassinated in 1963 it had become precisely that. Faced with a growing insurgency in southern Vietnam, Kennedy authorized a number of steps that made Southeast Asia the testing ground for his new approach to foreign policy and for his commitment to combat worldwide communism.

Kennedy's first year in office saw the situation in southern Vietnam deteriorate rapidly. The government of Ngo Dinh Diem, installed with the strong support of the United States in 1955, was not only failing to make political gains in the countryside but also alienating many potential constituencies, particularly students and Buddhist leaders in and around its urban strongholds, especially Saigon. Far from being the puppet of the United States he was often characterized as, Diem more often than not rejected efforts by his advisers from the Eisenhower and Kennedy administrations on such matters as political and land reform.[5] During this crucial period the NLF went from being a loosely affiliated network of political and military units nominally created in 1960 to a well-organized revolutionary force making huge gains in rural areas across central and southern Vietnam. The historian William Turley estimates that the armed forces of the NLF grew exponentially from about seven thousand to over one hundred thousand in 1960–64.⁶ Although he signaled increased American commitment to the Diem regime with an enhanced Program of Action for South Vietnam in May 1961, Kennedy also dispatched several advisers to Vietnam over the next several months to gain a better understanding of the situation.⁷

In May, Vice President Johnson flew to Saigon, followed later by the economic adviser Eugene Staley and, in October, Deputy National Security Adviser Walt Rostow and the military adviser Maxwell Taylor. Each of these missions confirmed that the situation was indeed getting worse and that the Diem regime was losing the battle against the NLF. What had been a minor concern in Eisenhower's outgoing security briefing with the new president—Laos was thought to be the most pressing crisis in Southeast Asia—suddenly became a major focus of the incoming administration and a test case for its new approach to foreign policy. James Brown of the United States Army Chemical Warfare Center at Fort Detrick in Maryland, a leading figure in the development and evaluation of the defoliation program who oversaw the early herbicide testing programs in Vietnam, accompanied Taylor and Rostow on their trip and helped make the case for the use of herbicides as a potential counterinsurgency tool. Brown's involvement with what would come to be known as the Taylor–Rostow report, which was instrumental in Kennedy's decision to increase American financial and military commitments to Diem, helped ensure that Vietnam would become the test case for herbicidal warfare.⁸ What made Southeast Asia a true test for Kennedy's vision, however, was that it offered a chance to take the best-equipped, most well funded military in the world into a situation defined by unconventional warfare and counterinsurgency. While U.S. forces were not particularly well prepared for such a task, the president felt this was the type of new challenge the country would have to meet in the developing world.

Kennedy's "flexible response" approach to foreign policy was designed first by Taylor as a response to what he saw as the flawed framework of the Eisenhower administration. Implemented in the Pentagon by McNamara, the new plan was designed to modernize the U.S. military to meet a more diverse series of threats and included an emphasis on chemical and biological alternatives to nuclear weapons. The architects of flexible response sought a range of approaches that made innovative uses of technology, a variety of "techniques and gadgets" that could reduce the NLF's advantageous use of the natural concealment of the forest and limit their food supply.⁹ Fortunately, the military had been working on this technology:

chemical herbicides that could be sprayed over large areas of forest to defoliate the landscape and deny sources of food to the NLF.

After seeing the domestic uses of chemical herbicides and pesticides, the U.S. military during and after the Second World War accelerated programs to develop delivery systems that could be used to clear vegetation around military installations and to control insects and thereby the exposure of troops to diseases like malaria. They could also be used to target enemy supply routes and food supplies.[10] As the historian David Zierler points out, however, this latter issue marked a major shift in thinking about the use of chemicals in a military context: "Whereas early research in plant growth manipulation required a cognitive leap to shift the field from growth promotion to weed killing, the idea that herbicides could become a military weapon necessitated a similar reorientation of the social function of plant physiology in a time of total war. Just as the idea to favor herbicides over growth promoters required new ways to unlock the potential of biochemistry, so did the notion of herbicidal warfare require innovative thinking about national security and the environmental dimensions of battle."[11] The line between military and civilian herbicides, however, was never clear-cut. Civilian scientists worked closely with both chemical companies and the military to research, develop, and test a variety of chemicals that had potential military uses.

An early report from the Advanced Research Project Agency (ARPA) noted that the choice of chemicals for such operations was somewhat limited but that the problem was largely one of testing, not availability. There were "a large number of compounds that could be screened," the report claimed. The Chemical Warfare Service (CWS) had "for years been screening new chemicals emerging from the laboratories through the United States, including industrial, government, and college sources." The CWS was in possession of more than twelve thousand of these new compounds and planned to closely coordinate the development and testing of them with the chemical industry.[12] Ultimately, more than sixty distinct chemical combinations would be used as herbicides, pesticides, fungicides, rodenticides, and riot control agents during the Vietnam War.[13]

In the late 1950s scientists had developed several herbicide combinations that had proven effective in defoliation and crop destruction. Most of these were based on the herbicides 2,4-dichlorophenoxyacetic acid (2,4-D) and 2,4,5-trichlorophenoxyacetic acid (2,4,5-T). Both of these herbicides were in wide use in the United States and around the world. Throughout the 1950s the domestic use of herbicides grew exponentially. Domestic production

TABLE 1.
PRIMARY RAINBOW HERBICIDES USED IN SOUTHEAST ASIA, 1961–71

MILITARY NAME	COMBINATION	PRIMARY YEARS USED
Agent Pink	60% n-butyl ester 2,4,5-T 40% isobutyl ester of 2,4,5-T	1961–64
Agent Green	100% n-butyl ester, 2,4,5-T	1961–64
Agent Purple	50% n-butyl ester of 2,4-D 30% n-butyl ester of 2,4,5-T 20% isobutyl ester of 2,4,5-T	1961-64
Agent Blue	100% sodium salt, cacodylic acid	1962–71
Agent White	80% triisopropanolamine salt of 2,4-D 20% triisopropanolamine salt of picolram	1965–71
Agent Orange	50% n-butyl ester of 2,4-D 50% n-butyl ester of 2,4,5-T	1965–71

of 2,4-D and 2,4,5-T increased from fourteen million to thirty-six million pounds and from virtually zero to ten million pounds, respectively, between 1950 and 1960. By the end of the fifties American companies were producing more than seventy-five million pounds of herbicides annually, bringing in well over one hundred million dollars per year. In 1959, according to the U.S. Department of Agriculture, American farmers treated more than fifty-three million acres with herbicides.[14] That same year the military began large-scale efforts to develop effective aerial delivery systems for herbicides.

The herbicide combinations produced for the military were variations on combinations of these and other chemicals. The six mixtures were given color-coded names by the military according to the color of the band placed on the drums in which they were shipped (table 1). Together, the six agents became known as the rainbow herbicides. Herbicides Purple, Pink, and Green were the first to be used in Vietnam. Agent Purple, the forerunner and closest chemical relative of Agent Orange, had been developed in the early 1950s for possible use in the Korean War. Agents Orange and White, introduced to Southeast Asia in 1965, would become the most heavily used chemicals for defoliation missions, while Agent Blue, a dessicant rather than a herbicide, became the weapon of choice for chemical crop destruction.

Before delivering them to Vietnam, teams of scientists, engineers, and military commanders working at a variety of sites, including Fort Detrick and Camp Drum, New York, experimented with multiple herbicide combinations and various aircraft and spray systems. As testing in Southeast Asia continued during 1961 and into 1962 the jury was still out on the tactical effectiveness of the herbicides, but an internal film produced for the military about the testing at Camp Drum stridently proclaimed how the tests were already contributing to the war in Vietnam. Against a backdrop of music like that used in newsreels, the film, entitled, "The U.S. Army Biological Laboratories Presents: Vegetation Control Testing, Vietnam," begins by showing the molecular makeup of 2,4-D and 2,4,5-T. "In a typical year, 1960," the voiceover notes, "fifty million pounds of these chemicals were sold."[15] Two case studies show how the compounds can be used to clear vegetation that was "hindering military efforts: Camp Drum and South Vietnam." The Chemical Corps flew more than thirty missions at Camp Drum, spraying more than four square miles with twenty-three hundred gallons of agents, which turned out to be mostly Agents Purple and Orange.[16] The film next turns its attention to southern Vietnam, noting that the dense vegetation there "retains its leaves the year round; consequently it affords ideal concealment to guerilla terrorists who sporadically attack the free people of Vietnam." Showing footage shot during the initial tests, largely of Pink and Purple, conducted by personnel from Fort Detrick in 1961 and 1962, the film highlights the areas where the tests took place, particularly around Bien Hoa and Saigon. Stressing the effectiveness of several chemical agents sprayed with several devices on a variety of species, "Vegetation Control Testing, Vietnam" juxtaposes a series of before and after shots of lush jungle and healthy trees and brown swaths of withering plants. As the film concludes by showing the first herbicide tests flown by the modified C-123s, the narrator comments,

> In summarizing the work in Vietnam, much worthwhile information was obtained, and the mission was successful, although conditions for operations were far from ideal. As a result of these tests, defoliation spraying was placed on an operational basis. Those chemicals already proven most effective as defoliants will continue to be used until faster acting types can be developed. It should be noted that the chemicals, when used appropriately, are harmless to man and do not render the soil unfruitful beyond one agricultural season. Defoliation operations are continuing on a daily basis in the Republic of Vietnam. Targets are vegetation along lines of communication, roads, canals, power lines and railroads. The number of ambushes in sprayed areas has dropped to practically zero. Military worth has thus been established.[17]

Internal reports from the same period were more guarded in their assessments of the herbicides' military worth, but the tone of the film mirrored the belief held by many in the Kennedy administration that the superior technology on display at Camp Drum and in the early tests in South Vietnam would be a major asset in the global war against communism. This would not be the last time the military value of herbicides was promoted with little evidence to support the assertion.[18]

As Zierler has persuasively argued, Kennedy's "authorization to launch herbicidal warfare against the National Liberation Front was a preeminent manifestation of 'Flexible Response' as its architects envisioned the strategy to contain Soviet influence on a truly global scale."[19] While the administration was divided over the use of chemicals for crop destruction operations (see chapter 2), nothing about defoliation programs held up the president's approval of the weaponization of herbicides, which came in September 1961 with his authorization of National Security Memorandum 115.[20]

By early 1962 the herbicide program was integrated with other counterinsurgency operations, but its effects on civilians received no sustained attention. William Buckingham argues that in these early years Ranch Hand operated largely ad hoc and was characterized by varying degrees of improvisation and evaluation, as both military commanders and scientific advisers refined spray mission procedures and reported back to Washington.[21] Between Diem, ARPA, the Chemical Corps, and the military chain of command, Kennedy was inundated with opinions on the effectiveness of the missions. ARPA's evaluations, coordinated by Brown, emphasized the limitations of herbicide use in the forests but indicated that the program could be tactically effective with some modifications and within the parameters he laid out.[22] Gen. Fred Delmore of the Chemical Corps offered his own report, focusing on problems with the spray systems in the C-123s but recommending that "the vegetation control program in South Vietnam be resumed immediately" after the minor technical adjustments were completed.[23] Buckingham has noted that in both the report and in his presentation to McNamara, Delmore emphasized the type of technical, data-driven approach for which the secretary of defense had a known affinity, and McNamara indeed appears to have been impressed with the report.[24]

Armed with data and enthusiasm, officials moved quickly to implement and expand the program. By the summer of 1962 the Military Assistance Command, Vietnam (MACV), which served as the headquarters of the U.S. military presence, and the Joint Chiefs of Staff (JCS) were on board, endorsing Delmore's approach and recommending to McNamara that the president authorize further herbicide operations. By November the president gave his approval, signing off on an expanded program of defoliation that allowed for the delegation of approval for defoliation but not for crop destruction.[25] Now in place, the plan was to spray the Vietnamese countryside with herbicidal agents designed to support the U.S. and RVN counterinsurgency in suppressing the Vietnamese revolution. The Kennedy administration was making good on its promise to combat communism in Southeast Asia, and herbicides were going to be part of that mission. How big a part it played remains a matter of some dispute among historians.

Kennedy, Zierler argues, always intended the herbicide program to be limited, remaining "confident that technology would effectively substitute for manpower." Zierler suggests even that Ranch Hand would likely not have reached the levels it did during the Johnson administration had Kennedy lived, given that Johnson never seemed to grasp that such technologies were meant primarily to serve as alternatives, rather than supplements, to large deployments of combat troops.[26] What is clear is that, in keeping with the larger goals of flexible response, many within the administration clung to the belief that technological innovations like herbicidal warfare could retard the spread of communism and revolution, controlling the population of South Vietnam in the same way the chemicals themselves controlled the growth of vegetation. These assumptions ultimately proved to be wrong, bringing with them considerable consequences, both intended and unintended.

The Global Reach of the Chemical War

Regardless of the policy choices made by Kennedy and Diem and of tactical choices made by military advisers on the ground, the chemical war could not have begun in earnest without a second key component. Any material to be used on a scale approaching that of the herbicides deployed in Vietnam requires a vast, global apparatus of testing, production, distribution, and storage; what makes this network unique is that years later, after the dangers of the dioxin found in many of these herbicides became well known, millions of people who had never set foot in Southeast Asia would find their lives touched, if not appreciably altered, by their encounter with the rainbow herbicides.

The path a typical barrel of herbicide took to reach Vietnam is instructive. While the journey varied depending on the specific agent and the year it shipped, a barrel of Agent Orange in 1967—when the use of herbicides in Vietnam reached its peak, with nearly two million acres sprayed—was produced at one of the plants operated by the eleven major chemical companies producing tactical herbicides according to military specifications. Since Dow and Monsanto together produced far more of these herbicides than the other nine companies combined, one might assume that a typical barrel was produced at one of those firms' plants. Among many potential sites, an average batch of Agent Orange in 1967 might well have been produced at the Dow factory in Midland, Michigan. Placed in a fifty-five-gallon drum of eighteen-gauge metal, in line with specifications issued by the Department of Defense (DOD), each barrel of Agent Orange was marked with a 7.6-centimeter-wide orange band to differentiate it from the other rainbow herbicides.[27]

Barrels of Agent Orange were shipped to a number of ports, including New Orleans, Baltimore, and Seattle, but after 1966 all herbicides sent to Southeast Asia were routed through either Mobile, Alabama, or Gulfport, Mississippi.[28] In 1967 the typical barrel of Agent Orange would have been loaded along with more than one hundred identical barrels and shipped by rail from Midland to Mobile. At the Port of Mobile the barrel would be placed directly onto a cargo ship, which would take approximately fifty days to reach Saigon.[29] Once it arrived in Saigon (approximately two-thirds of the tactical herbicides were shipped to Saigon and the remaining third went through Da Nang), the barrels were loaded into semitrailers operated largely by ARVN troops and distributed to various sites, notably the air bases at Bien Hoa, Da Nang, and Phu Cat, to supply Ranch Hand missions.[30] Once it arrived at its destination base, the herbicide would most likely have been transferred from the barrel, through a suction hose, to a trailer tank that could be used to fuel Ranch Hand aircraft directly. After 1968, however, herbicides at Bien Hoa were often transferred to one of several thirty-thousand-gallon holding tanks color coded (Orange, White, and Blue) to correspond to each of the major herbicides then in use (fig. 1).[31]

Given the large number of spray missions and corresponding herbicide supplies in 1967–68, however, the barrel might also have been stacked at the air base into one of several large storage pyramids. Even when their contents were transferred, the barrels were not emptied fully, as approximately one gallon (between two and five liters) remained in the drums. According to procedures, the residual herbicide was to be removed by placing the barrel on a draining rack to be emptied, cleaned, and used for defoliation around the perimeter of the base. Evidence from the period suggests that residual herbicide often remained in the barrel, and a study in *Nature* (2003) by Jeanne and Steven Stellman argues that the estimated residual herbicide in a given barrel contained an average of 5.96 mg of dioxin. By itself, that is not a significant amount, but multiplied by thousands of barrels, the residual dioxin eventually became a source of major contamination of the soil.[32]

Figure 1. Storage tanks for Agents Blue and Orange, Bien Hoa Air Base, 1970. (Courtesy National Archives, College Park, Maryland)

In the late 1960s, during the heaviest years of Operation Ranch Hand, Agents White, Blue, and Orange were stored in large holding tanks at Bien Hoa air base outside Saigon. Years later studies would confirm that storage sites such as these were contaminated with often dangerous levels of dioxin, far more so than areas that were sprayed repeatedly during herbicide missions.

Once the original herbicide was on its way toward its mission, the barrel itself and any residual herbicide trapped inside became a potentially useful piece of military hardware. Units from the United States, South Vietnam, Australia, New Zealand, and South Korea used the barrels in the construction of buildings, bunkers, and other fortifications. Enterprising local Vietnamese living outside of air bases like Da Nang regularly procured the barrels for their own use, including the transportation of fuel, which may have led to accidental defoliation in the urban areas of Da Nang and Saigon, as dioxin-infused fumes from motorbikes encountered nearby plants and trees. By 1967, according to stated policy, the typical barrel should have been "thoroughly cleaned, punctured, and flattened" and then buried in a landfill, but pictures from the period, reminders of the huge number of barrels passing through these bases from 1967 to 1969, beg the question of how often the procedures were followed.[33]

Internal army reports from the period confirm that drum disposal remained a problem well past 1967. An edition of MACV's internal series

"Lessons Learned: Vietnam" from 1968 noted that accidental herbicide damage in areas around air bases was often the result of residual herbicide leaking from improperly discarded barrels.[34] Investigating complaints from residents of Da Nang in early 1969, staff of the Civil Operations and Revolutionary Development Support (CORDS) reported noticeable damage to trees and crops in the area, believed to have been caused by the indiscriminate distribution of barrels, which the team found scattered throughout Da Nang city.[35]

Even when procedures were followed, leaks and spills were a common feature of the herbicide distribution process. Alvin Young has estimated that "about 10 out of every 10,000" drums shipped (0.1 percent) were defective and leaked. Others were damaged during loading and unloading.[36] Photos from the period show puddles and areas of stained ground around the storage areas, including the pyramids. The transfer from the barrels to trailers, tanks, and aircraft inevitably involved spills as well, the number of which increased after the large holding tanks were installed at Bien Hoa.[37] Accounts by Paul Cecil and Young stress the safety training and precautions undertaken by United States Air Force (USAF) crews, but they note that, once on the ground at ports of entry and air bases, the herbicides were commonly handled by ARVN personnel.[38]

The storage and inevitable spillage of herbicides are meaningful for the same reason it is important to retrace how the herbicides got to Southeast Asia in the first place. For all the concern about which regions were targeted and how many people lived in them, the areas of greatest impact on environmental safety and human health in Vietnam are not those that were sprayed but those where the herbicides were stored, particularly the air bases at Bien Hoa, Da Nang, and Phu Cat, which are the most heavily dioxin-contaminated sites in the country and among the most polluted in the world. For all the important work aimed at reconstructing flight paths through the use of the HERBS tapes, it is necessary to reconstruct how the herbicides themselves were used on the ground, particularly in regard to storage, barrel disposal, and the perimeter spraying at dozens of bases throughout central and southern Vietnam. A fuller explanation of the long-term environmental consequences of herbicide use and storage appears below (see chapter 5), but as an indicator of the risks involved the case of Da Nang is exemplary.[39]

As the second largest storage and distribution site of herbicides in Vietnam, behind only Bien Hoa, Da Nang has long been at the center of controversies over damage and pollution caused by Agent Orange and other chemicals. The number of complaints about accidental damage from herbicides increased markedly with the escalation of Ranch Hand missions in 1967–69.

In 1968, for example, representatives from CORDS, the United States Agency for International Development (USAID), and the Chemical Corps investigated locations in the vicinity of the air base at Da Nang. The areas that suffered the greatest damage were located immediately south and northwest of the air base, directly in the path of aircraft takeoff and landing patterns, respectively. The damage was believed to have been caused by a combination of ordinary leaks from aircraft containing herbicides and occasionally from so-called emergency dumps, rare but serious instances in which Ranch Hand planes dispersed a large amount of chemicals all at once.[40] A team from Fort Detrick, investigating several reports of herbicide damage in Vietnam months later, confirmed that four major dumps had occurred at Bien Hoa over the previous year. Around Da Nang the scattered damage was initially believed to have been caused by leaky valves in the C-123s, but later reports confirmed major dumps near the approach and takeoff points,[41] The final report from the CORDS/USAID investigation, prepared by the Chemical Corps, argued that "aircraft leakage is so minor it cannot be a contributing factor to the *extensive* damage throughout the city."[42]

The more likely culprit in the "extensive damage" at Da Nang appears to have been improper disposal of drums. The initial investigation into the damage of 1968–69 at Da Nang suggested that the supposedly empty drums were observed regularly lying around the areas of lighter damage, particularly to vegetable crops. "These drums were noted in the hamlets being used for trash containers and water barrels," the report observed, adding that since ARVN controlled the herbicide, "they also control the disposition of empty drums."[43] Later reports would follow up on the role of ARVN personnel in drum disposal. Interviews with local Vietnamese residents indicated that they were purchasing the drums for approximately three hundred piasters (around three dollars) and using them for various tasks.[44] Reporting on their trip to Vietnam in October 1969, the Fort Detrick personnel confirmed from additional complaints that barrel distribution was contributing to collateral herbicide damage near air bases. "Drums are used as containers for gasoline, diesel fuel, and water without complete removal of the residual chemical," the report noted. "The widespread use of herbicide-contaminated gasoline in motorcycles and other vehicles has undoubtedly contributed substantially to the herbicide damage caused by volatization [*sic*] from promiscuous storage of empty drums." The report concluded by recommending the implementation of a new drum disposal program. Unfortunately, these investigations and corresponding reports all offered different conclusions about what that disposal process should be. Some advocated burning the residual herbicide, others recommended alternative draining and cleaning procedures, and still others favored barring the sale of barrels to locals or removing ARVN personnel altogether.[45]

As much as American personnel wished to place the blame on ARVN, however, the size and scope of the problem at places like Da Nang implies that they were aware of the problem. Indeed, in many cases their use of the barrels was just as careless. A Chemical Corps investigation into tree and vegetation damage on the site of the Da Nang naval base found that a generator was regularly being filled with gasoline stored in Agent Orange drums. The report found the "volatilization of the herbicide in the generator" to be the likely cause of the damage and recommended that all herbicide barrels be removed from the base as a precautionary measure.[46] The policies for drum disposal were undoubtedly often followed carelessly and in some cases not followed at all. The long-term consequences of the scattering of residual herbicides are difficult to establish, but without question the former air bases and surrounding areas in central and southern Vietnam are among the most contaminated dioxin hot spots in the world.[47]

I assumed the typical barrel of Agent Orange in 1967 did, in fact, ship to Vietnam, arriving at places like Bien Hoa and Da Nang; had it shipped years earlier, it might not have. Buckingham has noted that in the early phases of Ranch Hand, up to 1964, the military consumed a very small portion of the total volume of herbicide production. Even in 1965, of the 3.4 million gallons produced in the United States, 2.8 million were used for domestic agricultural purposes, while only 400,000 gallons were purchased by the USAF. Prior to the use of Agent Orange in 1965 a variety of herbicides, some of which contained a similar combination of 2,4-D and 2,4,5-T and many of which contained significant levels of dioxin, were produced, distributed, and tested at sites around the world. Over time, as fears over Agent Orange have grown, people who worked and lived in these places have raised concerns about the effects of their possible exposure. The worldwide herbicide testing apparatus thus reveals how Agent Orange became a truly global concern. This network of military, corporate, and civilian actors grew in size and scope as a result of the escalation of the Vietnam War, but it was not created either for or because of that war.

Even after the initial spraying programs in Vietnam were under way, however, the testing network continued to grow, becoming even more global in its reach. In March 1962, for instance, the U.S. Biological Laboratories at Fort Detrick held a "bidders' conference" at which more than sixty organizations competed to be part of the herbicide testing program. The Pentagon awarded five initial contracts, the two largest of which went to the chemical industry giants Pennsalt ($464,000) and Monsanto ($460,000).[48] At several Defoliation Conferences held at Fort Detrick from 1963 to 1965, military leaders met with representatives of every major U.S. chemical company to enlist them in an effort to develop chemical-based solutions to the growing insurgency in southern Vietnam, share the current state of knowledge about vegetation control, and plan for the development of a military–industrial partnership that could deliver to the military the resources and infrastructure it needed to use these tools in Southeast Asia.[49]

Among the information revealed at the Defoliation Conferences is the widespread nature of herbicide and pesticide testing both by chemical companies and by the American military within the United States and around the world in the 1950s and 1960s. At the inaugural conference in 1963 Dow Chemical discussed its testing of Tordon-101, which would come to be known as Agent White, at sites in Davis, California, and Greenville, Mississippi. The Agricultural Research Service of the USDA offered its own presentation on its testing of Purple and White in Texas and Puerto Rico. In 1964 the Chemical Corps itself, the host of the conference, presented on its recent tests of Purple, White, and the combination that would soon be labeled Orange at sites in Tennessee, Georgia, and Maryland. At the time of the conferences similar tests were taking place in Thailand to determine if the results from elsewhere were repeatable in the climate and terrain of Southeast Asia.[50] The military–industrial partnership with the chemical industry continued to prosper, and the global testing apparatus expanded as the U.S. escalated the war in 1965. That year and the next saw augmented testing of Orange, White, and Blue, the three most commonly used agents in Vietnam, at sites from Arkansas to Hawaii, as well as ongoing tests in Puerto Rico, Thailand, and Canada.[51] By the time of the third conference, in 1966, both the military and the chemical companies were seeking more effective, more cost-efficient forms of vegetation control. As Ranch Hand reached its peak in 1967 and 1968, the worldwide testing apparatus spread even further.[52]

Laos and Cambodia were unwilling participants in the U.S. counterinsurgency experiments. Although data for Cambodia are predictably hard to come by since American operations there were illegal, some indications of the levels of herbicide missions directed at Laos exist. As one request for defoliation operations in Laos noted, "The legality of these out-of-country operations is uncertain." The Laotian missions, all supposedly approved by the U.S. embassy in Vientiane, posed logistical challenges for the USAF, which, departing either from Bien Hoa or Da Nang, had to fly over very hostile territory and along relatively high mountains to reach the targets. All of the requests included in memos now located in the National Archives targeted infiltration points near the demilitarized zone (DMZ) and along the Ho Chi Minh trail, probably the most common defoliation targets inside Laos.[53] What few records remain, however, often do not specify the locations of individual missions. The Stellmans' study found evidence to support the flight paths for more than two hundred missions, which sprayed about 1.8 million liters over Cambodia and Laos. Although they do not offer specific records, the Stellmans added that "[National Archives]–held documentation shows as much as 14% more herbicides as having been sprayed but no coordinates are given so that these data cannot be included in the revised HERBS file."[54] A memo of 1969 from MACV to the secretaries of defense and state contains a fairly thorough accounting of Laotian herbicide missions from December 1965 to March 1969. The memo, originally classified "Secret," indicates that defoliation missions sprayed over 109,000 acres, while crop destruction missions covered another 20,485 acres. Of the herbicides used, Agent Orange accounted for 291,015 gallons. With the addition of Blue (56,630), and White (14,500), the total sprayed is over 360,000 gallons, or about 1.37 million liters. If one adds some further missions, including those over Cambodia, the figure of 1.8 million seems reasonable.[55]

In December 2006 DOD released a study documenting the "testing, evaluation, and storage" of tactical herbicides from 1945 through the 1970s. The document identified forty distinct sites outside of Vietnam where testing or storage of herbicides had taken place, thirty-five of which were inside the continental United States.[56] These sites stretch from Hawaii to Thailand and from Florida to Oregon, but a few of them are notable because of their long-term meaning with regard to the Agent Orange saga.

Starting in 1961 Eglin Air Force Base in Florida was crucial to the training of pilots and to the development and refinement of the aerial spray systems that would be employed in Ranch Hand. Yet Eglin has become equally important as a test site for understanding the environmental disposition of herbicides containing 2,4,5-T after they enter the soil. Between 1962 and 1970 a test grid at Eglin of less than one square mile absorbed 4,440 gallons of Agent Blue, more than 15,000 gallons of Agent Purple, and nearly 19,000 gallons of Agent Orange.[57] One estimate that uses conservatively low TCDD levels for samples of Agents Orange and Purple is that the test grid received as much as one thousand times the amount of dioxin as an identically sized grid in Vietnam.[58]

Since the termination of Ranch Hand in 1971 this same grid, as the storage and loading area used for the drums, has been the site of extensive environmental monitoring. Various studies that have examined the effects of TCDD exposure on vegetation, animals, insects, and aquatic life around the base have shown that thirty years after its initial application TCDD was still detectable in the topsoil on the test grid and in the local wildlife, but that the overwhelming majority of the dioxin has disappeared and the remaining amount was concentrated near the surface.[59] Internal USAF studies showed not only that as much as 99 percent of the dioxin had been removed naturally, perhaps through photodegradation, but also that the overall amount in the soil had gone from 49.0 parts per billion (ppb) in 1964 to 0.3 in 1978.[60] Later studies from Vietnam and elsewhere complicated and in some cases undermined the findings at Eglin, but, given the massive amounts of 2,4,5-T applied to the test grid there, the findings are nonetheless noteworthy.

Another area affected by the long legacies of the chemical war is the Canadian Forces' Gagetown Base in New Brunswick, Canada. In 1966 and 1967 Agents Orange and Purple were tested at Gagetown, even after Purple had been discontinued in Vietnam.[61] Beginning in June 1966, the Chemical Corps ran tests in conjunction with Canadian Forces, spraying 172 gallons of herbicides containing 2,4,5-T that year (most of which was Orange or Purple) and another 12 gallons of Orange in 1967. After the issue of dioxin exposure reached the Canadian public and the media in May 2005, investigations revealed a more extensive herbicide testing program there, conducted not by U.S. forces but by the Canadians themselves. Well over a million liters (about 343,000 gallons) of Agents Orange, Purple, and White as well as other combinations of herbicides were sprayed at Gagetown from 1956 through 1984, when the Canadian government banned the use of 2,4,5-T.[62]

The issue of exposure of veterans and civilians who served, lived, and worked at Gagetown are as complex as any Agent Orange exposure scenario (see chapter 5), but the rash of investigations and publications—journalistic and governmental alike—about the Canadian herbicide case have reinforced several problematic misconceptions about Agent Orange, misconceptions that are all too common in similar events in other parts of the world., For example, *Blowback: A Canadian History of Agent Orange and the War at Home* (2009), a book written by the journalist Chris Arsenault, conflates Agent Orange and its associated dioxin, TCDD-2,3,7,8, referring to Agent Orange as "a deadly dioxin" rather than as a herbicide that contained varying levels of dioxin. Like Louise Elliot, the reporter who broke the story for the Canadian Broadcasting Corporation, Arsenault claims that Agent Purple was banned by the U.S. military in Vietnam because it was so much more toxic than Agent Orange, or, as Elliot put it, because "it was so bad."[63] Most scholars agree that Agent Purple was indeed, on average, more contaminated with TCDD than Agent Orange, but the United States never banned the use of Purple in Vietnam. Arsenault's arguments rely on the premise that military commanders and their civilian leaders knew about the toxicity of the herbicides containing 2,4,5-T, a premise for which there is, at best, scant evidence.[64]

In fact, DOD memos from the years prior to the arrival of U.S. troops in 1965 show unmistakably that the decision to replace Purple with Orange was driven by economics. Agent Purple is only slightly different from Agent Orange at the molecular level, but that small difference represented potentially large cost savings to the military as the use of herbicides increased with the arrival of U.S. land forces in 1965. Unlike Orange, which contained a 50-50 mix of the n-butyl esters for 2,4-D and 2,4,5-T, Purple was made up of 50 percent of the n-butyl ester for 2,4-D, 30 percent of n-butyl ester for 2,4,5-T, and 20 percent of the isolbutyl ester for 2,4,5-T. The isolbutyl ester of 2,4,5-T had been developed and patented by Dow Chemical, meaning that for every barrel of Agent Purple, Dow was getting additional revenue from the U.S. government.[65] In a series of follow-up memos in 1964, personnel at Fort Detrick confirmed the situation with Dow and advised substituting Orange for Purple immediately to achieve cost savings.[66] Nowhere in the memos is there any mention of dioxin, the toxicity of the herbicides, or their potential effects on humans, animals, and the environment. None of this explains why the Canadian government continued to spray any of the rainbow herbicides around Gagetown after their use in Vietnam was discontinued in 1971, but it does imply that the original decisions were not as clear-cut as Arsenault and others have suggested, and it divulges that worries about human and environmental health were in no way considered.

While the testing of herbicides continued at Gagetown and elsewhere around the world into 1967, Vietnam had become the primary theater of herbicidal warfare, a veritable laboratory, as the historian Marilyn Young has described it, for new counterinsurgency tactics and weapons being developed by the United States.[67] In that year alone nearly 18 million liters of herbicides were sprayed over 1.7 million acres. The heavy military use of Agents Blue, White, and especially Orange was beginning to strain the military–industrial partnership, as the Pentagon, the chemical companies, and American farmers and ranchers all began to chafe about possible shortages of 2,4,5-T. The U.S. military was now consuming a huge percentage of total herbicide production, and its projected needs for 1968 and 1969 threatened to far outpace the productive capacity of its corporate partners.[68] The shortage first became apparent in 1966, and throughout the next year the military worked closely with the companies, particularly Dow, to develop solutions. The military at one point floated the idea of placing directives on the chemical industry to make military production a priority, even at the expense of other markets, including domestic agricultural use. In early 1967 Secretary McNamara asked the director of the Office of Emergency Planning, Ferris Bryant, to coordinate this effort. According to Buckingham, Bryant "took steps to insure that the entire U.S. output of 2,4,5-T, the limiting component in the production of Orange, would be diverted to military requirements."[69] In response, Dow proposed a new combination it called Agent Orange Plus, essentially a combination of White and Orange. Dow offered even to ship extra supplies of White to Vietnam to be mixed on site, at no additional expense to the government.[70] The air force did, in fact, increase its use of White from 1967 to 1969, but the overall levels of spraying decreased during this period as a result of suspicions about the effects of 2,4,5-T on human health, thereby alleviating fears of domestic herbicide shortages.[71]

The question of a shortage of 2,4,5-T and therefore of Agent Orange was nevertheless a serious one, leaving a trace in the archival records of multiple countries. Investigations in Australia during the 1980s followed up on claims that Australian firms had produced Agent Orange during the war at the behest of the Pentagon. According to the final report of the investigation,

> In the latter part of 1967, the United States expressed interest in the purchase of 2,4,5-T from Australia and this was notified to Australian chemical manufacturers. Information regarding price and delivery details was passed back to United States procurement authorities but, in the event, no orders were placed because the United States decided to buy from their own domestic suppliers. No record has been found of any other contract or negotiation with any Australian Government Department, nor of any Australian supplier having made arrangements to sell 2,4,5-T to the U.S. government or U.S. defence forces. In the late 1960s, there were some commercial exports of Australian 2,4,5-T to American commercial interests, but there is no evidence that such supplies subsequently found their way into American military stocks. The same would apply to any 2,4,5-T of Australian origin re-exported from a third country (e.g. Singapore or Malaysia to which quantities of Australian manufactured 2,4,5-T were exported) to the United States.[72]

Another investigation, resulting in the "Report on the Use of Herbicides and Insecticides and Other Chemicals by the Australian Army in South Vietnam," (1982) "shows conclusively" that Australian troops used chemicals, believed to have been produced in Australia, in much the same manner as American forces.[73] Not everyone involved viewed the report as being conclusive. It did not show definitively either that the chemicals in question in fact included Agent Orange, or that the Australian chemicals were mobilized upon request from the Pentagon. It does underscore the uneasiness expressed in Washington about the potential shortage of 2,4,5-T in the late 1960s.

Similar investigations, with similar results, played out in New Zealand. In response to allegations made by the Australian journalists John Dux and P. J. Young in their book *Agent Orange: The Bitter Harvest* (1986), New Zealand launched a parliamentary inquiry into whether a local plant, partially owned by Dow Chemical, had produced Agent Orange for use in the Vietnam War. Community activists from the area around the plant and from the Vietnam Veterans Association of New Zealand (VVANZ) seized on the report. On March 13, 1989, the VVANZ sent a fax to the minister of defense saying they had "evidence that the defoliant Agent Orange had been manufactured in New Zealand during the late 1960s for use in the Vietnam War."[74] The concerns focused on the Ivon Watkins Dow (IWD) chemical manufacturing plant in New Plymouth.

Located on the western coast of New Zealand's north island, home of the only deepwater port on the west coast of either island, New Plymouth is the heart of the country's chemical and petroleum industries. Founded in 1944, the Ivon Watkins plant became Ivon Watkins Dow, Limited, in 1964, when Dow purchased a 50.1 percent interest in the plant. One of the many products the plant manufactured was 2,4,5-T, from 1950 until 1987. During this time it was the sole producer of 2,4,5-T in New Zealand, which used more of the herbicide per capita than any other nation. IWD maintained that it neither manufactured Agent Orange nor supplied chemicals of any kind to the Pentagon. "As part of IWD's overseas sales programme, small quantities of 2,4,5-T were sold to two United States companies during the time of the Vietnam War," the firm reported during the investigation in 1987. "The product was of a chemical composition that could not be converted for use in Agent Orange. Given the role of the Pentagon in setting market value for herbicides, it would not have made sense financially for IWD to have sold AO or 245T to pentagon [sic] for military use."[75]

The government inquiry found no evidence to contradict IWD's statements. The firm never denied making 2,4,5-T and dramatically increased its production and sales of the herbicide during the period. Between 1966 and 1969 total overseas sales of 2,4,5-T produced by IWD increased more than

240 percent, from NZ$88,858 to NZ$217,859 (table 2). During that period IWD shipped 180 tons of 2,4,5-T to two American companies, AnChem, in Fremont, California, and Miller Products in Portland, Oregon, neither of which ever produced Agent Orange.[76] Given the volume of herbicide in use by U.S. forces in Southeast Asia, it is not surprising that IWD's other overseas customers also increased their business at this time. Fiji and Malaysia, which consistently bought most of the 2,4,5-T produced at the New Plymouth plant, continued to make significant herbicide purchases throughout the years in question.[77] Documents from both the Customs Department and the Defence Ministry support the claims made by IWD. They also show that the Pentagon was indeed anticipating and planning for the likelihood of a 2,4,5-T shortage in the late 1960s, caused largely by its own growing demand for Agent Orange. While the U.S. government explored the idea of importing herbicides into South Vietnam from Australia, New Zealand, and Japan, there is no evidence that it did.

If the records show that the Pentagon approached the Australian government about the possibility of supplying herbicides to support Ranch Hand, New Zealand Defence Department correspondence with the American embassy from July 1967 reveals a request for defoliants, in response to a survey by the United States, should the United States have "the need." The embassy, however, indicated that "it was not desired to proceed to take up the New Zealand supply," mentioning Australian and Japanese companies as possible producers. While the New Zealand government was

TABLE 2. NEW ZEALAND EXPORTS OF 2,4,5-T, 1963–71

YEAR	TOTAL (NZ$)	TO US ($NZ)	TOP RECIPIENTS
1963	12,982	0	Australia/Fiji
1964	20,681	1998	Fiji/Malaysia
1965	36,271	0	Fiji/Malaysia/Australia
1966	88,858	0	Fiji/Malaysia
1967	112,958	6250	Fiji/Malaysia/Australia
1968	206,207	60,692	US, South Africa, Jamaica
1969	217,859	71,000	US/Malaysia/Fiji
1970	106,719	0	Fiji/Malaysia
1971	132,207	0	Fiji/Malaysia

Source: "Exports of New Zealand Products: Chemical Materials and Products—Weedkillers," ANZ, ABGX 4731, file B: Submissions to Inquiry.

willing to support the program, urging "the maximum effort by Defence to be made to get this project going and to assist in getting the material to South Vietnam," the logistical and financial constraints made the request less appealing. "There would be some problems," the embassy noted, "relating to the carriage of the material from New Zealand to Vietnam." The memos between the embassy and the DOD discuss specifically IWD's role. The company indicated to the New Zealand government that it could supply about twenty thousand gallons per year. The Pentagon "would not be interested" in such a minimal level, the embassy indicated, even though the cost of 2,4,5-T produced by IWD was nearly a dollar less per gallon than the product manufactured in the United States. Even when IWD proposed increasing production to eighty thousand gallons a year, "the problem would still be getting the stuff to South Vietnam."[78]

By mid-July 1967, as Ranch Hand missions neared their all-time peak in terms of volume and sorties, the New Zealand Defence Department decided the shipment of 2,4,5-T to Vietnam was simply not cost effective. Chief of Air Staff C. A. Turner wrote to the minister of defence on July 14 that the shipment would require an extra C-130 flight each month from New Zealand and thus, "from the point of view of economics, at US$6.50/gallon, the sale of 10 tons of chemicals [the monthly amount required] would produce NZ$9,350. The extra direct operating cost of a special C-130 flight to Vietnam and return, is of the order of NZ$9,000 of which the overseas content is approximately NZ$5,800. There would thus appear to be only a modest gain in overseas exchange of NZ$3,550 at a direct extra cost to defence of NZ$9,000."[79] Like so many aspects of the herbicidal warfare operations during the Vietnam War, the limitations here were those of cost, not of will.

An absence of evidence did not stop critics of IWD, of the war in Vietnam, and of the governments from devising elaborate conspiracies about how allied forces contributed to Ranch Hand. One of the most remarkable tales came from V. R. Johnson, the longtime president of VVANZ. During the parliamentary inquiry of 1989 into the manufacture of Agent Orange by IWD, Johnson submitted to the committee an "unsigned letter" sent to him anonymously. The author of the letter, supposedly a "major agricultural spraying contractor" based in Auckland, said he "was told to purchase chemicals from IWD in 1964–65" and then to place the chemicals in "Coca-Cola essence drums" and ship them, via Sydney, to the port at Cam Ranh Bay in Vietnam. The author claimed this procedure "happened on a regular basis during the war."[80]

The global apparatus for Agent Orange production, testing, and distribution did not include the Coca-Cola Company, but the lingering question of whether or not nations like Australia and New Zealand provided Agent Orange or its constituent components to the Pentagon during the war does reveal the global nature of the system that made Ranch Hand possible. The total global demand for 2,4,5-T was driven by the Pentagon during the Vietnam War, affecting producers, suppliers, and customers from Michigan to Malaysia. Even if, as the evidence suggests, IWD did not produce 2,4,5-T for use in Agent Orange, it dramatically increased production and overseas sales to meet the rising demand. This point was not lost on critics, who attempted to link the increases to complicity in the herbicidal warfare program and thus to the ongoing effects of Agent Orange and its associated dioxin in Vietnam and among New Zealand veterans. In his submission to the inquiry, B. R. Thomas argued that "politically, New Zealand has been a full participant in the war in Indochina and must inescapably share full responsibility for the use of chemical warfare agents and, in particular, of Agent Orange." Thomas concluded that it was "an academic issue" whether IWD "actually exported Agent Orange for use in Vietnam. All production here would, at the least, have aided diversion of materials from other sources. For a time the military were taking all the Agent Orange they could get."[81]

The question of a global shortage of 2,4,5-T is about more than simply who produced what for whom and for what purpose. By the mid-1960s the major chemical manufacturers of the herbicide were well aware of the problem of dioxin contamination during the production process. Documents made public during the legal battles of the 1980s revealed that Dow had learned that limiting the temperature at which the herbicide was produced, through a slower reaction, could limit the amount of TCDD produced. Dow attempted to make this information available to other manufacturers, but there is no evidence that they informed the Pentagon.[82] If indeed there was pressure on chemical companies around the world to increase the volume and speed of production of 2,4,5-T in the late 1960s, much of the herbicide produced during that period, particularly by companies other than Dow, likely had dioxin levels above historical norms and above what came to be defined as safe limits. The global scale of the legacies of the chemical war becomes both difficult to imagine and easier to understand in light of these worldwide networks.

Environmental Warfare and the Illusive Dream of Control

Agent Orange has long since become the symbol of the environmental devastation wrought on Southeast Asia during the Vietnam War. Understanding it in its fullest historical context requires situating Agent Orange and the other rainbow herbicides as one part of a larger chemical war, one driven largely by the elusive dream of control by the White House and the Pentagon, including control over nature itself.

As James Gibson argues in his discussion of what he calls the technowar waged by the United States, counterinsurgency in the Kennedy years was conceived "as the use of unconventional *techniques,* rather than as social and political mobilization." In the hands of the best and the brightest Vietnam became a laboratory for a variety of scientific and social science approaches to warfare and pacification.[83] This embrace of technical expertise manifested itself in a number of ways in Vietnam, from the strategic hamlet program, in which the United States and the RVN attempted to control the NLF's access to the population by relocating villagers in camps surrounded by barbed wire, to the so-called McNamara Line, an electronic barrier designed to track troop and supply movements along the Ho Chi Minh trail.[84]

Herbicidal warfare fit easily into this framework. Vegetation control, as ARPA and industry experts often called it, was an outgrowth of the belief that by deploying the proper technology and expertise humans could understand, harness, and ultimately control the power of nature. In his discussion of the history of weed control, Zierler notes that during and immediately after the Second World War industry and the government promoted the value of plant sciences as a means to expand the growth potential of various plants and, later, to inhibit such growth. Zierler quotes the plant scientist George McNew, who claimed, "The person who can control the activities of the living cell without destroying its life can determine the ultimate fate of the individual plant." E. J. Kraus, a key figure in the development of 2,4-D, claimed that through the process of weed control "we are going to make plants grow taller, if you wish them taller, and shorter, if you wish them shorter . . . when I start prophesying the sky is the limit."[85] Once the United States began to escalate the war, the Kennedy administration applied not just the same attitude but the same weedkillers, writ large.

Herbicides were only part of the effort to impose order on the unfriendly terrain and unruly environment of South Vietnam. As the situation grew increasingly out of their control, American military commanders continued to draw on other government experts and the arsenal of chemical and technical knowledge and resources they represented to develop even more innovative ways of negating the natural advantages of the NLF. By transforming the battlefields of Vietnam into landscapes more hospitable to American goals, they sought to use technology to achieve what American forces could never accomplish politically.

Malaria was nothing new to the U.S. military. During the Pacific campaigns of the Second World War more American casualties resulted from malaria than from enemy forces. During the Korean War the military contributed resources to fighting malaria as well. Vietnam was no different. After the introduction of major ground combat forces to southern Vietnam by the United States in 1965, the initial rates of hospitalization for malaria were approaching those of the Pacific war.[86] In response the military launched Operation Flyswatter, a coordinated effort involving each branch of the armed forces but ultimately delivered by Ranch Hand through the 12th Air Commando Squadron.[87] During 1965 and 1966, the army and navy began experimenting with insecticide spraying by helicopter, which achieved limited success around secure areas and base perimeters but was subject to NLF ground fire in more contested zones. Even prior to 1965 the air force was working on modifying C-123s for insecticide missions, but these plans were initially put into practice elsewhere, including Afghanistan and Iran to assist in the suppression of locusts.[88] By 1967, however, in light of the increasing number of American forces in-country, Flyswatter began running regular spray missions that used modified C-123s to target mosquitoes with malathion. The planes used for Flyswatter, which came to be known as the Bug Birds, were identical to those used for Ranch Hand missions, except that the camouflage coating had been removed, apparently in the hope of convincing enemy forces to hold their fire since they, too, would benefit from the spraying. Leafleting and loudspeaker operations by Psychological Operations (psyops) preceded most Flyswatter missions over both friendly and hostile areas, informing those below that malathion was harmless to people, animals, and crops and asking them not to fire on the aircraft. The efforts did little to deter ground fire, however. Unlike Ranch Hand missions, Flyswatter sorties involved extended, slow flights close to the ground and were not normally accompanied by armed support from fighter jets.[89] While NLF forces may have stood to gain from antimalarial missions, they had more to gain from downing an American aircraft (figs. 2 and 3).

People on the ground, whether civilian or military, Vietnamese or American, might not have easily distinguished between Flyswatter and Ranch Hand missions. Unlike Ranch Hand, Flyswatter targeted bases and urban

Figure 2. C-123 aircraft used in Operation Ranch Hand. (Courtesy National Archives, College Park, Maryland)

Most herbicide missions in Vietnam utilized modified C-123 aircraft. After 1967 these aircraft were also used increasingly for pesticide missions like Operation Flyswatter, which sprayed large amounts of malathion. Many U.S. veterans who believe they were sprayed directly by Agent Orange were, in fact, possibly sprayed by malathion.

Figure 3. C-123s in spray formation over A Luoi valley, 1967. (Courtesy National Archives, College Park, Maryland)

Contrary to many accounts from the period, the spray from Ranch Hand aircraft appeared as a white mist and did not correspond to the color of the agent. Agent Orange, like the other rainbow herbicides, was named for the 7.6-cm band wrapped around the barrels in which they were shipped. Ranch Hand missions normally flew very low and close to the ground, often eliciting enemy fire.

areas directly and made no attempt to clear or avoid friendly forces or civilians on the ground. Flyswatter aircraft regularly sprayed malathion over nine large U.S. bases in I, II, and III Corps and their surrounding areas, including urban areas such as Da Nang and Bien Hoa.[90] Missions over these primary targets were normally repeated about every two weeks from 1967 to 1971, a schedule that overlapped with the heaviest operational years of Ranch Hand, 1967 to 1969. During its five-year run Operation Flyswatter flew more than thirteen hundred missions, spraying approximately 1.75 million gallons of malathion. The total amount of malathion sprayed over the course of the war is estimated to be more than double this number.[91]

Despite the differences between mission procedures and formations and aircraft appearance and despite the psyops efforts, those exposed to the spray often confused the two and assumed the worst. During the crop destruction incident in 1969 in Da Nang, discussed earlier, the witnesses on the ground described seeing silver planes fly directly over the area where their crops were located. That the color of the planes was silver points to Flyswatter aircraft, and investigations on the ground discovered patterns that were inconsistent with direct spraying of herbicides.[92] Other cases reflect similar incidents of mistaken identity, noting the direct spraying of crops and troops by silver planes that were most likely Flyswatter aircraft rather than Ranch Hand. Given the similarities in the aircraft, the overlap in mission, and the more common occurrence of direct spraying of U.S. and allied troops as well as civilians, it is plausible that on the basis of archival sources much of the anecdotal evidence from veterans about being directly sprayed by Agent Orange and other herbicides while serving in Vietnam from 1967 to 1971 describes exposure to malathion from Operation Flyswatter.[93] Such confusion is even more likely to have occurred in 1970–71, when insecticide missions continued while Ranch Hand missions were curtailed and eventually suspended.

While the battle against malaria was a modestly updated version of earlier battles against nature, other attempts by American forces at manipulating and controlling the environment in Southeast Asia were more direct in trying to impose order through chemicals. The most radical form of the chemical war was the repeated use of forest fire as a military weapon. A combination of herbicides, napalm, and airstrikes, these operations reveal officials' unwavering faith in chemicals and technology for overcoming a multitude of problems facing U.S. forces in Vietnam. They also demonstrate the ultimate folly of that reliance and thereby the larger failings of the United States in the Vietnam War.

In March and April 1968, near the end of the dry season in Southeast

Asia, a huge forest fire raged out of control in the U Minh forest, approximately 150 miles southwest of Saigon. Described by the *New York Times* as the worst to hit the area in more than twenty years, the fires were partly owing to a remarkably dry season that triggered similar blazes in Cambodia and Thailand.[94] The exact origins of the U Minh fires remain unknown, but, as the *Times* noted, there had been "similar fires three years ago, when American B-52 bombers dropped incendiary bombs on the forest" in an attempt to drive the Front from their concealed bases. The peninsula, one of the most impenetrable NLF strongholds of the Mekong Delta region, had concerned the U.S. military for some time. No direct evidence proves that U.S. forces started this fire, but they did everything they could to exacerbate it. For more than a month beginning in late March the United States saturated the forest with grenades, incendiary bombs, white phosphorous, and napalm to encourage the inferno to spread, while naval vessels in the Gulf of Siam harassed NLF forces working to contain the fire and remove ammunition caches in its path.[95] By the end of April the monsoon season began and the fire gradually dissipated, but some 1,150 square miles of the forest were destroyed. U.S. military records later revealed that "the benefits of such a vast fire became increasingly obvious."[96] These benefits, however, had not been detected by James Brown. In a report in 1962 on the effectiveness of early herbicide missions, Brown noted that large-scale burning of the forest was unlikely to be successful given the climate, conditions, and plant species in the region. Pointing specifically, and somewhat ironically, to a fire that burned but did not spread after the crash of a C-123 in February, Brown argued that forest fires would not be effective and would lead to negative perceptions of the larger herbicide program.[97]

Ignoring experts like Brown, military commanders decided to pursue the weaponization of forest fire. Three distinct yet related operations, Sherwood Forest, Hot Tip, and Pink Rose, exemplify how American war planners viewed the landscapes of Vietnam, Cambodia, and Laos as subjects to be mastered by technology. Unsatisfied with the ongoing defoliation programs, the military sought to ignite huge areas of tropical jungle, burning vegetation to make it easier to locate and attack guerilla forces. In the end, the three operations were failures; much like the guerilla forces it shielded, the canopied forest proved largely impervious to the superior technology of the United States. The jungle consistently refused to burn, symbolizing the larger futility of the military effort. Some scholars have held that the United States began to view the forest itself as an enemy, a somewhat misleading position.[98] War planners and military commanders treated the natural environment less as a combatant to be destroyed than as an object to be pacified and controlled. In other words, they treated the forest less like the NLF than like the southern Vietnamese population the United States was ostensibly trying to protect. As in the case of the civilian population, this American endeavor to control a foreign, often hostile host through technology resulted in a type of widespread destruction that not only wreaked havoc on the land and people of Vietnam but also, in doing so, undermined the stated objectives of the United States in Southeast Asia.

Operation Sherwood Forest began in March 1965, just before the arrival of hundreds of thousands of additional U.S. troops, although the project was the culmination of several months' work. The target was the Boi Loi Woods just west of Saigon, an NLF stronghold that ARVN had essentially abandoned by the end of 1964. Beginning in December 1964 U.S. forces carried out a combination of defoliation operations and preparatory bombings, along with psyops to warn the civilians to flee the area, but some within the USAF continued to push the idea of igniting the forest.[99]

Disregarding the seeming consensus that such efforts would fail, MACV approved the request in March, and U.S. planes dropped diesel fuel, napalm, and incendiary cluster bombs over thousands of previously defoliated acres in Boi Loi. The official MACV history of Sherwood Forest extolled the virtues of the mission, calling it "reminiscent of a World War II B-17 bombing strike" and arguing that the operation was "probably unique in the history of air combat." "Unfortunately," the report concludes, "a huge rain cloud formed over the target area," and the resulting rain "did not let the fire spread as expected."[100] Later analysis showed that the rain had little part in the failure. As the official USAF historian of Operation Ranch Hand later wrote, "The U.S. military had to try several times before learning" that the ecological conditions made it "almost impossible to set a self-sustaining forest fire in the jungles of South Vietnam."[101] The unit history for Sherwood Forest suggests such a mindset: failure aside, it noted, "The Boi Loi woods would not furnish the same safe sanctuary to the Viet Cong that it did before 1965."[102] The overwhelming failure of Sherwood Forest notwithstanding, the JCS in December 1965 asked the secretary of defense to approve research to determine if forest fires could be used as a military weapon, leading ARPA to contract with the U.S. Forest Service (USFS) to explore that possibility.[103] ARPA initiated Project EMOTE (*E*nvironment *MO*dification *TE*chniques), a wide-ranging program that studied forest fires as well as other forms of environmental modification, including the alteration of rain and weather patterns in Southeast Asia.[104]

The ironies of these missions abound, not least the active involvement of the USFS, which over the previous several years had advised and assisted the South Vietnamese in the development of their lumber industry.¹⁰⁵ The USFS had long supported the use of herbicides in both domestic and foreign vegetation control scenarios as a safe, efficient, and cost-effective means of making forests more "productive."¹⁰⁶ Its close consultation with Operation Ranch Hand, whose motto, "Only You Can Prevent Forests," openly mocked the USFS and its beloved mascot, Smokey the Bear, marked a departure for the agency: assisting in the destruction of huge swaths of forest with no clear environmental or economic goals in mind.

Aided by USFS experts, Operation Hot Tip began in March 1966, aimed once again at igniting a large-scale forest fire, this time targeting Chu Pong Mountain in the central highlands near Pleiku. Even if the fire was ineffectual, according to military intelligence, the "defoliation and burning should hamper PAVN/VC activities in the area." A civil affairs or psyops component to precede the mission was omitted for the mission "since the only 'inhabitants' of the area are VC/PAVN soldiers and their laborers."¹⁰⁷

As Operation Hot Tip began, like Sherwood Forest, with dry season defoliation followed by napalm and incendiary cluster bombs, everything seemed to go according to plan. On February 16 and 17 the initial napalm strikes and applications of Agent Blue were carried out. After several weeks of evaluating defoliation and initial burn effects and closely monitoring local weather patterns, officials deemed the mountain ready. At two p.m. on March 11, 1966, the Second Air Division attacked Chu Pong Mountain with eleven additional napalm strikes and a heavy load of M-35 bomblets delivered by B-52s. "Heavy flames" were observed from above, and the strike produced a massive convection column that burned significant portions of the forest canopy.¹⁰⁸ Early assessments characterized Hot Tip as a success, both operationally and technically, but ARPA's summary report did not determine the "total effectiveness" of Hot Tip because additional data were needed.¹⁰⁹

When the smoke cleared, literally and figuratively, from Operation Hot Tip, its total effectiveness was easy to discern: the fire failed to achieve self-sustainability. Nonetheless, later reports described the results as inconclusive and counseled that the idea warranted further study. In Hot Tip not recent rain but previous rain caused the failure. The fuel for the fire, particularly the twigs on the forest floor, was simply too wet to "carry fire into and through the canopy."¹¹⁰ ARPA saw this as simply another natural obstacle to be overcome through superior knowledge and firepower. Along with its USFS allies, ARPA used the results of Hot Tip to prepare new "guidelines for chemical treatment and incendiary operations" to "modify the moisture content of these critical stem and twig fuel components." The new formula for igniting a forest fire called for early and repeated applications of Orange and White followed by application of Blue for "maximum desiccantation [sic]." Once the moisture and distribution of foliage were controlled, the USFS argued, the forest fire would, in the right weather conditions, achieve self-sustainability. In other words, more chemicals, more firepower, and a little help from Mother Nature would allow ARPA, the USAF, and the USFS to impose their will on the landscape of Southeast Asia. Throughout the late summer and fall of 1966 USFS personnel worked closely with ARPA to devise a plan to apply the lessons of Sherwood Forest and Hot Tip. This "definitive test" became Operation Pink Rose, the test plan that even ARPA admitted would "probably be the final test of the feasibility of destroying forest/jungle growth by fire."[111]

From September 1966 to January 1967 Ranch Hand aircraft sprayed over two hundred thousand gallons of Agents Orange, White, and Blue over the three targets, two in northern Tay Ninh province, near the Cambodian border, and one in southern Phuoc Long province, about one hundred kilometers east of Saigon. From early January through early April 1967, B-52s carpeted the targets with napalm and incendiary devices, while analysts monitored rainfall and wind data, observed the huge convection columns that formed over the blazes, and collected soil and brush samples from cooling areas where the forest refused to burn, again demonstrating its resilience. The photos accompanying the final appraisal of Pink Rose show that the problem of Hot Tip had indeed been solved: the "ground litter," the fuel that failed to ignite on Chu Pong Mountain, was successfully attacked in Pink Rose, but this did not create a self-sustaining fire. To the degree that forest canopy burned at all, it was ignited by direct contact with the M-35s, many of which "activated prematurely." "Although the ground litter burned," the photo captions read, "the forest did not."[112]

The final report on Operation Pink Rose seemed to express the frustration of the military: "Fire has been used successfully for centuries by native people employing 'slash and burn' agricultural techniques." During wartime, however, the slash component of that process, that is, the widespread clearing of the trees in the area, "is not feasible in insecure areas."[113] The report went on to claim that the problem was primarily one of insufficient desiccation; even with the wide application of Agent Blue, the chemical resources deployed simply could not dry out the forest enough to make it burn. In the final analysis, even the normally (and often wildly) optimistic internal reports of ARPA and the USFS concluded that Pink Rose, at long last, "gave conclusive evidence that tropical vegetation was not appreciably altered by fire under existing conditions."[114]

Even in the face of complete failure, some in the U.S military were determined to locate a silver lining. In the preliminary memorandum prepared for General Westmoreland in February 1967, W. G. McMillan in the Office of the Science Adviser for MACV summarized the dismal results of the multiple, concerted attempts to ignite large forest fires for military purposes, noting that there was no prospect of developing an "operationally effective method of jungle burning in South Vietnam." However, McMillan went on, "in our disappointment over not being able to produce on call a spectacular jungle firestorm, there is a danger of overlooking . . . [the] ancillary military effects," above all in the realm of psychological warfare: "There may be a psychological factor in adding incendiaries to all the other weapons systems the enemy has to suffer. He might well ask himself: 'Good Grief! What next?' The psychological effect would, of course, be greater if some of the enemy were actually killed or injured, which stresses the importance of trying to get some feedback from captives or Chieu Hois [NLF soldiers and supporters who defected as part of the Chieu Hoi, or Open Arms, amnesty program] who may have experienced or witnessed these events."[115] Adding these "not inconsiderable potential ancillary military effects" to the additional "desired technical data," the report recommended proceeding to the next phase of the operation, Pink Rose III. That phase was never implemented, but ad hoc disparate burning continued.

Even as the zeal for forest fires abated and Pink Rose was terminated, American war planners did not abandon their desire to alter the landscape to better serve their strategy and tactics. In 1966 and 1972 the USAF ran Operation Popeye, an attempt to manipulate regional precipitation patterns to extend the length of monsoon season in hopes of causing flooding, landslides, and erosion around roadways to inhibit the movement of troops and supplies. U.S. aircraft would seed clouds with silver and lead iodide to increase rainfall. Beginning in 1966 the navy began testing these so-called weather modification techniques over the Laotian panhandle and determined that cloud seeding could be effective along supply lines from northern Vietnam through Laos. Reporting to Congress on the program in 1974, Lt. Col. Ed Soyster noted that intelligence indicated there would be "no significant danger to life, health, or sanitation" in the targeted areas because the local population was "very experienced in coping with seasonal heavy rainfall conditions. Houses in the area are built on stilts, and about everyone owns a small boat."[116] At the same hearing Gen. Ray Furlong of the USAF testified that he was aware of a related operation that would drop "emulsifiers" on the Plain of Jars area of Laos to erode part of the Ho Chi Minh trail and make it slippery and difficult to navigate.[117] And until it was ended in early 1971 Operation Ranch Hand continued to spray more than seventy-two million liters of chemical agents over central and southern Vietnam.[118]

The USFS maintained its role in Vietnam during these years, assisting Ranch Hand in the evaluation of herbicide missions while advising the South Vietnamese timber industry, mostly with the goal of supplying the U.S. military. As James Lewis has shown, USFS personnel, working under the auspices of USAID, trained ethnic minority Montagnards on "logging, milling, and restoration efforts." In addition to providing U.S. and ARVN forces with local lumber supplies, the USAID plan hoped t make the Montagnards economically independent and presumably less willing to collaborate with the NLF.[119] As Lewis shows, this program, too, was a "political and economic failure":

> The Viet Cong demanded bribes from loggers and infiltrated operations. The best sawmill operator [USFS project leader Jay] Cravens trained turned out to be the leader of the local Viet Cong unit. Although the United States military provided logistical support and military protection for the foresters as they flew around South Vietnam to advise on logging operations and set up sawmills, the military also continued its defoliation and bombing missions, often near the proposed logging operations. Cravens visited all forty-four provinces of South Vietnam while there, and recalled that everywhere he went the country reeked of herbicide. Damage to vegetable, fruit, and rubber tree farms angered farmers, and shrapnel in tree trunks wreaked havoc with saw blades at the lumber mills. Instead of aiding the Vietnamese, the U.S. alienated them.[120]

None of these operations proved measurably more successful than forest fires in directly aiding U.S. military efforts, but they did succeed, along with the rest of the American war effort, in further alienating the population and altering the environment with consequences both intended and unintended. It is not surprising that the U.S. military devoted resources to exacerbating the fires in U Minh in 1968, aiding in the destruction of millions of acres of land; U.S. forces proved themselves incapable of starting a deadly blaze on their own, but they were only too happy to enhance an existing forest fire. If they could not beat nature, they could join it.

The failure of these operations proved to be no impediment, as the

Pentagon clung to its resolve to weaponize the weather and pacify the natural environment. At a Senate hearing in 1972 on weather modification and warfare, the commanding officer of the Naval Ordnance Laboratory in California rejected congressional attempts to ban environmental modification: "Primarily, the work is aimed at giving the U.S. Navy and the other armed forces, if they should care to use it, the capability of modifying the environment, to their own advantage, or to the disadvantage of the enemy. We regard the weather as a weapon. Anything one can use his way is a weapon and the weather is as good a one as any."[121] But the weather, like so many other potential instruments of war in Vietnam, proved to be an ineffective weapon for the United States.

Conclusion

U.S. forces used everything at their disposal, including artillery, herbicides, the weather, and even the forest itself, as a weapon in Vietnam in their attempt to turn the tide of a war that was running against them. As Buckingham has written, the substitution of technological solutions for manpower represented a stubbornly persistent pattern for the military in Vietnam.[122] Operation Ranch Hand and Agent Orange are the most notorious and most recognized aspects of these efforts, but they were smaller components of the larger chemical war waged in Southeast Asia, a war that sought technological solutions not only as a substitute for manpower, but also as an answer to political problems. When a political solution in southern Vietnam failed, the United States turned to traditional military solutions. When traditional military approaches failed, the United States sought to use unconventional weapons. And when these failed, war planners convinced themselves that the technological superiority of such weaponry would be a powerful psychological weapon against a technologically primitive opponent. How could a supposedly ragtag group of guerillas not be intimidated by the weaponization of nature itself?

Through it all, policymakers in the White House and military commanders in Southeast Asia continued to rely on the most important illusion of all: the illusion of control. The Kennedy and Johnson administrations in particular never came to terms with the limits to American power—military, political, economic, and technological. The United States could not control the French, so it ultimately took over the fight against the Vietnamese revolution. The Eisenhower administration could not control the course of the revolution, so it actively assisted and supported the creation of the Republic of Vietnam and the ascension of Ngo Dinh Diem as premier and president. President Kennedy could not control Diem and ultimately allowed the coup that led to his assassination. His administration could not control the population of southern Vietnam, so the military rounded up civilians and placed them in strategic hamlets to isolate them from the NLF. Once the war began in earnest, Johnson and his administration could not control the actions of the North or the NLF, so they attempted to bomb them into submission. Military leaders could not fight the land and air wars they wanted, so they sought to exert greater control over the landscape, but they could no more control the landscape than they could its human inhabitants. Ultimately, they also could not control the effects of the chemicals they let loose on that landscape and its inhabitants, leading to the long-term environmental and human costs of the Vietnam War.

CHAPTER TWO

HEARTS, MINDS, AND HERBICIDES
The Politics of the Chemical War

In the late summer of 1962 Edward R. Murrow was concerned about crop destruction. Writing to National Security Adviser McGeorge Bundy in August, Murrow, then serving as the director of the United States Information Agency, expressed his skepticism about the ability of the United States to "persuade the world—particularly that large part of it which does not get enough to eat—that defoliation 'is good for you.'" Among the many issues Murrow raised in his memo was that the *New Yorker* magazine had just run a series of pieces that called attention to the ecological consequences of insecticides. The articles, written by Rachel Carson, became the basis for *Silent Spring,* the book that helped usher in the modern environmental movement in the United States.[1] Two years later Carson died, but the ideas she popularized would fundamentally change the way people thought about chemical herbicides and insecticides. They would also complicate the crop destruction program in Vietnam, well under way at the time of her death.

Often concerned more with the global perception of the chemical war than with the chemical war itself, the Kennedy and Johnson administrations consistently fretted over the politics of defoliation and crop destruction as they escalated the war in the mid-1960s. Throughout this period the military stubbornly clung to its belief that the military effectiveness of herbicide use outweighed any possible negative political consequences. From the Pentagon down to field commanders, it became a given that Ranch Hand missions made life more difficult for the NLF without having overly adverse effects on local populations. But in the face of mounting outside pressure to justify this increasingly controversial program in an increasingly controversial war, the internal debates over the pros and cons of the chemical war took on added significance during the critical year of 1967. That year the RAND corporation, the Santa Monica–based think tank closely associated with the military and intelligence communities, released two scathing reports that called into question the herbicide program in general and crop destruction in particular. These reports, "An Evaluation of Chemical Crop Destruction in Vietnam," by Russell Betts and Frank Denton, and "A Statistical Analysis of the U.S. Crop Spraying Program in South Vietnam," by Anthony Russo, offered by far the most critical assessments to date of herbicidal warfare programs from within the military–industrial complex itself and touched off a serious debate within the Pentagon.[2]

The origins of this policy debate lay in how the political battles over the herbicide programs had played out in multiple settings, from Washington to the Vietnamese countryside, and how the chemical war shaped and was shaped by the battle for hearts and minds. Most of the literature on the chemical war has notably overlooked such aspects. Military histories like William Buckingham's *Operation Ranch Hand* and Paul Cecil's *Herbicidal Warfare* summarize and too often simply reconstruct the official military line on herbicides without interrogating the assumptions on which those policies were based; neither takes seriously the possibility that Ranch Hand was undermining support for the war in the countryside. Other works on the use of herbicides in Vietnam have traded advocacy for historical context, collecting firsthand accounts from veterans and other victims of Agent Orange without considering the ways in which people encountered the herbicides. Yet most of these works focus on Agent Orange at the expense of other chemicals, such as Agent Blue, thereby ignoring the vital component of crop destruction altogether.[3]

By focusing on crop destruction as it is addressed in previously ignored military records, I demonstrate that the political and psychological aspects of the chemical war were far from ancillary components of the military campaign. Rather, both the revolutionary forces of Vietnam and the U.S.-led forces saw the propaganda battle over the program as a central front in the war. In this battle there was no clear winner: both the U.S.–ARVN forces and the NLF struggled to make their case to Vietnamese villagers, and both drew their share of blame from local populations. There were, however, clear losers: Vietnamese civilians were, as one villager put it, like "a fly caught between two fighting buffaloes."[4] In rationalizing and justifying the use of herbicides and other chemical agents in Vietnam, U.S. policymakers consistently distinguished between civilian crops and those controlled by the NLF, distinctions which proved even less effective than similar differentiations between combatants and noncombatants in the American war of attrition.[5] The account I give here reveals that the basic American assumptions about the military effectiveness of herbicide programs were based, at best, on limited evidence, while far more substantial proof of the political costs was readily ignored. As in so many other areas of the war, the United States sought in vain to reconcile its political and military objectives in Vietnam while villagers were forced to live in fear of bombs, bullets, and the spray. The extent to which these debates were grounded more in politics than in ethical or moral concerns can be measured in the decision to extend the chemical war to crop destruction in 1962.

To Win the People

As we have seen, the idea of herbicidal warfare resulted from a combination of the approaches the Kennedy administration took to the Cold War. But if the military's view of the problem was framed largely by the new administration's flexible response to the global war against communism, the policymakers also realized that the political consequences of their actions would be framed through the lens of the Cold War and decolonization. Concerns about charges that the use of herbicides and other chemical agents would open up the United States to accusations of violating international prohibitions against chemical and biological warfare were present from the earliest discussions of the program. By the early fall of 1961, when the first herbicide test missions were under way, the American embassy in Saigon raised concerns that the introduction of the aircraft and the herbicides would be seen as a violation of the Geneva Accords of 1954 and likely result in accusations of chemical warfare from the North Vietnamese.[6] Worries over the political fallout, however, were normally raised by the embassy or the Department of State (DOS), and as the vulnerability of Diem's regime became more apparent toward the end of Kennedy's first year in office, the president increasingly relegated political and diplomatic considerations to a status secondary to military necessity. During this period the public remained almost completely in the dark about the chemical war. Only in the late 1960s, amidst revelations by scientists and antiwar activists about the potential dangers of dioxin, would the full scale of the program become public knowledge.

Two central questions framed these early debates: first, did the military use of herbicides constitute a form of chemical warfare prohibited by international law? and, second, regardless of the legal definitions, what was the political cost of the herbicide program in terms of both international public opinion and the political struggle in southern Vietnam? The first issue was dismissed without much debate; the consensus of the administration was that the program was not tantamount to chemical warfare. This did little to assuage the uneasiness of many advisers about the potential political fallout, however. One DOS memo, recalling "the propaganda circus created by the communists on alleged U.S. use of 'germ warfare' in Korea on the basis of fabricated evidence," noted that the Communist bloc would take advantage of any such opportunity to launch a great propaganda offensive.[7] In a memo to the president in late 1961 Secretary of State Dean Rusk argued that although the use of herbicides was not specifically prohibited by international law, such use would open the administration to charges of biological and chemical warfare.[8] Even Secretary of Defense Robert McNamara expressed concerns about the political impact of the program, suggesting that Diem make a public statement clarifying that the herbicides were not harmful.[9]

Early on, however, the political costs of herbicidal warfare were phrased almost solely in terms of world public opinion or Communist propaganda. Both the Pentagon and the White House feared that herbicides would be seen by others around the world as a violation of prohibitions against the use of chemical weapons, but outside of the DOS, policymakers rarely, if ever, discussed the political costs of such programs to the battle for hearts and minds.[10] Writing to the director of the United States Information Agency, a member of the Far East division of the DOS noted the preoccupations of many at State: "I am no military strategist or tactician, although I did learn a few things about chemical or bacteriological warfare (most of it harassing) at the Air War College. Perhaps defoliation can be a critical factor in exposing Viet Cong strongholds and destroying Viet Cong food supplies. If it is, and must be used, we can take the psychological bumps which are certain to be dealt to us. But the spectre of charges that 'U.S. imperialists are waging germ warfare on Asians' haunts me."[11] In the end, President Kennedy determined that the program was worth the risks and approved major defoliation operations that would later include crop destruction.[12]

The documentary record shows that both military and civilian advisers within the Kennedy administration were operating on the assumption that the herbicides were not harmful to people or animals. Their concerns about chemical warfare were consistently political rather than moral in nature and relied on the fact that the same herbicides were being used domestically in the United States. Under Secretary of State George Ball noted that while the administration should make clear that these initial operations were not aimed at human targets, herbicides should be deployed in a "low-key" manner, "since defoliant is harmless to personnel and animals."[13] In a memo prepared for Deputy National Security Adviser Walt Rostow, Robert Johnson of the National Security Council argued that to get out in front of any charges of biological and chemical warfare, the United States should make the program "as open and above board as possible," emphasizing "the fact (I believe it is a fact) that the chemical agents involved are the same kind that are used by farmers against weeds." Johnson suggested it might be advisable to get the International Control Commission, the international team charged with overseeing the implementation of the Geneva Accords, "to examine every drum of the defoliant mixture to determine that it is what we say it is." Regardless of the course of action, however, Johnson argued that the informational efforts were urgent: "Otherwise we may pay many of the political costs while reaping no military advantages."[14]

Throughout 1962, the first major year of operations for Ranch Hand, the military constantly evaluated those supposed advantages, measuring the effects of the herbicide program on the vegetation itself as well as on the NLF. The "Review and Evaluation of the Defoliation Program" of 1962, performed by ARPA, strongly supported continuing and expanding the program but made specific recommendations for minor adjustments, such as modifications to the dispersal equipment in the C-123s and continued assessment of whether the herbicides then in use (largely Agents Purple, Green, and Pink; Orange had yet to be introduced) were the most effective means of vegetation control.[15] The report targeted the destruction of "Viet Cong crops" as the primary area for expansion. As Buckingham has observed, Ngo Dinh Diem had long advocated expanding the herbicide programs to include crop destruction. ARVN troops had long been at work manually destroying crops, and Diem saw herbicides as a "cheaper and more efficient" method. Kennedy had seen crop destruction as a potentially more troubling form of herbicidal warfare, however, and was initially reluctant to employ it. The president was not alone in his misgivings. "Advocates of crop destruction," Buckingham notes, "would have to overcome strong opposition from the State Department."[16]

In the summer of 1962 these tensions came to a head, as the DOS and DOD squared off over the crop destruction program. From the beginning the central issue was military utility versus political consequences. Within this larger debate, both advocates and critics regularly came back to two issues: first, how could U.S. and ARVN forces distinguish civilian crops from those controlled by the NLF? and, second, for the program to be effective, would not the civilian population already have to be isolated from the NLF? Within the DOS Roger Hilsman was one of the most vocal critics of the proposal. "Destroying crops will inevitably have political repercussions," he wrote to Assistant Secretary of State Averell Harriman in July 1962, noting the potential fallout within South Vietnam and in the international arena. If the program could indeed force the NLF to focus on generating and transporting additional sources of food, "rather than fighting," he went on, "the political price may be acceptable," but only "after the Viet Cong have been isolated from the peasants and driven into well-defined areas of concentration."[17] The early response from MACV and the JCS to such concerns was that the initial crop destruction targets, located in the central highlands, were essentially areas the Montagnards, a local ethnic minority, had abandoned to the NLF. Additionally, they observed, the destruction of crops had a powerful psychological effect on all Vietnamese: "An interesting side effect in such an operation as crop destruction is that because of the superstitious nature of the rural peasant in Vietnam, the ability of the GVN [Government of Vietnam] to kill large areas of vegetation 'magically' makes a deep impression on him. During the Mangrove defoliation operation, one hundred and twelve VC [Viet Cong] surrendered when it was publicly announced that additional defoliation operations would be conducted."[18] The JCS realized that not all crop destruction targets would be limited to the remote highlands. Eventually the program would require the destruction of crops in close proximity to civilian populations. As this memo makes clear, however, in addition to the psychological benefits of herbicides, another key factor was that crop destruction would support and strengthen pacification efforts, especially the strategic hamlet program.[19] From the outset the military considered herbicidal warfare in general and crop destruction in particular as part and parcel of the larger pacification effort; that attitude resulted in a key difference of opinion in the debate: the DOS argued that if crop destruction was to be effective, the population would have to be separated from the NLF; supporters of the program, chiefly in the military, argued that crop destruction should assist directly in the forcible relocation of civilians.

As the MACV proposal for crop destruction worked its way up toward the White House in August 1962, the debate between DOS and DOD intensified. On August 8, McNamara sent his recommendation to the president, arguing that crop destruction would effect a "substantial military advantage." He agreed with MACV and the JCS that the program should be seen as supporting pacification programs and that the technical resources for the program would be an efficient, effective means of improving crop eradication efforts already under way. "The only possible drawback anticipated is in the psychological area," he noted, again concurring with those who suggested that this could be dismissed as Communist propaganda.[20] Nowhere in the memo does McNamara discuss the potential political impact among the civilian population of South Vietnam.

Secretary Rusk attempted to reframe the issue in his own recommendation, which went to President Kennedy on August 23. While the proposed course of action would likely result in increased propaganda, he remarked, the larger worry should be over the political fallout among Vietnamese civilians. "The way to win a guerilla war," Rusk asserted, "is to win the people. Crop destruction runs counter to this basic rule."[21] Rusk's explication of the argument would prove to be prophetic: "The problem of identifying fields on which the Viet Cong depend is hardly susceptible to solution so long as the Viet Cong and the people are co-mingled. The Government will gain the enmity of people whose crops are destroyed and whose wives and children will either have to stay in place and suffer hunger or become homeless refugees living on the uncertain bounty of a not-too-efficient government." Rusk disputed the psychological arguments about the supposed magic of herbicides that had gained currency among advocates in previous months: "Other people, who merely sympathize with [the NLF], will also hate the government for crop destruction. The use of strange chemical agents to destroy crops strikes at something basic implanted in human beings (even if people do not—as many will—fear that the chemical agents are also directly harmful to people)." Rusk went on to outline potential benefits as well but ultimately recommended that the program be terminated.[22]

As President Kennedy mulled over his decision, the debate between the DOS and the DOD continued. At a meeting at the Pentagon on August 24 the staff of the JCS and the East Asia desk at State went back and forth, talking point for talking point, about the central issues Kennedy was considering: Would crop destruction be effective only after other pacification programs had succeeded in isolating civilians from the NLF? or could it be an effective weapon in that effort? Would crop destruction force the NLF to redirect energies and resources that might otherwise be used in military and political efforts? or would the hardships of food denial be passed on to the local population? Would the program result in worldwide condemnation?

or had the United States already weathered the storm of criticism, which might be written off as Communist propaganda? And, finally, would the program further alienate the rural population from the Diem regime and its U.S. sponsors? or would it help provide security from the NLF and be a potential ally in the battle for their hearts and minds? During one tense moment the DOS representative, Deputy Under Secretary Alexis Johnson, commented that "the program posed great psychological problems even though the chemicals to be used were obtainable at a hardware store. They were in fact similar to materials we use on our own lawns," to which Gen. Lyman Lemnitzer of the JCS staff replied, "It is strange that we can bomb, kill, and burn people but are not permitted to starve them."[23] The two sides were not going to come to any consensus.

President Kennedy's decision to approve limited crop destruction operations, as Buckingham has argued, was ultimately shaped not just by the internal debate between State and Defense, but also by a meeting he had with RVN Foreign Minister Nguyen Dinh Thuan in mid-September. Thuan was in Washington for meetings with the International Monetary Fund but met with various constituencies in the crop destruction debate. At a meeting at the DOS on September 19, Thuan assured those in attendance, among whom were some critics of crop destruction, that local province chiefs could help distinguish between civilian and NLF fields, although his reasoning—that the Montagnards put huts in their rice fields and the NLF do not—was not very convincing.[24] A week later Thuan met with President Kennedy and repeated his earlier assessment. The president asked why the Viet Cong could not also build huts in their rice fields if this was indeed the primary way to distinguish civilian and military crops; he too was apparently less than convinced by Thuan's reasoning.[25] Still, Thuan insisted that crop destruction could be effective on a limited, targeted basis. On October 2, 1962, the president approved operations on exactly those grounds.

As crop destruction was put into action over the next several months, State continued its attempts to rein in the program, insisting on prioritizing political and psyops efforts to inform the civilian population about the purpose and the effects of the spray. The department developed strict guidelines for all defoliation missions, including a role for the embassy in approving requests. This role was based in part on requests from MACV and the JCS that approval for missions be delegated down the chain of command to allow for more rapid response, an approach Kennedy resisted throughout the remainder of his presidency. Under President Lyndon Johnson crop destruction and defoliation missions increased in direct proportion to the escalation of the war. Any remaining misgivings about political fallout trumping military necessity were marginalized. "In 1964," Buckingham writes, "the restraints placed on chemical crop destruction by Washington officials, fearful of the potential domestic and international outcry again against the tactic," slowly crumbled.[26]

Even as the situation in southern Vietnam deteriorated in 1963, however, the political fallout should have been increasingly clear to officials. The Communist bloc was no longer alone in drawing attention to the herbicide program. The *New York Times* had begun reporting on crop destruction and the "sensitivity [of the United States] to the possibility that accusations would be made that Americans took part in chemical warfare." Citing the *Times* piece in a letter to President Kennedy in March 1963, Congressman Robert Kastenmeier of Wisconsin noted Franklin Roosevelt's refusal to engage in crop destruction against Japan during the Second World War and castigated the president for "our present starvation program in Vietnam."[27] In his reply to Kastenmeier, though, Deputy Assistant Secretary of Defense William Bundy argued that "the use of chemical and biological weapons has not occurred, and the compromise of moral principles has not been at issue." "Chemical warfare as defined by international law," Bundy noted, involved the use of chemicals on "the physical person of the enemy," and the commercially available "weed-killers" being deployed by the United States and its South Vietnamese allies did not fit that definition.[28]

The debates among policymakers and military leaders over herbicidal warfare and crop destruction centered around a number of false distinctions, each of which was magnified in the environment and historical moment the United States found itself in in southern Vietnam in the 1960s: that between the political and the military in a counterinsurgency, counterrevolutionary situation; that between civilian noncombatants and military combatants; and that between "the physical person" and the larger environment in which that body was located. The first two represent a failure by war planners to grapple with the political and military realities of their counterinsurgency efforts in Vietnam. The last, however, is more complicated. In hindsight, it seems absurd to believe that policymakers could separate the "physical person" of the enemy or, for that matter, of civilians from the physical environment in which they lived. How could the sprays being used to defoliate forests and destroy rice crops and fruit trees be considered any more separable than civilians and combatants in a guerilla war?

In the early 1960s this bodily divide was not simply a distinction relied upon by military commanders and the National Security Council. It was a common assumption about the relationship between body and nature held by many Americans at the time—before *Silent Spring* had helped draw attention to the dangers of pesticides and accelerated the spread of environmentalist thinking in the United States. This framework was under assault in the United States and elsewhere as a nascent environmental movement began to complicate common understandings of nature, the human body, and the effects of pesticides and herbicides but had yet to enter the mainstream.[29] The herbicidal warfare waged by the United States in Southeast Asia would become part of this discussion, demonstrating the futility of such distinctions and also the ultimate futility of the Vietnam War.

Negotiating the Spray

We have seen how the politics of the chemical war were viewed from Washington. Understanding the full political stakes of crop destruction and the larger chemical war is impossible without knowing how local villagers and the NLF encountered and negotiated the spray. While much recent work on Agent Orange has used oral history and ethnography to reconstruct these early encounters, no studies to date have employed existing firsthand accounts from the period.[30] Among the most valuable and underutilized sources are a series of interviews conducted by the RAND Corporation during the war.

Between 1964 and 1968 RAND employees interviewed over two thousand captured NLF soldiers, refugees, and participants in the Chieu Hoi (Open Arms) program.[31] An entire series of these interviews focused on reactions to the herbicide programs. While the statements of the interviewees do not constitute a random sample of Vietnamese villagers or of members of the NLF, they do constitute one of the few written records about the reactions of these populations to the defoliation and crop destruction programs. The sources are potentially problematic in other ways, not least because many of the interviewees were prisoners being interrogated, and many others, especially participants in the Chieu Hoi program, were relying on securing protection and jobs from the United States and RVN.[32] Nevertheless, the interviews suggest a number of similarities in the reactions to the chemical war of populations in three distinct regions of South Vietnam.[33] The RAND interviews were very structured and focused on specific areas, including the effects of defoliation missions on local crops and vegetation; the immediate effects of the spray on NLF activity in the area; whom the villagers blamed for the spray; and whether the subject and, in the subject's view, the local population considered the spray more or less dangerous than bombs and bullets. In the responses to these and other questions a picture of the reactions to the spray begins to emerge, one which mirrors many of the underlying tensions of South Vietnamese life during the period, namely, that the villagers of provinces throughout the country remained trapped in a war being fought ostensibly for their future. The interviews also help reveal the futility of many of the distinctions made by the U.S. military and make clear that the political problems created and exacerbated by the chemical war would not be easily overcome.

The interviews, along with captured documents, reveal the extent of the NLF's preparations and precautions vis-à-vis the herbicide programs. Members regularly carried nylon masks or surgical masks to cover their mouths and noses. When available, similar masks or nylon sheets were distributed to villagers as well, and some units even had glasses to cover their eyes. The masks were normally "padded with absorbent cotton" and lined with "crushed charcoal," but when the supply of masks was low the members regularly followed the instructions given to villagers: they would urinate on a nylon sheet or a towel and cover their head with it. Sometimes a family covered individual trees with sheets and towels and placed lids on wells to prevent the contamination of water supplies.[34] Captured documents reveal plans for developing rudimentary operations to decontaminate soil and water exposed to chemical agents.[35] The NLF took seriously the potential threat to soldiers and civilians of exposure to these chemicals. While they may have been uncertain at first about the nature of the spray and its immediate impact, NLF leaders provided a variety of means of dealing with exposure, ranging from protective gear to various folk remedies for inducing vomiting if herbicides were consumed directly.[36]

While the NLF was seemingly effective in getting word out about possible precautions, the prescribed methods of dealing with chemical exposure were based largely on experience and anecdotes rather than on scientific data.[37] One NLF memo from 1962 encouraged local cadre to experiment with various remedies among the local population to determine which were most effective.[38] The RAND interviews not surprisingly reveal major differences in how various villages reacted to exposed crops, water, and animals, largely, it would seem, because of confusion in the NLF's propaganda regarding the potential harm of the chemicals. In some areas rice was eaten soon after the attacks. One subject in Binh Dinh province noted that the rice "tasted funny" but elicited no harmful side effects.[39] Another, from Ca

Mau, claimed that while no one in the area became sick from the chemicals, the villagers and the NLF took far greater precautions after attacks. Local villagers would seek out alternative water supplies, and chickens that were directly sprayed or that drank exposed water would be slaughtered immediately as a precautionary measure. According to two separate reports, at least one village in Ca Mau province regularly destroyed and replanted rice crops and manioc after attacks.[40] A captured member of the NLF main force operating in Ba Ria province revealed that cadres in that area instructed members and locals that food and water exposed to the spray could be fatal and should be thrown away. Rice fields in Ba Ria were left to recover for one season, and some orchards were left unattended for a year.[41]

The various ways in which the NLF prepared local populations for the attacks reveal their understandable confusion about the nature of the chemicals and their possible effects. According to several interviewees, cadres claimed that if locals did not cover their mouths and noses they would have difficulty breathing and might even die, a claim villagers became skeptical of as the war went on. As one subject put it, "People worried about the chemicals until they saw no one died."[42] Several captured documents refer to the people being poisoned by the spray, and the RAND interviews describe some anecdotal evidence of minor irritations after exposure, but there is no evidence of the kind of radical health effects claimed by much of the early NLF propaganda.

One of the most immediate goals of the RAND interviews, and one of the major projects of RAND more broadly, was to evaluate the effects of chemical missions on the NLF's activity, exploring in particular the military effectiveness of defoliation missions in altering troop movements and of the crop destruction programs in destroying local food supplies. The interviews offer competing visions on this point. Many local farmers and lower-level NLF members argued that the missions had little effect on the NLF. "There are a lot of trees in the jungle," one respondent replied. "The Front simply found new bases." Others described minor disruptions, such as forcing units to move at night instead of during the daytime because of greater exposure from defoliation.[43] Several interviewees noted that if the food supply of a village was disrupted, the NLF would simply move on to the next village in the province. One local farmer reported that occasionally the NLF would even donate rice to local villages when food supplies were threatened. In some instances the NLF also lowered or temporarily ceased collecting local taxes on rice.[44] The overall picture offered by lower-level operatives is that herbicide missions only intermittently strained access to food supplies and to tax revenues. Local families and villages, in these accounts, suffered far more from the effects of the spray than the NLF did. As one "economic cadre" put it, the chemicals were not effective in destroying the food supply or in forcing relocation of units. They were, he claimed, an effective source of NLF propaganda because of their drastic impact on civilians.[45]

Higher-ranking cadres and NLF officials offered a slightly different view, describing the threat posed by crop destruction as a potentially serious concern of the NLF. One "education cadre" revealed extensive knowledge about the chemical operations and reported that high-ranking party officials believed that if the United States "used chemicals on a wide scale, that is to say, if they destroyed a large part of the crops in South Vietnam, this would be an immense danger to the Front's achievement of victory in this war." The "American imperialists wouldn't do this," they believed, because it would undermine their political objectives in the region.[46] Still, captured documents and intelligence reports do demonstrate that crop destruction was having an impact on food supplies, especially following the escalation of missions after 1965. Most often, descriptions of food shortages involved production and distribution, but in several instances existing problems in these areas were exacerbated by crop destruction. A report from August 1967, for instance, notes that NLF units operating in the central highlands were experiencing food shortages for a variety of reasons, including "chemical defoliation." The report notes that some additional rice was purchased from the local population but that if defoliation missions continued to destroy crops, agricultural production would continue to have problems supplying adequate food levels and manpower, and that would likely require raising additional taxes from local villages. Another set of captured documents from 1969 describes food shortages among units in Phu Yen province owing to "natural disasters, *enemy sabotage,* and ineffective production plants." A local commander in Phu Yen wrote to Hanoi shortly after that memo was written, "I have heard that our crops were mostly destroyed by recent defoliation activity. This made me anxious because our unit is now confronted with many difficulties in the economic field."[47] Even if the difficulties posed by crop destruction were tertiary, however, such evidence supports claims made by early advocates of crop destruction that the program would, if nothing else, force the NLF to divert resources to agricultural acquisition, production, and distribution.

At best, the evidence in the reports and the interviews suggests that the military effects of herbicide programs were mixed, a view confirmed by other recent studies. In his recent social history of the southern Vietnamese revolution, David Hunt made use of other RAND interviews, concluding that the immediate negative health impact of crop destruction was minimal, but that villagers bore a disproportionate share of the burden when the crops were destroyed.[48] In his massive, detailed study of the Vietnamese revolution in My Tho province, David Elliot found similar results, the minor military disruptions limited by larger political concerns. Writing of cadres operating in "minibases" in and around My Tho, Elliot notes that ARVN forces, "despite heavy defoliation and extensive clearing, simply could not level the entire village."[49] Even in cases in which the NLF was hampered in part by chemical crop destruction, military utility cannot be considered separately from the political consequences of those programs. If the NLF was diverting not only its manpower from fighting to farming but also food from civilians to soldiers, the problem was both military and political for the NLF and for U.S./RVN forces. As was true of so many aspects of U.S. strategy during the war, military tactics were incompatible with and often antithetical to political and ideological goals. Just as search-and-destroy missions undermined the battle for hearts and minds, so the spray helped destroy the socioeconomic base of village life throughout southern Vietnam in the early to mid-1960s. For the chemical war to have succeeded, the devastation would have had to be far greater even than it was in 1966 or in 1970. As in the case of the air war and the village war, drastic increases in levels of both defoliation and crop destruction were required to bring about the desired effect, arguments that General Westmoreland and others made repeatedly in 1967–68. What they failed to see, however, is the impact such an escalation of herbicidal warfare would have had on Vietnamese civilians.

Regardless of the specific message, the NLF made heavy use of the spray as a propaganda tool, and, at least in the early stages of Operation Ranch Hand, villagers needed little convincing about whom to blame. Some changed their view as the war dragged on. Early on, unsurprisingly, most locals blamed the Americans or the government or, occasionally, the nationalists, the two latter terms both meaning the RVN government. They realized that only those forces had the aircraft capacity to undertake such missions and that the destruction of crops was aimed at limiting the resources of the NLF. Any uncertainty about this conclusion, however, was likely aided by extensive NLF propaganda. The NLF regularly proselytized in villages after attacks, explaining the motivations for the chemical war and encouraging villagers to petition the RVN to stop the attacks, a method rarely, if ever, pursued except by individuals seeking indemnification claims. Failure to formally petition the government did not mean that villagers did not blame the RVN and its

American sponsors. Both civilian farmers and cadre explained that the spray increased local "hatred of the Americans."[50] In some areas the NLF used the spray to explain the link between the United States and the RVN. As for what this link was, the explanations ranged from the understandable ("to destroy the VC economy") to the arguable ("to make the people hungry and miserable") to the fantastic ("The Americans used the chemicals so that [they] could send their own people here to live").[51] The NLF also used the spray to argue that the U.S./RVN forces were desperate, that the Americans had turned the spray against the people because the NLF "regularly defeats ARVN on the battlefield."[52] A grammar school text captured in a "VC-controlled area" in 1966 included a nursery rhyme with the lines, "We children hate Americans who are cruel. They scatter poison to destroy our paddies and vegetables."[53]

The biggest potential threat to the NLF posed by the chemical war came not when its food supply began to erode, but when the basis for its propaganda about the program began to wane. Over time, villagers in several provinces began to be aware of two things: that the spray was not as dangerous (at least in the short term) as the NLF claimed, and that the presence of the NLF in their area was exposing local villages to repeated chemical attacks, undermining their way of life, their food supply, and their economic lifeblood. Just as the presence of the NLF in an area would result in patrols by U.S./ARVN troops through local villages and increase the likelihood of air attacks and civilian casualties, that presence could also draw the spray, increasing the likelihood of crop destruction. As the chemical war escalated after 1965, local villagers became more willing to blame and occasionally speak out against the NLF.[54]

As a civilian rallier operating near My Tho put it, "When the chemicals were first sprayed, the people were all excited and frightened. Thus, when the Front called on them to protest to the GVN against the sprayings, they agreed and responded enthusiastically." But later on, the same figure notes, "as they saw that nothing happen [sic] and nobody was going to die, they stopped thinking about the chemicals."[55] The education cadre familiar with the program concurred: "Before the aircraft sprayed chemicals, the people had heard a lot from the VC about how the Americans sprayed chemicals to kill the people and destroy the crops. After the sprayings the people found that only the crops were destroyed and that the people weren't affected. They realized that the VC's propaganda was only partly true; that is, the chemicals only destroyed the crops and didn't kill the people."[56]

The explanation offered by local NLF leaders was not that they had intentionally misled the local population about the possible health effects of the chemical agents but that their studies had led them to believe the spray could have potentially deadly effects. "The documents which we studied said the chemicals could be deadly for the people," explained the education cadre operating around Long An and Hau Nghia provinces. "In our studies, we learned that the chemicals were very dangerous to our health and that they could kill people. But actually I didn't find them very harmful to men and animals."[57] The NLF performed extensive analyses of the chemicals, regularly sending in "decontamination teams" and "observation cells" organized by People's Chemical Teams shortly after herbicide missions.[58] As early as 1962 one NLF directive correctly identified the two primary components of the herbicides in question as 2,4-D and 2,4,5-T but suggested that 2,4,5-T was both less effective and less poisonous than 2,4-D. The report does not specify how the team distinguished between the two, but the fact is the opposite is true: the production of 2,4,5-T is the primary means of generating the dioxin found in Agents Pink, Green, and Purple, which were in use at the time of the report. Captured documents later revealed that the NLF had taken note of the introduction of Tordon, the commercial name of Agent White, a combination of 2,4-D and picloram, but remained confused about the exact nature of the agent. Initially, at least, the NLF was unaware that White contained 2,4-D, its analyses showing it to be "much stronger than 2,4-D or 2,4,5-T." The analysis also suggested that Agent White, like Agent Blue, contained arsenic, which it did not. The suggested response to this new agent was to increase precautionary measures and not to eat or drink food and water exposed to Tordon.[59]

The teams going into villages were finding that the immediate effects of the chemicals were not always in line with the predictions laid out in NLF directives. After a chemical mission in Long Dinh province in 1963, the NLF brought in a medical team to study the impact: "The doctors said that the chemicals only destroy vegetation and that they have no effect on people's health. But they also said that the Americans have other types of chemicals that can kill human beings."[60] Initially, the cadres "lost face when the chemicals didn't turn out to be as deadly as we had told the people," but explaining the existence and use of multiple types of chemicals was seen as a way "to protect the prestige of the front." According to the education cadre, most of the people believed the explanation, but the fear motivating much of the earlier propaganda seems to have been partially undermined.[61]

Perhaps the most intriguing set of answers from the RAND interviews was to the question, "In your opinion, which is more dangerous, the chemical spraying or bombings?" Given the level of bombing that many of these areas experienced during this period, at the height of Operation Rolling Thunder, this is not an insignificant comparison.[62] Despite the claims made in most of the interviews that people had come to believe that the sprays were less harmful than they were originally told, the majority of the interviewees said that the spray was more dangerous than bombings. In this sense, the larger uncertainties about the effects of the spray in many ways mitigated any erosion of support for the NLF.

Regardless of whether the interviewees found one more dangerous than the other, they articulated two important considerations: that the spray could have long-term implications for the exposed areas and that the use of chemicals, particularly in crop destruction missions, was as much an economic weapon as a military one, a weapon that affected civilians disproportionately. An interviewee from Ba Ria province found the spray "much more dangerous than bombs and bullets. You can avoid bombs and bullets by hiding in underground trenches, but you cannot protect yourself against the spray, especially when the operation is carried out at night. Furthermore, the spray destroys crops and causes much damage. Bombs and bullets can't destroy crops to that extent."[63] A member of the Binh Dinh local forces responded, "We would rather be killed by bombs than be poisoned by chemicals or starved to death." In his opinion, "spraying is like a bomb—a bomb with a delayed fuse."[64] A sergeant from the Binh Dinh forces who had joined the Viet Minh in 1954 stated, "The spray is an economic weapon, a slow killer."[65]

As the war dragged on, villagers became more concerned about repeated disruptions to their food supply and their increasingly precarious position, caught as they were between the NLF and U.S./ARVN forces in all aspects of the war. The people in a village in Binh Dinh province "blamed the VC for the spraying because they had lived with the GVN before and they knew how peaceful it used to be. When they found out that the VC only lied (about the impact of the spray), they became resentful." When the people complained to the cadres, "The cadres waited until the women—because only the women dared to say things—stopped complaining, and then they'd come and softly explain that they'd come to liberate the people. . . . The women told the cadres to fight hard and get it over with quickly so that people could live and work in peace."[66] "I blame the Americans," noted another civilian farmer. "They spray the VC area and it is understandable; but why do they also spray areas where people are living?"[67] A member of the NLF local forces in Binh Dinh province summed up the villagers' overall reaction to the spray and to the war in general when he replied to the question asked of every interviewee, "Whom did [the villagers] blame?": "They blamed both sides. They blamed the VC for their own presence, which brought damage to the crops. They also blamed the GVN for not taking people into the consideration before spraying. They, the innocent villagers, were caught between and were hurt."[68] He concluded, "I think the bombs and shells hurt the VC forces most. Most of the victims of the chemical spraying were innocent people. Myself, I'm scared of them all."[69]

For both the NLF and the local population from which it drew support and resources, the chemical war had an impact on everyday life. As Bui Lan puts it, "While Washington was worried about the global disdain for chemical warfare, its enemy in the rainforests was busy trying to identify the type of chemicals being sprayed, and to educate their troops and local inhabitants with precautionary measures against what they called poisonous substances."[70] The NLF sought scientific answers about the nature of the spray but was more often than not forced to rely on experiential and anecdotal reports from its cadres and local villagers. The more they learned about the spray, it seems, the less they knew, and the less they could convince locals about its potential dangers. Even in the absence of concrete data, however, and even when villagers and soldiers learned the spray was less harmful than claimed, several sources indicated that they were still fearful of its potential effects.

The seemingly innate fear of the effects of chemicals that were destroying trees and crops is a reminder that the military and political impacts of the chemical war cannot be separated. Just as the disruptions to the daily life of the NLF confirmed predictions of military advocates of crop destruction, so the fears about the spray confirmed those made by their critics in the DOS. The military's persistence in dismissing these concerns before, during, and after the years of herbicidal warfare made the uphill battle that faced the United States in winning over villagers even steeper.

Hearts, Minds, and Herbicides

From the inception of the chemical war, most U.S. officials, irrespective of their thoughts about defoliation and crop destruction, realized it would have political consequences. Because of the requirements State built into the program at the outset, U.S. and ARVN forces attempted to integrate political efforts with the use of herbicides. From pacification to payments to propaganda, the RVN and its American sponsors tried to make the chemical war more palatable to Vietnamese civilians between 1964 and 1970. But policy often failed to translate into practice. When examined closely, each of these areas reveals the fundamental weaknesses of the programs themselves as well as the larger contradictions of the Vietnam War. Sheltering civilians from the spray, allegedly educating them about the benefits of the chemical war, and recompensing them for damage to their crops and property proved to be no more effective than other, similar military programs.

During the debates in 1962 over the authorization of crop destruction, advocates and critics debated how herbicides should be considered relative to larger pacification efforts, especially the strategic hamlet program. Military officials believed that defoliation and crop destruction would support the relocation of the population, while the DOS believed such programs would be effective only if the NLF was already effectively isolated from the population. The strategic hamlet program itself, however, was based on the same faulty logic that held that RVN officials and their U.S. advisers could determine who was and was not sympathetic to the revolution.[71] In practice, the relationship between pacification and herbicides was never truly defined, resulting in an ad hoc approach that once again caught Vietnamese civilians in a tangled web.

One such example of the muddled relationship between pacification and defoliation is the case in 1964 of the forced relocation of several hundred villagers living just south of the Long Dinh district of Dinh Tuong province. For several years the NLF had fought fiercely to gain control of this area of the Mekong Delta region (called My Tho province until 1954 and renamed Tien Giang after the war). Located about forty miles southwest of Saigon and ten miles northwest of My Tho, the district was not far from the infamous battle of Ap Bac in January 1963, in which the NLF proved its ability to confront ARVN directly on the battlefield.[72] On April 7, 1964, U.S. advisers from the Chemical Corps were called in to evaluate an "urgently requested defoliation target": the canal that bordered Long Dinh and Phu My districts. The area in question, the advisers reported to the observation team, had become "fortified by the VC as a main arsenal and ordnance supply point" and was largely concealed by the "dense, green shrubbery" along the canal, which was lined with fruit trees, rice fields, and garden crops. Many local families, the request noted, had already fled the area because of operations and "appeals to leave the area," but more than "400 VC families" remained. Lt. Col. Peter Olenchuk, the chief of the Chemical Section, recommended that "the defoliation of the canal, as a matter of tactical urgency, begin without delay."[73]

The purpose of the mission was "to enable friendly operations to proceed along the canal with greater visibility and therefore more security." That security was not simply a question of the enhanced visibility that would come from defoliation; it also included relocating hundreds of villagers. Having declared the area tactically urgent and a major security risk for ARVN troops, U.S. advisers decided to proceed with the forced relocation of the remaining population, about eight hundred people, in the vicinity of the canal. These "VC families" would be moved to the same New Life hamlet where the previous evacuees from the area were placed, in nearby Long Dinh district.[74] Prior to the removal the families in the area received a variety of propaganda pieces about the proposed defoliation and the reasons they were being forced from their homes. Only two days after Olenchuk's recommendation the RVN province chief sent word to the families: "For security reasons, request all people living within 4,000 meters along both sides of the Commercial Canal, from Ba Beo, Hung Thanh My Village, Long Dinh District, to the canal Cong Tuong, Phy My Village, Ben Tranh District, to hastily move away from this area to take shelter in the safe areas of Long Dinh and Ben Tranh District. Ten (10) days after this date, any family that is still located in this area, their property and lives will not be guaranteed. The local District Chief has been ordered to help any relocating family, if asked."[75]

On April 15 another round of notifications came, along with another note from the province chief explaining the impending evacuation. The claim in this letter, that "taking advantage at the dense vegetation close by the hamlets and villages of the populace, the VC come back at night time to kill the populace and to implement their illegal tricks," must have struck the locals as strange, given that they were considered by all accounts to be supportive of the NLF. "Dear Compatriots," the chief's letter concluded, in order "you will bravely move out from that area and return to stay in the Hamlet of New Life."[76]

Within a month the defoliation request had been approved and the district had been declared a free-fire zone, which was not an unusual procedure when defoliation targets were located in known populated areas and when it was obvious that civilian crops would be destroyed. Anyone left in the village was considered to be a member of the NLF and susceptible to enemy fire or defoliation. The evacuation did not go exactly as planned. Beginning on May 12 about one hundred families were forced out of their homes and into the hamlets. As they left their village, ARVN forces, accompanied by U.S. advisers, burned the houses along the canal, often in sight of the villagers who had just abandoned them. Reports from that day describe "extreme hostility" from the residents, some of whom were taken into ARVN custody as suspected NLF agents. Others included a woman and the baby to whom she had given birth just the day before. The villagers wondered if they were being taken to prisons and if they would be given food. One noted that "to a Vietnamese, no other things are more barbarous than to burn a house of other [sic] however small it is even a hut. Such an act is very horrible."[77] Many of the villagers blamed the Americans who were present (figs. 4a and 4b). As the provincial representative to the U.S. embassy observed in his formal report to the U.S. mission in Saigon:

> The question here is not whether or not it was right to move the people. The question is the manner in which it was accomplished. Women and children were literally snatched up during the morning and held until about two PM when they were put on *American* helicopters, flown by *Americans,* and taken to a field and dropped off. There they were met by young USIS [United States Information Service] men who did their best to tell them they would not be hurt, but to no avail. They were hungry and thirsty. They left behind practically all their belongings. . . . We missed the boat for it turned out to be a black mark against the Vietnamese and American governments.[78]

On the basis of reports like this one, several later accounts tried to explain how the evacuation was handled, particularly the accusations of the villagers that they were removed so that the area could be defoliated. The evaluation Olenchuk sent to the operations unit of MACV stressed that while the process might have been handled better, the operation was driven by military and security needs. Defoliation, he claimed, was a minor concern in the process. The evacuation, he argued, "was in no way precipitated by the planned defoliation of the area, which was a subsequent supporting action." Yet the materials Olenchuk included in his final assessment, including the initial reconnaissance reports and propaganda materials described above, belie that claim. Those documents show that the mission was designed to provide "greater visibility and therefore more security" and that the villagers were told they were being evacuated because they would "obstruct" the defoliation of the canal.[79] As was often the case, herbicide operations and pacification were seen by the military as being closely interrelated and mutually reinforcing: defoliation provided increased visibility, visibility provided security, and security would lead to greater control of the area. And yet when those assumptions were challenged by people on the ground, the military consistently denied in public what was considered a central operating assumption behind the scenes: that herbicidal warfare was as much a political and psychological weapon as a tactical one. This fact becomes even clearer in the U.S./RVN propaganda about the program.

Figure 4a. Long Dinh resettlement hamlet, 1964. (Courtesy National Archives, College Park, Maryland)

Figure 4b. Resettled refugees at Long Dinh hamlet, 1964. (Courtesy National Archives, College Park, Maryland)

In the early phases of the war, the Pentagon and Department of State disagreed over the relationship between herbicide programs and other pacification efforts, resulting in a muddled policy that enacted its heaviest toll on rural Vietnamese villagers. In the spring of 1964 ARVN troops and their U.S. advisers forcibly relocated more than one hundred families living along the major canal in My Tho province into a "New Life" resettlement hamlet in Long Dinh. After many of the families' houses were destroyed, the areas adjacent to the canal were defoliated to provide greater security to ARVN patrols. The operations failed to improve security in the region, and the relocation of families and the destruction of their houses increased hostility toward the Americans and the Saigon government.

As part of the lengthy approval process required for herbicide missions, each formal request was required to include one section that addressed the possible effects on the morale of NLF/PAVN in the area and another that detailed the "civic" or "psywar" actions recommended in conjunction with the application of herbicides. Even when civilian population figures were negligent, spray requests often came with a recommendation for standard psywar/civil affairs programs to deal with damage to civilian crops and property. While the NLF spent much time explaining to villagers why the chemicals were harmful to them, the United States and the South Vietnamese regime claimed to spend at least as much time explaining why there was nothing to fear from the spray.

Early in the war U.S. and ARVN forces entered targeted villages fairly regularly to offer educational programing about the herbicides. They explained that the chemicals were being sprayed to deny food and shelter to the NLF and that civilian property and crops would not be harmed, a claim most civilians rightfully became skeptical of as the war went on. The village education teams went beyond making claims about the safety of the chemicals: ARVN soldiers applied Agents Blue and Orange to their skin in small amounts to show that the chemicals would not harm humans.[80]

Among the tools used to spread the word about the benefits of herbicides were cartoon leaflets featuring "Brother Nam," who learns that, contrary to Viet Cong propaganda, the chemicals are harmless (fig. 5). A typical leaflet, "Brother Nam has questions about chemical defoliants," tells readers that Nam and his family had been regular victims of Viet Cong ambushes, on the road as well as on water. Against images featuring buffoonish-faced NLF agents strangling and stabbing innocent victims, the caption reads, "Besides stealing property, the VC killed Brother Nam's brother on the spot." Clean-cut, friendly looking government agents in civilian attire spray the brush with backpack-mounted tanks labeled "defoliant," responding to Nam's questions about the safety of the chemicals: "Nice to see you, Brother Nam. The only effect of defoliant is to kill trees and force leaves to wither, and normally does not cause harm to people, livestock, land, or the drinking water of our compatriots.

"Do you see how well I look, Brother Nam? Each day this is my responsibility and I normally inhale defoliant chemicals into my lungs, but do I look ill to you?" Nam is not convinced: "So you say that, but what if my crops are damaged?" "If you are unfortunate enough to have your produce damaged by the effect of chemicals," the agent replies, "then government will compensate you." "From now on," the cartoon concludes, "Brother Nam like

Figure 5. "Brother Nam" defoliation pamphlet. (Courtesy National Archives, College Park, Maryland)

While the NLF used fears about the herbicide program for purposes of propaganda and recruitment, U.S. and ARVN forces attempted to use tools such as this pamphlet to assuage fears about the effects of the spray. In this cartoon, "Brother Nam" learns that, contrary to "Viet Cong Propaganda," the chemicals are safe for humans and that the government will compensate his family for any damage to their crops. Despite such efforts, villagers often remained afraid of the effects of herbicides and were continually frustrated by the destruction of their crops and by the corrupt and inconsistent indemnification program.

everyone else is calm and does not listen to the distorted propaganda of the Viet Cong." Nam himself reassures readers, "Now I understand thoroughly and am no longer concerned about defoliant chemicals." In another variation on the cartoon, Nam concludes by saying, "Now I fully understand and have no questions about defoliants. I now resolve never to listen to Viet Cong Propaganda."[81]

Although the individual issues of the "Brother Nam" series varied slightly, they are nearly identical in their presentation of the safety information and especially of the compensation claims for lost crops readily available to villagers. A standard indemnification letter, distributed in An Xuyen province in 1964, reads as follows:

> *Dear Fellow,*
>
> *In civilized countries the defoliation chemical is generally used to clear wooded areas so that the culture may develop easily. In our country, the government used the above chemical to turn wooded areas into open fields where mosquitoes are eradicated and communication by road and waterway will be improved, and your rural life is also more insured. The defoliation chemical only kills vegetable [sic] right in the defoliation area. It does not cause damage to adjacent areas. The defoliation chemical is dropped on [ARVN troops], but the health of those soldiers is still normal. Otherwise, at any defoliation area, the peasants can cultivate effectively.*
>
> *VC are very afraid of defoliation operation because they will lose good locations where they can hide and terrorize you people easily.*
>
> *Dear Fellow, welcome defoliation plan, and don't listen to subversive propaganda.*[82]

The experience of most Vietnamese farmers was that Brother Nam's claims about indemnification were greatly exaggerated.

Vietnamese civilians were free to file damage claims against both the RVN and the United States. Claims against the United States could also be made by other entities, including private firms and foreign governments, under the Foreign Claims Act. Through the end of 1969 the U.S. Foreign Claims Commission in the RVN received forty-nine claims, totaling US$66,816, related to herbicide damage and, with the exception of a controversial Cambodian rubber plantation incident, denied every one.[83] The denials were based on findings that the damage in question was either not caused by defoliants or was "combat-related" and thus not covered under the Foreign Claims Act.[84] In response to two incidents that occurred in 1968,

MACV issued a stern memorandum ordering American advisers to avoid encouraging Vietnamese civilians to file damage claims (herbicide-related or otherwise) in instances that resulted from combat. Promises made by U.S. personnel that conflicted with the language of the Foreign Claims Act, the memo noted, "were unauthorized and are of no force or effect."[85] Given the military's justifications for the herbicide program, however, it was relatively easy for the United States to argue that nearly every defoliation incident was ultimately related to combat. In the summer of 1965, for instance, sixteen farmers in Tay Ninh province filed claims against the United States for damage to their crops, most of which were rubber plants in close proximity to the perimeter of a major airfield in the province. In its verdict, issued on May 8, 1967, nearly two years later, the commission expressed its deep regrets about the delay and noted that it was unable to process the claims because "defoliation complaints are of military character" and "beyond [our] jurisdiction." According to the principle given in this ruling, not one necessarily followed in every instance, all claims related to herbicide damage were combat-related and hence not subject to indemnification payments.

Formal claims against the U.S. government were the exception, not the rule. The vast majority were made through the RVN, and because most indemnification claims were handled at the district level, records on actual payments are spotty, above all before 1967. Only in that year did the indemnification funds appear as a distinct category in the RVN defense ministry budget. According to the Herbicide Policy Review of 1968, coordinated by the U.S. embassy in Saigon, the RVN paid RVN$35 million to 5,848 claimants during the calendar year; RVN$14 million of this amount was paid out in December alone as part of a "clean-up" of long-standing unsettled claims. The amount for herbicide damage claims represents less than 10 percent of the RVN$381,004,000 for other war-related damage paid out the same year.[86]

After 1967 the RVN put in place new accounting procedures, and CORDS began to pay greater attention to the claims, resulting in clearer documentation. Under the new program guidelines, claims for up to RVN$100,000 could be processed and awarded at the district and province levels. Claims in excess of that amount, up to the RVN$500,000 maximum, had to be forwarded to the Ministry of Defense for approval. Between 1967 and 1970 the RVN paid out nearly RVN$400 million in herbicide damage claims, suggesting that the new procedures resulted in increased claims and payments.[87] In light of the regular references to the high rate of denial, however, it is difficult to know whether this amount reflects a significant number of claims. If, as reports suggest, most claims were handled on the district level, then the average claim would have been less than RVN$100,000. If the average claim was RVN$20,000, the amount dispersed during this period, the heaviest of Ranch Hand activity, could represent as many as twenty thousand claimants. If the charges of corruption among RVN officials are to be believed, one wonders how much of the total amount actually made it to the claimants. The RAND studies of 1967 argued that "the incidence of GVN aid to people affected by crop spraying was very low" and often went to wealthy, well-connected landowners. In contrast, the NLF aid to affected villagers "commonly attested to," according to the RAND reports, normally "took the form of arranging for redistribution of rice from those with a surplus to those in need."[88]

In 1967 a CORDS study documented significant levels of corruption within the process, including one province chief who "announced to the people that whoever wanted a quick compensation for his crop damage should hire someone to write it for him," since many of the villagers in that hamlet could not write. A typical fee for such a service, the report noted, was relatively inexpensive, around RVN$300. If the claim was awarded, however, the hamlet chief would then exact an additional RVN$300 "tip" for each major plot of land for which the farmer received compensation. Commonly, the claims were not awarded or at least did not get to the farmers themselves. In 1966 more than one hundred families in that same hamlet filed indemnification claims through this process, none of which were resolved in the villagers' favor.[89] An internal MACV memo from late 1967, prepared as part of the run-up to a total review of the herbicide program in 1968, confirmed this view, noting, "We also continue to receive unsubstantiated reports that the indemnification program is fair game for corrupt officials at local and higher levels of the GVN."[90] The final report of the Herbicide Policy Review the following year reinforced these findings but cloaked the charges of corruption in bureaucratic language, describing the program as inefficient, inconsistent, "insufficiently responsive to the claimant," and, above all, "subject to administrative irregularities."[91]

The overwhelming picture of the compensation program presented in the limited documentary record is one of dismal failure, one that failed to offset the political fallout from herbicide programs. Even claims that were reportedly paid quickly and fully demonstrate that such was the exception rather than the rule. The accidental defoliation of Cha La, in An Xuyen province, is a case in point. On April 22, 1964, Ranch Hand missions targeting waterways controlled by the NLF accidentally sprayed fruit trees adjacent to the strategic hamlet in Cha La. A wire story picked up by the *Washington Post* called the damage "one of the more tragic mistakes of the war."[92] MACV investigations challenged many of the facts in the article but concluded that the mission had indeed resulted in some damage near the hamlet. Regardless of the veracity of the claims, the military realized it needed to respond to them and urged RVN officials to issue prompt payments for all indemnification requests, even those considered suspicious or fraudulent. As Buckingham observes, the fifty-seven residents, who received payment within two months of the incident, "were reportedly highly satisfied with their government's handling of the situation," and MACV noted that "*in contrast with their past reluctance to make such payments,* South Vietnamese officials had paid for defoliation damage at Cha La promptly and completely."[93]

The failures of the indemnification process, from corrupt officials to fraudulent claims, obscure the underlying meaning of the program, namely, that civilian crops were regularly, consciously targeted by the United States. While the standard psyops program was theoretically carried out even in areas sympathetic to the NLF, the reports included in such instances reveal the underlying contradiction of nearly every crop destruction mission: the impossibility of distinguishing civilian crops and crops "controlled" by the NLF. Despite assurances that defoliation and crop destruction missions were targeted, accurate, and generally confined to nonpopulated areas, reports from areas with overtly pro-NLF populations reveal their inherent lack of precision.

As part of the effort to defoliate canals and other waterways around An Xuyen province in 1964, the psywar annex report comments that "numerous homes and crops" were located in the target area. "Spraying defoliation chemicals on these people," the report adds, "would probably increase resentment against GVN." But since "all eligible male members of these families were reported to be active Viet Cong," the request recommended proceeding with the attack. To add to the effort, however, the psyops program was amended to inform the locals that "if Viet Cong terrorists continue to murder government troops who are trying to protect the people, the government will be forced to defoliate all canals," adding that unless they "prevail upon terrorists to stop ambushes and mining incidents, their crops may be accidentally destroyed."[94]

That indemnification claims were included as a central component of nearly every herbicide mission implies that the military was well aware that civilian crops and property were regularly exposed to herbicides. Yet American evaluations of the program continually stressed that the impact on civilian attitudes toward the United States and the RVN was minimal, downplaying frequent incidents of allegedly accidental crop destruction of so-called civilian crops. In some cases, reports argued that the program was, albeit unintentionally, helping drive village support away from the NLF. Official military reports on Ranch Hand prior to 1967 often employed twisted logic to demonstrate how the herbicide program was an asset in the battle for hearts and minds.

A report from the Combined Intelligence Center, Vietnam (CICV), though supplemented with captured documents, was largely based on the same sources RAND used for its report on NLF morale in 1966. The CICV report offers little evidence of the military effectiveness of crop destruction, save a few reports of intermittent food shortages in isolated NLF units and the temporary reassignment of some personnel for crop production. In lieu of documenting sustained food shortages or increased disruptions to unit activities, the report rests on the specious claims that the military effectiveness of the program can be judged by the increased ground fire directed at Ranch Hand aircraft and by the increase in propaganda about the program: "The recent intensity of NVA [North Vietnamese Army] and VC propaganda directed at herbicide operations, although principally designed to influence world opinion, probably also indicated the general effectiveness of crop destruction operations."[95] Historians of the program have echoed this line of reasoning. In his *Herbicidal Warfare*, the former Ranch Hand pilot Paul Cecil argues that even in 1963 "an indication of the effectiveness of earlier herbicide missions was the increasingly strident tone of communist antiherbicide propaganda."[96] There appears to be some truth to the claims of sporadic food shortages and strains on personnel, but, given the certainty the military continued to express in defending the tactical effectiveness of herbicide operations, it is striking how little direct evidence is offered in the report to support such claims. This absence of evidence leads to three conclusions: first, the U.S. military was not particularly interested in gathering evidence to support the tactical and strategic value of herbicidal warfare; second, in the absence of such evidence, there was never any real basis for the claim that the military utility of herbicides outweighed their potential political costs; and, third and perhaps most glaringly, there appears to be little, if any, concrete evidence that herbicidal warfare was in fact militarily effective.

The startling lack of evidence to support assertions of the military effectiveness of herbicides paled in comparison to the logic employed in describing the effects of the program on the morale of the NLF and Vietnamese civilians. Citing a claim that had by then become well known among U.S. military leaders, the CICV report stated that NLF soldiers who had directly encountered herbicide attacks expressed little fear of the spray, believing their countermeasures to be effective. Nevertheless, NLF soldiers who had not witnessed a Ranch Hand mission firsthand remained fearful. From the perspective of MACV, the fears were largely a function of the NLF propaganda that stressed the physical danger of the chemicals. These "misconceptions," according to CICV, stoked the fears of NLF soldiers. They also played into the psychology of the village war.[97]

The CICV report was adamant that despite the supposedly severe losses to the NLF caused by crop destruction, civilians in targeted areas suffered far more, a remarkable claim given the paucity of evidence of broad military effectiveness. The basis for this contention seemed to be that, unlike the NLF, which could simply move on to the next village, local farmers and residents had limited alternative supplies of food. One of the unmistakable consequences of crop destruction operations was that they increased the likelihood of civilian relocations to areas controlled by the RVN. As one informant described it, crop destruction, while not the sole factor in relocation, "tipped the scales": "The truth is, if these people moved to the GVN-controlled areas, it was not only because their crops had been sprayed with chemicals; because since their areas had been hit by bombs and mortars, they had already had the intention to leave; and they would probably have done so, had it not been for the fact that they could not decide to part with their crops. Now that their crops were destroyed by chemicals, they no longer had any reason to be undecided."[98] The question of whom to blame for such hardships remained difficult to answer: "From their sadness and futility, bitterness and hatred often spring. The direction which this bitterness and hatred take is by no means uniform, but is influenced to a considerable extent by misconceptions and confusion, and by preexisting loyalties or inclinations. Misconceptions concerning the effects of herbicides, apparently attributable to intensive VC propaganda, sometimes cause subsequent illnesses and misfortunes to be attributed to chemical spraying. The natural result is to enhance any belief of cruelty by the GVN and its American allies."[99] The RAND interviews do provide some indications that as the war went on resentment against the NLF for drawing the chemical attacks increased in some areas. But the attribution of civilian animosity toward the United States and the GVN, who all villagers realized were ultimately responsible for the attacks, to "misconceptions" about the potential hazards of the chemicals raining down upon their houses and farms, reinforces the fact that there was no practical way to distinguish between civilian crops and those designated as "VC-controlled." The report notes that while some confusion persisted among the civilian population about why the United States and RVN were destroying their crops, "most people in the target area do have at least a rudimentary understanding of the purpose of the spraying. Herbicide operations are in fact commonly directed at civilian crops, although the ultimate target is the VC. Therefore, it is understandable that many people fail to understand the subtleties of the program."[100] Perhaps no statement in any U.S. military document better encapsulates the Americans' inability to grasp the significance of the chemical war at the village level. For the rural Vietnamese who were historically, spiritually, and economically tied directly to their land and their crops, there were likely no "subtleties" to be found in the face of chemical crop destruction.

Interpretation aside, the effects on civilians in targeted areas were clear: they suffered enormous hardships. Rather than seeing such distress as an obstacle in the battle to win hearts and minds, however, the CICV report dismissed it: most of the people in the target areas "already support the VC."[101] Like those living in areas targeted by B-52 carpet bombings and in villages designated as free-fire zones, the Vietnamese citizens living in areas targeted by herbicides had essentially been written off by the U.S. military: their hearts and minds long since lost, they were now simply bodies, numbers to be accounted for in the American war of attrition.

Finally, the report went on to describe the effects of crop destruction on civilians in nontargeted areas. As the CICV itself noted, even if the program disrupted the NLF food supply, the cadres simply moved on to obtain crops in another location. Such increased demands on the local population, the report concludes without citing a single supporting piece of evidence, "tend to foster resentment toward the VC. On the other hand, effective VC propaganda concerning herbicide operations promotes fear and hatred of the U.S. and GVN. These factors probably cancel one another, but the precise effect cannot be observed."[102] At best, then, the official military line on crop destruction in the summer of 1966 remained the same as throughout the previous five years. Despite the scarce evidence and the confusing logic of the report, the recommendation of CICV was to expand the crop destruction and defoliation programs because the advantages "substantially outweigh the disadvantages."[103] The CICV report demonstrates as unmistakably as any evidence from the war that the U.S. military was simply not interested in addressing the political costs of the program. Relying instead on specious claims of military effectiveness, the chemical war dragged on.

While domestic critics in the United States increasingly challenged the program in 1966 and 1967, the central assumptions underlying it went unchallenged within the military.

No Way to Make Friends

In 1967 these assumptions were brought into sharp relief by the release of two studies by the RAND department. Just a year earlier a RAND study had focused attention on the morale of the NLF, noting that, according to information gathered primarily by interviews of defectors and prisoners, the NLF "continued to lose the sympathy of the rural population."[104] While the morale study did not emphasize the herbicide program, it did describe the fear that the increased air war by the United States was instilling in both Vietnamese soldiers and civilians. The study claimed that despite limited evidence of the overall military effectiveness of herbicide operations, fear of the chemical spray was "widespread" and "reinforced by the nature of Viet Cong propaganda, which stressed the alleged toxic nature of the spray."[105]

The larger point of the study was that villagers saw themselves as being trapped between the two sides of the war. With amplified ground forces and air attacks came a marked increase in civilian victims and further disruption of daily life. As the war escalated, the authors note, "the villager seems to have acquired the feeling that he cannot afford to choose sides.... Military activity and fear of attacks tend to disrupt farm work and reduce the villager's economic interest in continuing to work in his fields, since he is less able to earn a living form his labors."[106] Despite a lack of data supporting claims of military effectiveness and concerns about the political fallout from the chemical war, the study offered the same recommendations as the CICV report: strengthen, expand, and accelerate the herbicide programs.

Just over a year later, in 1967, the assessments from RAND began to diverge from the official military line. RAND had been active in Southeast Asia throughout the period of U.S. military involvement, producing hundreds of reports related to military operations, intelligence, and especially the attitudes and capabilities of the revolutionary forces of Vietnam. The overwhelming majority of RAND reports, however, supported existing U.S. policies and procedures. In her recent study of RAND in Southeast Asia, Mai Elliot argues that "[its] research rarely advocated drastic policy changes, and RAND followed rather than led, and was usually behind the curve rather than ahead of it." More commonly, policymakers and military leaders used

RAND reports to show that a particular issue had been given serious consideration: "At its most influential, RAND's research reinforced what policymakers were already inclined to do, encouraged them to believe that they were on the right track, and motivated them to persist in doing what they were doing or to do more of the same. The research also served as an additional arrow in their quivers to persuade the president or their colleagues that the course of action they proposed or were undertaking was correct and would produce results."[107] Part of what made the two reports on the chemical war so controversial was how sharply they broke with existing policy.

In their report Russell Betts and Frank Denton attacked head-on the myth of military effectiveness of the chemical war and placed the program squarely in the context of pacification efforts and the battle for hearts and minds. The summary offered a basic but pointed observation: "A nonhostile populace would seem desirable if the GVN ever hopes to make progress in its pacification objectives, and destroying a farmer's source of sustenance is not a way to make friends."[108] Whereas just a few years earlier RAND had been somewhat more optimistic in its findings of isolated food shortages among the NLF, it now argued that they were the exception rather than the rule. "The crop destruction program has not in any way denied food to the VC," which appeared "to have been able to transfer most of the deprivation burden to the local peasant." The report pinpointed the most basic contradiction in the crop destruction efforts: at the local level any distinction between civilian and military crops was arbitrary at best and imaginary at worst. Even when missions succeeded in attacking food supplies that were useful and convenient to NLF units, they would simply pass along the hardship to the local population.[109]

Furthermore, the report detailed, the propaganda battle was failing to pay dividends. Despite the emphasis in mission requests and reports on informational programing about Ranch Hand programs, the RAND report indicated "an almost total absence of efforts by U.S./RVN to educate people about herbicide use or to assist those who have been affected." "These peasant perceptions," the report continues, "appear to contribute to a temper of mind which is receptive to the Viet Cong propaganda."[110] Why, the authors asked, did the spray seem to "stir up so much hostility?" The answers are telling: "First, crop destruction strikes at the very heart of the peasant's existence—both his food supply and his handiwork. Second, the civilian population generally lacks knowledge and understanding about both the nature and the purpose of these operations. And finally, the hostility is due to what the peasant conceives to be a lack of GVN concern for his welfare, combined with an active and generally effective Viet Cong effort to exacerbate his already intense feelings."[111] Regardless of the NLF's role in passing along the burden, the RAND study argued that peasants failed to comprehend why the GVN was spraying their crops. In a response the authors describe as representative, one interviewee said, "Even under the French nothing so awful has ever occurred."[112] RAND found major gaps in the indemnification payments, noting that "the incidence of GVN aid to people affected by the crop spraying was very low" and that those who did receive compensatory claims were more likely to be wealthy landowners. In contrast, nearly a third of respondents reported receiving compensation and surplus rice from the NLF.[113]

Despite what U.S. and ARVN forces were telling villagers about the herbicides, the RAND data showed that around three-quarters of the interviewees believed exposure to the spray caused adverse health effects. The report stopped short of demonstrating direct health effects, claiming only that 70 to 80 percent of the subjects "indicated that people who were significantly exposed to these chemicals could expect, at least, runny noses, nausea, cramps, and diarrhea for several days." It is telling that the report does not show that the respondents themselves experienced such effects, only that the respondents indicated that others could and only if they were "significantly exposed," although the authors do not define what they mean by *significant*. Regardless, if this was indeed the perception among a high percentage of Vietnamese villagers, the same precept would hold true: if destroying a farmer's source of sustenance is not a way to make friends, neither is providing them with a steady diet of health problems. That a high percentage of civilians believed the herbicides to be harmful to humans would, at minimum, suggest that U.S. propaganda was failing to convince them otherwise. In short, RAND saw the propaganda campaign as a failure. At the village level, according to the study, the NLF was superior to the U.S./RVN forces in warning villagers about the dangers of herbicides, providing assistance for those affected by the spray, and ultimately arousing the anger of the local population and directing it at the government, if not always in direct support of the NLF.

Anthony Russo's report reinforced every major theme and finding of the first article. On the military front, his analysis offered a dismal view of crop destruction. "No significant relationship" could be found between the percentage of rice lands sprayed and daily rice rations to NLF soldiers. Yet, according to Russo, the effects on civilians were considerable: "The civilian population seems to carry very nearly the full burden of the results of the crop destruction program; it is estimated that over 500 civilians experience crop loss for every ton of rice denied the VC." As the report went on to detail, one ton of rice was equivalent to the yearly rice ration of four NLF soldiers. In order to have an impact on NLF rice rations, Russo estimated, the United States would have to "destroy large portions of the rural economy—probably fifty percent or more." Russo concluded by offering that "the returns from the program" were "insignificant at best, and the costs of the villager seem disproportionately high." While admitting shortcomings in his model, Russo said he believed the program should be discontinued immediately.[114]

Given how closely RAND had supported the U.S. mission in Vietnam to this point, the RAND studies touched off a furious debate inside the Pentagon. The reports had not simply offered seemingly overwhelming evidence of the political costs of the program, but had also called into question the central premise of MACV: that the herbicide program was militarily effective. An unlikely adversary had put the military on the defensive. Given their long-standing claims about the tactical and strategic value of the chemical war, however, it should not have been difficult for Westmoreland, McNamara, and the JCS to mount an effective defense of the program. The military response, instead, reveals just how anemic the evidence supporting their case really was.

Shortly after the release of the RAND reports, the assistant secretary of defense for systems analysis reviewed the data and issued a report to Secretary McNamara confirming that crop destruction was counterproductive and harmful to U.S./RVN goals.[115] On November 21 McNamara ordered the chairman of the JCS to initiate a review of the crop destruction program. Over the next several months staff members from the Pentagon, MACV, and RAND engaged in a lengthy, line-for-line, note-for-note debate about the effectiveness of crop destruction. Science advisers from ARPA and General Westmoreland's staff challenged Betts's and Denton's claims, particularly with regard to the degree of civilian alienation caused by the program.

The actual evidence offered by the military in support of its program, however, did little to discredit this crucial area of the studies, arguing for the effectiveness of crop destruction by repetition rather than by a thorough presentation of evidence. Westmoreland's memo to the Pacific commander (CINCPAC) in November laid out the central talking points that would be echoed by a variety of military voices in the debate. Beginning with a summary of the goals of the program ("to deny food to VC and VC sympathizers in the area, to place an added burden on his logistical system and to weaken VC strength, resolve and morale"), Westmoreland noted that these goals were being met with varying degrees of success. The major shortcoming of the program, in his estimation, was the lack of enough spray capacity to respond quickly to requests by field commanders. In other words, the primary reason crop destruction was not more effective was that there was not enough of it. The military response made no attempt to take seriously the idea that the program was damaging the United States and RVN politically among the Vietnamese. "The best measure of the utility of crop destruction," Westmoreland concluded, "is its value to the military commander."[116]

Instead of demonstrating that information and propaganda campaigns were countering the negative effects of herbicide missions, military advisers relied on problematic assumptions about the effectiveness of the efforts. David Griggs, Westmoreland's science adviser, wrote, "RAND reported that the civilian populace lacks understanding of the nature and purpose of the crop destruction operation, *whereas I am informed* that there is an extensive GVN leaflet program" explaining the program. On the matter of indemnification claims Griggs said, "The RAND authors do not seem aware of the fact that a guarantee of indemnification for crop loss to all friendly people is a mandatory *condition* for each mission."[117] Griggs and others who reviewed the program for Westmoreland and McNamara offered no new evidence to support these claims, only their belief that such conditions were followed closely in the field. Whereas RAND sought to show that the programs were problematic, Westmoreland's staff continued to assume that educational and compensatory efforts were consistent and effective. This continued pattern implies that the military was uninterested in addressing such questions. If the population was being alienated by the failure to distinguish civilian from combatant in the chemical war, what would that say about the conventional war? The answers to such questions would have cast doubt on the militarization of herbicides in Vietnam, but in addition would have given the lie to the entire American military presence in Southeast Asia: the United States was alienating the very people it was ostensibly protecting.

Russo himself found out as much when he went to Saigon in hopes of briefing the military on his findings but was allowed to meet only with Griggs. As Mai Elliot describes the encounter, "Griggs told Russo that the program worked because 'he had flown over' the VC areas up in the mountains and had seen 'how spraying was wiping out the rice.' According to Russo, Griggs 'never sat down, he just took a copy, threw it down, and said, "in World War II, we had some good operational research, but this is not good research, this is crap."'"[118] The exchange confirms that Griggs had, at best, a "fly-over" knowledge of the program and no real sense of the effects crop destruction was having on the NLF or on civilians.

The military response, from the field commanders to Westmoreland to the JCS, relied on two areas they believed RAND had overlooked. First, they argued that most of the RAND data came from respondents who had not been active militarily since 1966. The increased efforts of Ranch Hand during 1967, the military claimed, were more effective than those of previous years and included an increased emphasis on informational campaigns. While the number of missions and sorties increased during 1967, whether there was renewed emphasis on education is unclear. In its review, conducted in accordance with the JCS request, the scientific advisory group of the Pacific Command gave no indication of increased psyops efforts in 1967. It simply noted that enemy propaganda continued to be employed "to spread fear and arouse hatred." The report concluded that such efforts were unlikely to succeed because since 1961 defoliation campaigns had been conducted "*without any adverse effect whatever* on either civilian or military personnel in the affected area."[119] As late as 1967, then, the official military line was not only that the tactical utility of herbicides justified the political tradeoffs, but also that the operations were having no adverse effect on civilians.

Intelligence and interrogation reports after 1967 confirm little improvement, belying military claims. The reports, in fact, follow the model of earlier ones and of the RAND interviews. A report of June 1967 by the RVN National Interrogation Center (NIC) summarizes the effects of defoliants and other chemicals in Quang Ngai and Binh Dinh provinces, indicating that the operations "caused food shortages and demoralized both the troops and the civilian population," but not giving any evidence to support the latter claim. "Since the people were dependent upon their crops for a living," the report says, "they became destitute when these crops were destroyed, and they began to lose faith in the VC propaganda."[120] Why the authors correlated the destruction of civilian crops with declining support for NLF propaganda is unclear, and reports from other places refute that pattern. After crop destruction missions were carried out in Hau Nghia province in May 1967, for instance, a survey team sent to Mui Lon hamlet returned reports of what it termed "Anti-Americanism," in which several interviewees described their frustration with the United States. "The Americans come to Viet Nam to fight the war, and they look for battles to fight," one of them is quoted as saying. "But why must they spread defoliants? We don't see the Viet Cong die; we only see honest people suffer when their crops are damaged."[121]

Other documents confirm the pattern of political blowback against the United States. A MACV intelligence summary dated January 5, 1968, describes papers captured from NLF units operating in the central highlands indicating that perhaps as much as 30–40 percent of local crops had been destroyed by herbicide missions during 1967, but also suggesting that the units were able to recoup their losses from other villages in the region.[122] A report by the Defense Intelligence Agency (DIA) on the interrogation of a witness to a defoliation mission in Binh Duong province noted not only that the missions allowed the NLF greater mobility in partially defoliated areas at night, but also that crop destruction was fomenting anti–United States sentiments in areas where civilian crops had been destroyed; that sense was bolstered by several other reports from different provinces during the same period.[123]

The second point highlighted in the military response to the RAND report was that RAND analysts had failed to "consider the needs of the military in the field." Given the focus of the RAND report, this seems like a specious claim, but Westmoreland used the opportunity to allow the commanding officers in different combat zones to make their case that crop destruction was indeed hampering the military abilities of the revolutionary forces. If ever there was a place Westmoreland's field commanders could demonstrate persuasively the military effectiveness of herbicides, this was it. Yet again the evidence was thin.

The commanding general of the Marine Amphibious Force, which operated mostly around Da Nang, began with the caveat that while "empirical data is not available," there were several "indicators of the success" of crop destruction. He cited one captured enemy document to show the "seriousness of the enemy food supply in I Corps." The document addresses planning for *potential* food shortages but does not indicate that NLF or PAVN units in the area were suffering from shortages, let alone that such shortages were caused by Ranch Hand missions. In fact, the one specific mention of crop destruction in the passage cited could easily be read as highlighting the political, rather than the military, effectiveness of the program. "At areas where the crops are seriously destroyed by the enemy," it reads, "we must motivate a unification and mutual help program."[124] Other documentation, discussed earlier, implies that food shortages were causing difficulty for some NLF units in some areas of operation. But if, as Westmoreland argued in his cover memo, the best indicator of the success of the program is the commander in the field, why would he point to this document, which offers proof neither of actual food shortages nor of enemy crops being destroyed, as evidence of field expertise? And why was that single specious report used as one of Westmoreland's examples to be forwarded to the Pacific commander and ultimately to the JCS? Where was the overwhelming evidence of military utility?

The report uses one additional document to demonstrate qualified support of a cause-and-effect relationship between the missions and military food shortages, a document which it says is one among "numerous prisoner interrogations." Even granting the accuracy of this assessment, however, the general made a massive leap of logic in his final point: "While no positive evidence exists that food shortages cause specific enemy operations to be cancelled, it can be assumed that food shortages have degraded enemy capabilities. It is strongly recommended that the crop destruction program be continued. It is believed that enemy food shortages will become more acute by continuation of these operations with further degradation of his combat potential."[125] Those supposedly degraded capabilities would be on full display throughout the provinces of I Corps only weeks later, when the revolutionary forces of Vietnam launched the Tet Offensive on January 31, 1968.

The case made by the commanding general for all of I Corps was not much stronger, relying on generalizations about intelligence documents without including specific references to them. Breaking down the degree of success in limiting food to NLF units operating along the coastal versus those in the central highlands, the general asserted that the objectives of the program were being met, while acknowledging that "crop destruction does not deny completely food to the VC/NVA except in certain local situations." Recommending expansion of the program and thereby providing the language Westmoreland would use in his cover memo, the general stated, "While empirical data on its effects on VC/NVA are lacking, current intelligence reports establish the validity of the program."[126]

The idea that "insufficient statistical data" at MACV precluded an "empirical evaluation" is striking, in light of the data-driven management style of the McNamara-era Pentagon. More striking still is that the military's defense of crop destruction criticizes the RAND study for its lack of data from 1967, but they themselves had no data, from that year or any other, with which to counter RAND's arguments or to shore up their claims of improvement from that year. If MACV, ARPA, and other Pentagon entities had any empirical data whatsoever in support of military claims of the effectiveness of herbicidal warfare, presumably the place to exhibit them would have been in the response to the scathing reports from RAND; the memos in question, after all, were internal documents, not public statements.

In fact, one long-classified document raises the question of whether the U.S. government *ever* had any empirical data to back its contentions about the effects of herbicides on NLF activity. Produced by the Remote Area Conflict Information Center (RACIC) as part of Project Agile, the report, entitled, "Defoliation-Incidents Correlation Study," begins by noting, "Some studies have been conducted on the psychological reactions of

VC to defoliation operations and on the attitudes of the inhabitants of the Vietnamese villages as they relate to these operations. However, no extensive look had been taken at possible changes in VC activity in the defoliated areas."[127] As the summary indicates, ARPA contracted with RACIC to explore precisely these questions. Using information on the date, location, and "relative intensity" of NLF-initiated actions supplied by the DIA, data from ARPA about the date and location of herbicide missions, and activity data generated from daily situation reports from MACV, RACIC generated a complex computer program to compare the levels of NLF activity before and after defoliation missions in the southern third of the RVN from 1963 through 1966. Although the analysts at RACIC found that NLF activity declined in amount and intensity following defoliation missions they added in no uncertain terms that correlation did not amount to causation: "Since a decline in activity follows the spraying, one might be led to believe that the spraying had caused the decline in activity, if it were not for the fact that the same decrease can be observed for the corresponding months before defoliation." The reason? NLF activity and the number of spray missions coincided with seasonal variations in foliage growth; the forces made maximum use of the forest cover until Ranch Hand missions depleted it. Then, as the RAND interviews made clear, the NLF simply went elsewhere. The report also noted that the seasonal decline was reinforced by the fact that defoliation missions were usually followed by increased activity of U.S./RVN forces in the targeted area: "It is quite likely that defoliation makes air support more effective by improving target visibility, thus causing the VC to reduce their actions in the defoliated regions. However, the evidence is strong that there was increased presence of friendly forces, and this factor may have been the primary cause of reduced VC activity." The RACIC study showed the impossibility of separating these factors and determining the relative impact of defoliation missions.[128] A generous reading of those findings might hint that defoliation was effective as part of a well-coordinated strategy to limit NLF activity, but when considered alongside the larger lack of evidence in support of effectiveness, such an argument becomes difficult to sustain, as the military was finding out in 1967 and 1968.

What the RACIC study found is not nearly as meaningful as the larger issue it lays bare: that the U.S. military had almost no evidence, even in 1967, other than the limited qualitative information stemming from reports by prisoners and defectors, on which to base its claims that the use of herbicides for defoliation and crop destruction missions was tactically and strategically effective against the NLF. This should not come as a surprise: in a war in which enemy "body counts" were regularly inflated by U.S. forces and the relative security of an area was defined by the data-driven Hamlet Evaluation System, the quantitative, managerial approach to the war taken by the United States consistently sought to explain away political realities with military and technical solutions. For all of the supercomputers, analysts, and piles of data, neither policymakers nor military commanders could ever answer fundamental questions about whether or not the tactics with which the United States was prosecuting the war—from search-and-destroy missions to carpet bombings to crop destruction—were proving effective. Even less surprising is that they were also unable to determine how those tactics were impacting the civilian population. Yet, considering McNamara's emphasis on data and empirical analysis, the near-total lack of evidence of the military effectiveness that the field commanders, MACV, and the Pentagon were able to produce for their data-driven boss is nothing short of remarkable.

By the end of 1967 the JCS had responded not only by attempting to discredit the RAND studies but also by offering a broader defense of crop destruction and the chemical war in general. The official JCS response to McNamara claimed that the RAND studies were of "questionable validity" because of their "limited scope" and "questionable time frame." "Crop Destruction," the chiefs concluded, "is an integral, essential, and effective part of the total effort in South Vietnam; program objectives are being met; and no program changes are necessary at this time."[129] Even in the face of questions raised by the new reports, the military response invariably relied on the stale assumptions that Ranch Hand missions operated in practice exactly as they operated in theory and that the military utility of the herbicide programs continued to outweigh any adverse political consequences. The results of the review left Denton nonplused. Responding to the Pentagon's findings on his study in February 1968, he remarked that the arguments could be reduced to the following: on the one hand, "there is very strong empirical evidence to support the thesis that the operations significantly alienate (as well as impose suffering on) the non-VC portion of the SVN population." In contrast, however, to the degree that the military acknowledged negative effects, "It is admitted that the spray alienates people, but these people are VC supporters anyway."[130] The herbicide program would continue to be criticized and reviewed, but within the military there was no official doubt about its overall effectiveness.

Given the evidence available, the RAND reports of 1967 likely represented a shift in the politics of the war within and among the RAND analysts more than a policy revelation. The data available to compile the reports were not in any meaningful sense different from those used in RAND reports that supported crop destruction just a year earlier. Indeed, over the next year Russo in particular would grow more frustrated with the war itself as well as with RAND's general support of it. By the end of 1968 he had been fired by RAND and, as he would later tell Elliot, had also become "100 percent radicalized" against the war, which he saw as being "based on lies."[131] Russo continued his new friendship with Daniel Ellsberg, who had joined RAND just before Russo was fired. Together, the two would conspire over the next two years to copy and leak to the press the internal DOD study that came to be known as the Pentagon Papers.[132] Regardless of the motivations, however, Russo's criticisms brought the politics of herbicidal warfare to the forefront of internal debates on military policy for a brief period from late 1967 to early 1968. Russo was not the only one whose stance on the war had shifted. McNamara's recommendations from the summer and fall of 1967 reveal his growing sense that the war was doomed. There is no evidence that the crop destruction debate played a major role in his resignation as defense secretary in November 1967, but the timing of his departure may explain the apparent absence of any response by McNamara to the military's reaction to the issues raised by RAND.[133]

When Ellsworth Bunker was appointed ambassador to the RVN in 1968, he ordered a full overview of the program, which became known as the Herbicide Policy Review. Coordinated by the embassy, the review drew on a variety of military, civilian, and scientific records and experts. The final report relied on familiar military assessments of the effectiveness of the program, but also took seriously the economic and political costs of the program among the civilian population, particularly in the realm of crop destruction. After noting that crop destruction had been militarily successful "to an undetermined degree," the reports noted that "at the same time, the program has had significant but again undetermined adverse political, psychological, and economic impacts on civilians in VC controlled areas."[134] A later section, however, in which the review did offer a refutation of MACV's claims about the increased psyops efforts after 1966, stated that although "the psychological costs of the program have not been serious or unmanageable," "effective psyops" had not yet been put in place and should be considered a priority in moving forward. In the final analysis, the review provided the same answer as always to the long-standing question of military utility versus political costs: "In weighing the overall costs, problems, and unknowns of the herbicide program against the benefits, the committee considers that, on balance, the latter clearly outweigh the former and that the program should be continued and refined in accordance with the findings and recommendations contained in this Report."[135] Despite offering the most thorough and thoughtful analysis of the program from within government channels to that point, the Herbicide Policy Review was simply the latest in a series of analyses that was willing to trade "undetermined" military claims of success for what should have been seen as easily determined and unmistakably negative political costs.

Conclusion

By the time the Herbicide Review Study was released in May 1968, the situation in Vietnam had changed dramatically. The Tet Offensive, which began as the herbicide review was just under way, exposed how far away the "light at the end of the tunnel" remained. The attention to herbicides brought on by the international scientific community was causing an ever-greater uproar over crop destruction and defoliation. As a result of concerns about the safety of the chemicals in the United States as well as in Southeast Asia, President Nixon ordered his military commanders to phase out all herbicide operations in Vietnam by late 1970. Agent Orange was the focus of this decision and the first chemical agent to be prohibited. By early 1971 Ranch Hand had flown its last official mission, and the crop destruction program was discontinued as well. The suspension went against the wishes of the military, which fought the decision every step of the way.[136]

In the final formal defense of the chemical war, the Special Report on herbicide operations of 1970 offered a defiant justification:

> From the most basic of beginnings, this new phase of warfare—the use of nonlethal chemical agents—has progressed to an important role in the prosecution of the war. Politically handicapped from its inception due to the propaganda value associated with plant destruction and the use of chemical agents of any kind, this new concept has been subjected to the closest scrutiny and not found wanting. It is impossible to say how many American and Allied lives the defoliation operations have saved. This alone provides a moral justification for the use of nonlethal chemicals. The use of chemicals requires a strong and constantly active psywar program to counter propaganda, and the United States should resist the temptation to abolish herbicides just because of attacks upon the program.[137]

In a nutshell, this passage effectively summarized the basic, flawed assumptions about the chemical war. Questionable assertions about military effectiveness continued to trump concerns about political tradeoffs, which themselves were dismissed by reference to a dubious, sometimes nonexistent psychological warfare program.

When the RAND corporation broke with the herbicide programs in 1967, the handwriting was on the wall for those willing to read it. A close look at the evidence on which the RAND reports rested as well as at the military responses to them reveals that despite the contrasting views about how the chemical war was shaping the battle for hearts and minds, the U.S. military was unable or unwilling to grasp the contradictions at the center of its crop destruction and defoliation programs. Even if villagers became frustrated with the NLF's appropriations of their crops and of the way in which the front's presence attracted the spray, the uncertainties and fears that the war in general and the chemical war in particular had made commonplace in their lives would not easily be overcome by leaflets, cartoons, empty assurances, and in-person demonstrations.

The RAND evaluations made a major understatement in pointing out that "destroying a farmer's source of sustenance is not a way to make friends." But the questions raised in the documents discussed here call into question more than crop destruction and defoliation programs. They reveal the ways in which those same contradictions exposed the larger futility of the Vietnam War. Even if the propaganda programs had been more consistent and effective, even if the indemnification claims program had been insulated from the rampant corruption of the RVN, the war itself—the application of military technology to address a fundamentally political problem in southern Vietnam—would have undermined the very objectives such policies were designed to mitigate. All the psyops in the world could never have resolved the fundamental contradiction that the chemical war exposed: that to destroy the ability of the revolutionary forces of Vietnam to continue their war it was necessary for the military largely to ignore such distinctions as civilian and noncombatant, farmer and guerilla, upon which the battle for hearts and minds was supposedly based. This was as true for crops as it was for carpet bombs, as true for herbicides as it was for hamlets. The erosion of those distinctions in Vietnam, however, has clarified the distinctions relevant to a historical verdict on the war: those between just and unjust and between a winnable war and one that was, from the beginning, doomed.

CHAPTER THREE
INCINERATING AGENT ORANGE
Dioxin, Disposal, and the Environmental Imaginary

On April 15, 1970, the secretaries of agriculture, interior, and health, education, and welfare announced at a White House press conference that the U.S. government was suspending registration of the herbicide 2,4,5-T. That statement made it effectively illegal to sell or transport products containing the compound for most domestic purposes while continuing to allow them to be sprayed along roadsides and on nonagricultural ranchlands. Although known by few Americans at the time, 2,4,5-T was a common ingredient in many commercial grade and household weedkillers. It was also one-half of the chemical mixture constituting Agent Orange. In a separate press conference minutes later the Pentagon announced that it, too, despite the objections of military commanders, was suspending most uses of herbicides containing 2,4,5-T. Amid growing health concerns about the presence of dioxin in herbicides used in the United States as well as in South Vietnam, President Nixon overruled his military commanders, ordering them to phase out all herbicide operations in Vietnam by the end of 1970.[1]

Nobody told Russell Bliss. Bliss was a waste oil hauler operating in rural Missouri, about thirty miles west of St. Louis. Part of his business was to collect various forms of waste oil, which he then reprocessed as a spray to be used as a dust suppressant in rural areas. Around the same time Operation Ranch Hand was being phased out, Bliss contracted with the Independent Petrochemical Corporation to remove "still-bottom" waste from reactors at a trichlorophenol plant in Verona, Missouri. Although in 1971 the plant

[97]

was owned by Independent Petrochemical, some of the still-bottom waste had been in the tanks since the early 1960s, when the plant was owned by Hoffman-Taff, one of nine major chemical companies contracting with the U.S. government to produce Agent Orange. Although Hoffman-Taff produced the lowest overall volume of Agent Orange of all the companies, its dioxin content was among the more concentrated mixtures.[2] Over time, the dioxin content in these still-bottoms had risen to levels of hundreds, perhaps thousands, of ppm, far higher than the normal concentrations found in the herbicides used in Vietnam, which are estimated to be somewhere between three and thirteen ppm.[3] Over the next several months Bliss mixed the still-bottoms from Verona with other waste oils, processed them, and for the next three years sprayed them over horse stables, farms, and dirt roads in and around Times Beach, Missouri. In late May 1971 Bliss sprayed Judy Piatt's horse stables. Within a few days the farm was covered with dead birds. Within a month dozens of horses, cats, and dogs had died as well, and several local children had been admitted to the hospital for a variety of illnesses. The spray Bliss used on Piatt's stables contained between 306 and 356 ppm, a total of between twelve and fourteen pounds of heavily concentrated dioxin. By 1974 the Centers for Disease Control (CDC) had been called in to investigate the cause of the human health problems and the deaths of dozens of animals in Times Beach. They initially found soil samples containing well over thirty ppm dioxin.[4]

By the time of the Times Beach disaster the central problem surrounding Agent Orange had shifted considerably. Concerns about its military effectiveness and political consequences in Southeast Asia were quickly replaced by a growing emphasis on the effects of herbicides containing dioxin on humans and the natural environment. In the mid- to late 1970s, however, before lawsuits by veterans of the Vietnam War from the United States, Australia, and New Zealand put the herbicide, the government, and the chemical companies that produced Agent Orange on trial, questions about the effects of Agent Orange focused largely on the issues of how best to dispose of it. During this period the U.S. government was forced to confront the dilemma of disposal through Operations Pacer IVY and Pacer HO and to confront a major dioxin disaster at home in Times Beach. The government in New Zealand followed suit, dealing with the growing preoccupations of its citizens about the possibility of dioxin contamination near the Ivon Watkins Dow plant in New Plymouth.

Operation Pacer IVY and Pacer HO as well as the episodes in Times Beach and New Plymouth reveal how various constituencies negotiated shifting, contested concerns about environmental safety, public health, the body, and the role of the state in the 1970s and beyond. By forcing the question of how best to deal with the vast remaining stockpile of chemicals and their residual effects on storage and production sites, these events offer a unique opportunity to explore the evolving global environmental consciousness taking hold in the early 1970s and the ways in which the legacies of the Vietnam War contributed to this sensibility. As the U.S. military continued to cling to its assertions about the relative safety of Agent Orange, it was forced throughout these operations to deal with the newly implemented environmental regulatory apparatus in the United States, particularly the EPA. While the bulk of this story focuses on the United States, the disposal of Agent Orange is also a useful case study because it is global in scope. From transporting the stockpiles through the Panama Canal and across the Pacific Ocean to debating if Agent Orange could be reconstituted to be sold in the global marketplace, to the final incineration of the herbicide reserves at sea in the South Pacific, the U.S. military was forced throughout the seventies to deal with an increasingly complex matrix of environmental attitudes, laws, and regulations that grew out of both the early environmental movement of the period and the Vietnam War itself. During this period the USAF, which continued to deny that Agent Orange was harmful to soldiers and civilians, was required to write lengthy, detailed Environmental Impact Statements (EIS) documenting the safety procedures to be used in the operations, the hazards posed by the chemicals themselves, and the overall potential impact of the procedures on everything from drinking water and air quality to the growth of coral reefs and the migratory patterns of rare birds.

In examining these events and the reactions to them by various local actors, one can see how they contribute to what Christopher Sellers and others have termed the "environmental imaginary" that emerged during the 1960s. In this formulation, citizens and activists used local-global geographic conceptions and a focus on the human body as environmentally threatened to develop various forms of ecologically oriented politics. This social formulation had become fairly stable by the end of the 1960s, yet the legal and policy formulations for which environmentalists had fought came to valorize objectively measurable scientific effects on nature over subjective and experiential effects on the body. Later events like the Love Canal disaster in New York, Sellers concludes, helped the public reimagine a "naturalistic human ecology beyond the pale of the biologically provable."[5] In exploring the apprehensions about environmental and human health raised during Pacer IVY, Pacer HO, Times Beach, and New Plymouth, one can see not only several examples of Sellers's formulation—as the immense volume of quantitative data on ppm of dioxin in water, soil, and animal tissues became the standard for determining human risk and safety while providing little comfort or knowledge to the human actors involved—but also how affected populations began to frame these formulations in response to the perceived inadequacies of scientific knowledge and of the state authorities through whom that knowledge was disseminated, a tendency that would firmly take hold in the lawsuits, benefits, and regulatory battles of the 1980s. As we will see, the same uncertainty over the precise effects of dioxin exposure that led many individuals and communities to declare, Not in my backyard! (NIMBY) when it came to issues of disposal and incineration would later be used by the state, the courts, and the chemical companies to limit both government and corporate liability to those who claimed damages as a result of exposure to Agent Orange.

Pacer IVY and the Dilemmas of Disposal

As opposition to the war in Vietnam grew during the mid-1960s, American scientists came to play a major role in making known to Congress the environmental impact of U.S. policies in Southeast Asia. As David Zierler has shown, even when scientists checked their politics at the door (and largely because they did so), their efforts to end the program of herbicidal warfare were instrumental in bringing about the end of Operation Ranch Hand and, ultimately, drawing down the American war in Southeast Asia.[6] During the late 1960s the American Association for the Advancement of Science (AAAS) attempted to turn the Republic of Vietnam into a laboratory to study the effects of herbicidal warfare in much the same way the Pentagon had turned it into a laboratory for counterinsurgency. In multiple trips and studies from 1967 to 1970, members of the AAAS documented the effects of Agent Orange and other chemical agents on the landscape and, to a lesser extent, on the people of Vietnam. Throughout the sixties the U.S. government had ignored, dismissed, and, in the case of the study in 1968 by the Bionetics Research Laboratory showing the teratogenic effects of 2,4,5-T on mice, even covered up information about the potentially dangerous effects of Agent Orange and TCDD. By 1970, however, in the midst of growing dissent over the continuation of the war and greater knowledge about the effects of 2,4,5-T, both Congress and the White House, if not the Pentagon, became more willing to listen to the claims of scientists. This readiness became very clear during the debates in 1970 over the Geneva Protocols on Chemical Weapons, which, as Zierler shows, were in many ways a forum on herbicidal warfare and U.S. policy in Vietnam. For at least a brief period in the early seventies scientists were able to make the case that while the effects of 2,4,5-T, Agent Orange, and dioxin on humans and the environment were still not entirely understood, the uncertainty over those effects was enough to curtail Ranch Hand and lead to the suspension of the use of Agent Orange in Vietnam and of 2,4,5-T in the United States.[7] But, as the USAF was about to find out, scientists and other critics of Agent Orange were not the only ones preoccupied about the effects of dioxin. As the war in Vietnam slowly drew to a close in the early part of the decade, concerns about the safety of Agent Orange and TCDD were framed largely by the question of what to do with the nearly 2.4 million gallons of Agent Orange still in possession of the USAF and the ARVN.

While the United States continued to use other herbicides throughout central and southern Vietnam, the permanent suspension of Agent Orange missions posed a major dilemma to the military. At the time of the announcement more than 1.5 million gallons of Agent Orange were sitting in fifty-five-gallon drums at several bases throughout Vietnam. Another 860,000 gallons were awaiting shipment to Southeast Asia from the Naval Construction Battalion Center (NCBC) in Gulfport, Mississippi.[8] These chemical agents, which had for years been seen as vital to the war in Vietnam, now became a logistical thorn in the side of the military. Faced with the reality of the permanent ban, the USAF confronted a plethora of options for divesting itself of the stockpiles. Every option, however, carried with it a set of problems that ranged from the practical and political to the legal and ecological. In stark contrast to the largely unregulated use of the chemicals during the war, the process of divestiture and disposal would be closely monitored by the recently created EPA.

During the summer of 1970, CINCPAC, the JCS, and MACV all lobbied the Pentagon and the White House to lift the ban on Agent Orange use in Vietnam while moving forward in dealing with the challenges posed by the remaining stocks of the herbicide. The first response from MACV, aside from continuing to allow the use of Agent Orange by U.S. and ARVN forces, was to "Vietnamize" the herbicides, in keeping with the larger policy promoted by the Nixon administration of turning the entire war effort over to the Republic of Vietnam. Under this plan the United States would simply transfer ownership of the remaining inventory to the RVN, thereby circumventing U.S. regulations and domestic political battles. MACV also considered the possibility of incinerating the remaining stocks inside Vietnam, since burning the chemical was a proven way to destroy the dioxin present in the herbicide. When the Pacific commander forwarded this list to the JCS, however, he elevated the status of "Option III: Retrograde from Vietnam" as the next best option to continued use, removing the plan for Vietnamization altogether. By December the administration affirmed that the ban on Agent Orange in Vietnam would continue, informing the DOD that Nixon would have to approve personally any use of herbicides outside of the remote areas to which they had been restricted and any Vietnamization of the remaining supplies.[9]

During the spring of 1971, U.S. Senate hearings on updating the Geneva Protocols to include riot control agents and herbicides offered a chance for critics of the chemical war in Vietnam to air their views. Scientists such as Matthew Messelson of Harvard University, the author of the Bionetics study, along with DOS officials, used the opportunity to deal a major blow to any hopes among the military that defoliation and crop destruction missions would be extended in some form.[10] By mid-March even the JCS seemed to have realized that the herbicide program was over for good and that attention needed to be refocused on the disposal dilemma. An internal memorandum summarizing a JCS meeting on the issue noted that the "assumption of lifting the ban on Agent Orange is unrealistic" and that "the JCS assertion that reinitiation of orange would produce military benefits that outweigh political risks is both unreal and unfounded." Instead, the notes indicate, "the military should begin to address the potential domestic market for these herbicides."[11] According to William Buckingham, Secretary of State William Rogers convinced President Nixon that domestic political concerns over the program now trumped the military benefits of continuing the program.[12] This was a marked shift from the previous year, when the military clung, without providing much evidence, to its belief that the military effectiveness of herbicides outweighed any political fallout. Some of this shift is surely owing to the fact that the administration was facing domestic political difficulties over the program, as opposed to international condemnation or criticism from Vietnamese civilians. But the change also reflects the growing and shifting knowledge about the dangers of dioxin and its effects on humans and the environment, worries that could no longer be easily dismissed, even in internal military communications.

The coalescing of domestic political opinion and environmental consciousness helps to explain Nixon's role in this process. Unlike the Kennedy administration, which in 1962 could only speculate about the impact of Rachel Carson's *Silent Spring* on domestic policy and its crop destruction program, Nixon came into office as environmentalism was nearing its modern apogee, just prior to the first Earth Day celebration and a few years before the disasters at Love Canal and Three Mile Island would bring the dangers of industrial pollution into sharp relief.[13] Yet diplomatic, political, and even environmental historians have continued to downplay or ignore Nixon's important legacies in this area. Many recent biographies of Nixon contain no references to either environmental politics or the establishment of the EPA, and a reader in modern American environmentalism published in 2007 gives Nixon only two brief sentences and no primary documents.[14] The author Rick Perlstein has recently argued that Nixon's environmental efforts were little more than crude political calculations, pushing environmental regulations to capitalize on the growing popularity of such measures among the American public and establishing the EPA as a way of further consolidating power within the executive branch.[15] Regardless of his motives, Nixon played a central role in advancing environmental politics and policy in the United States. His administration's record on environmental regulation is without question the most progressive in the second half of the twentieth century. To be sure, his environmental policies, like most of his decisions, were based on a desire to maximize political capital while crafting policies that would leave visible legacies of his presidency. The creation of the EPA, the passage of landmark legislation such as the Clean Air and Clean Water Acts, and the decisions on 2,4,5-T and Agent Orange allowed him to do both.

Amidst the swirling winds of environmentalism at home and the war still raging in Southeast Asia, the DOD reached its final decision on the Agent Orange inventory in September 1971, recommending, for both logistical and political reasons, the immediate return of all herbicides from Vietnam to the United States. In his memo, Secretary of Defense Melvin Laird referenced the environmental concerns about the herbicide, noting that "all stocks, both RVN and CONUS [Continental United States] with unacceptable levels of impurities will be incinerated. Options for possible use of remaining stocks would be considered."[16] Eventually, the entire stockpile of Agent Orange was determined to be unacceptable; the immediate issue, however, was to locate, consolidate, and relocate the remaining stocks. By this point supplies of the agent had been removed from nearly all field units and were located at four storage sites: Da Nang, Phu Cat, Nha Trang, and Bien Hoa (fig. 6). Bien Hoa housed the majority of the drums.[17] Before they could be shipped, however, thousands of deteriorating, rusted, and leaking

Figure 6. Agent Orange stocks in need of transfer to new drums, Da Nang air base, 1969. (Courtesy National Archives, College Park, Maryland)

When President Nixon discontinued the use of Agent Orange in 1970, the military was faced with the challenge of what to do with the nearly 2.4 million gallons of the herbicide still on hand. The first task was to deal with the inventory of barrels, many of which were leaking and rusted. The massive redrumming operation, known as Operation Pacer IVY, took place at the air bases at Bien Hoa, Tuy Hoa, and Da Nang.

barrels needed either to be repaired or have their contents transferred to new drums, that is, be "redrummed." Thus began Operation Pacer IVY.[18]

By the end of 1971 the remaining stocks of Agent Orange had been moved to three locations that would serve as the staging points for Pacer IVY: Bien Hoa, Tuy Hoa, and Da Nang. The Seventh Air Force was charged with overseeing Pacer IVY, but ARVN personnel did the bulk of the work handling the drums. According to Richard Carmichael, a USAF engineer, hundreds of local Vietnamese women were also employed to assist with the operations. In an interview with Maj. Alvin Young, USAF (ret.), Carmichael recounted that he "had verbally expressed concern to the contractor because the boots, aprons, and gloves issued to the Vietnamese women were too large for them to wear. As a result, many of the Vietnamese women wore sandals and handled the herbicides without gloves."[19] Their direct, repeated exposure to large amounts of herbicide raises uncertainties about the potential long-term health consequences for the women involved in the operation, but it also indicates the changing attitudes about exposure to the herbicides in general. In internal memos, the USAF was still disputing many of the criticisms about the effects of Agent Orange, yet during Pacer IVY air force personnel were questioning the lack of protective gear among the female contractors.

According to Young and Carmichael, at least half of the inventory at the three bases was in need of redrumming. Photographs from Da Nang reveal extensive leaking and spillage during Pacer IVY, complicating issues about dioxin levels in the soil and surrounding area. Unusable barrels from the redrumming operation were to be crushed and buried in local landfills, while usable ones were rinsed, cleaned, and turned over to ARVN forces. Carmichael, who was involved in the inspection of redrumming operations, reported to Young that very little of the proposed cleanup at Da Nang was completed and that substantial contamination of the soil had occurred at the base. Warren Hull, another USAF officer, confirmed similar results from Pacer IVY operations at Tuy Hoa airbase, a site he later described as "heavily contaminated with Agent Orange."[20] All told, more than twenty-five thousand drums of Agent Orange, half of which were in their original containers and half in new drums, left Vietnam in March and April 1972, when the *M/T TransPacific* picked them up at the three ports and set sail for the U.S. military installation on Johnston Atoll in the South Pacific. The barrels arrived there on April 19, 1972.

The supplies in Vietnam were only part of the story of Pacer IVY. The other 860,000 gallons then in control of the United States were at the NCBC in Gulfport, which had been used as a storage and shipment site since 1968. Gulfport, like Da Nang, Bien Hoa, and other bases in Vietnam, has long been a focus of environmental and health concerns. In August of 1969 Hurricane Camille struck Gulfport with sustained winds of over 160 miles per hour, one of the strongest storms ever to make landfall in the United States. Much of the Gulf Coast experienced serious damage, and the NCBC was no exception. More than 1,400 drums of herbicides scheduled for shipment to Vietnam were blown into the water and scattered around the port by the winds. Young reports that 412 of these were located and shipped to Vietnam; most of the others were recovered from the water. But as many as 240 drums of Agents Orange and Blue were never located and became a focus of future inquiries and investigations about environmental contamination at Gulfport.[21]

Over the next several years, as Pacer IVY was under way, the USAF considered no fewer than thirteen options for what to do with the massive stockpile of herbicides. By the time the remaining supply of Agent Orange had been delivered to Johnston, the Air Force Logistics Command had narrowed the options and began a formal study of five primary disposal scenarios: incineration, soil biodegradation, fractionation (converting the herbicide to its acid components through distillation), chlorinolysis, and industrial reprocessing. The Scientific Advisory Board for the USAF focused on these options only after eliminating such alternatives as deep ground burials, deep well disposals, microbial reduction, and return of the herbicides to the manufacturer.[22] Throughout the exploratory process the USAF encountered reservations and regulations from local governments and constituents about the possible effects of Agent Orange disposal in their areas. The similarities in these local responses show, years before Love Canal and Three Mile Island, the growing anxieties about the dangers these dangerous chemicals posed to the health of humans, animals, and the natural environment, most of which revolved around the ultimate uncertainty about the effects of dioxin exposure.

When Utah was proposed as a candidate site for herbicide disposal through soil biodegradation, the state's governor, Calvin Rampton, directed the State Department of Health to investigate the issue. The department's report indicated "universal agreement" among various state agencies that the information about the disposal process was "simply too sketchy" to support the proposal. While inconclusive biodegradation studies had been completed on acidic soils, most notably in Oregon, no data were yet available on the persistence of dioxin in alkaline-laden arid landscapes like Utah's. The report further noted that the "major push" on disposal initiatives by the USAF at the time was a combination of several factors, including the ongoing cost of storage and maintenance of herbicide barrels and the "emotional problem surrounding its use in Vietnam." Local press reports described the effort as an attempt to dump dioxin in Utah, leading to public outcry. On the basis of the report's conclusions and despite the "close and friendly" relationship between the state and the USAF, Rampton felt it was his "duty to resist the proposal."[23]

Other candidates for disposal included Deer Park, Texas, and Sauget, Illinois, both home to the type of large-scale chemical facilities required for destroying the dioxin in Agent Orange through incineration. The State of Texas rebuffed proposals for the Deer Park facility, owned and operated by Rollins Environmental Services, citing "possible harmful effects to the area." The primary objections raised by the Texas Air Control Board were that the additional contamination caused by the incineration of Agent Orange would compound an already serious air pollution problem in the area, which was located just outside of Houston and was home to a number of industrial plants, including a major Shell Oil processing plant.[24] The Sauget site was even more problematic. Originally named Monsanto, Sauget had been the original home of Monsanto Chemical Company since the early twentieth century. The Sauget plant produced Agent Orange during the war as well as DDT, plastics, and other chemicals. Sauget would later be identified by the EPA as the largest producer of polychlorinated biphenyls (PCBs) in the world and would eventually be declared a Superfund site. When worries over PCB contamination became an issue in the early 1970s, the unincorporated town of Monsanto changed its named to Sauget but remained squarely on the radar screen of environmental groups and regulators.[25] When the USAF named Sauget as a possible destination for the Agent Orange supply, the EPA was quick to take note. The appeal of the site seemed to be its unincorporated, seemingly isolated status, although that assessment failed to take into account the town's immediate proximity to St. Louis, Missouri. The matter of proximity was not lost on the residents of Missouri, however. In response to an invitation to submit public comment on the draft proposal of 1972 for incineration at either Deer Park or Sauget, citizens wrote to the USAF and Congress to protest the plan. Douglas Thornberry, a resident of St. Louis, wrote a lengthy, detailed letter to Secretary of the Air Force Robert Seamans criticizing the air force for "the insinuation that Sauget, Illinois is just some small place, where no one has ever heard of, and probably would assure it is located out in the back country. If you would consider a 15 mile radius circle drawn about Sauget, Illinois, you would discover a metropolitan area with a population of much more than a million people."[26] Anticipating the potential release of as much as five hundred pounds per day of hydrogen chloride emissions based on the data provided by the USAF, the EPA's reply to the initial EIS noted that the normal wind patterns around Sauget would blow those emissions directly over downtown St. Louis, located directly across the Mississippi River from the facility. While it was "impossible to accurately determine the effect of these emissions on the surrounding community," the reply concluded, "it is safe to say that such an amount of emissions over such a long period of time could represent a potentially serious condition." In the end, the EPA recommended pursuing the construction of a new incinerator in a "remote region."[27]

For all of the discord about the final resting place for Agent Orange, the proposals that generated the most attention were those involving the reprocessing of the herbicide for domestic use. Although virtually identical to domestic herbicides in use at the time, Agent Orange was technically designed specifically for the military according to specifications provided to the DOD. Because of these specifications and because use of the agent predated the creation of the EPA, the tens of thousands of barrels of Agent Orange awaiting their fate were not registered for domestic use in the United States. Given the heightened anxieties about Agent Orange, dioxin, and the recent ban on 2,4,5-T, it was unlikely that the herbicide could be successfully registered at the time. Still, given the historically heavy domestic use of 2,4-D and 2,4,5-T, timber and forestry interests, ranchers, and other advocates expressed a great deal of interest in the supply of Agent Orange then in storage.[28]

"We're being besieged by both sides," an EPA spokesperson told the San Antonio *Light* in April 1972, "from ranchers who want us to get it for them, and from environmentalists who object to plans to pay for it." Ranchers and many of their congressional representatives advocated that the supply of Agent Orange be declared military surplus so that it could be sold to them at greatly reduced costs, but the EPA noted that this was unlikely, given the presence of "impurities" in the supply. In fact, military procurement guidelines for contractors specifically stipulated that "any contractor inventory dangerous to public health or safety shall not be donated or otherwise disposed of unless rendered innocuous or until adequate safeguards have been provided."[29] Many advocates of domestic herbicide applications would have disputed the dangers of dioxin, but convincing the EPA and the public that 2,4-D and 2,4,5-T in any form were "innocuous" and safe for local use remained a major obstacle.

Throughout 1972 and 1973, while other constituencies balked at having Agent Orange buried or burned in their backyards, advocates of herbicide use lobbied the government to allow its use both at home and abroad. Michael Newton, a forest ecologist at Oregon State University, wrote to Young, then a captain in the USAF, requesting that they work together to "shake loose the juice," as Newton put it, allowing the USFS or the Bureau of Land Management to use the herbicides for brush control in forests. Newton was interested in comparing the effects of clean and dirty, heavily contaminated Agent Orange. Forwarding the request, Young wrote a note on the memo, calling it an "excellent proposal for the price of five drums. Recommend we send him the Agent!"[30] Newton and Young were not the only ones who saw the destruction of Agent Orange as a waste of resources. In a letter to Rep. Robert Sikes of Florida, who represented the area surrounding Eglin Air Force Base, a major test site for military herbicide use, Col. G. C.

MacDonald of the USAF wrote to protest "the utterly pointless destruction of 2,300,000 gallons of military herbicide Orange, which to date has cost the taxpayer about $20,000,000." Citing previous domestic use by the USFS and the results of long-term testing at Eglin, MacDonald dismissed the recent furor over 2,4,5-T by claiming that "much of the scientific community has no intention of being either honest or objective as regards herbicides or the war in Vietnam, for that matter." He recommended a number of potential uses, including by the USFS and for "several major road building projects in Central and South America."[31]

There was indeed interest by some Central and South American constituencies, most notably the government of Brazil. Working through USAID and the Blue Spruce Chemical Company in New Jersey, the Interior Ministry of Brazil in April 1972 requested "shipment of certain chemicals for use in Brazil to update cattle grazing lands, thereby helping to increase the output of meat and milk commodities." The chemicals in question were in fact 2,4-D and 2,4,5-T, which the ministry and its advocates understood to be surplus materials, the application of which "would have a most beneficial impact on the Brazilian economy." While both USAID and Blue Spruce asserted that the herbicide formulation was legal in Brazil, Blue Spruce nevertheless planned to set up an elaborate front corporation based in the Bahamas to avoid undue regulatory difficulties and to allow for distribution not only to Brazil but also to a number of countries in the Amazon basin, including Colombia and Venezuela. The plan was eventually nixed when the USAF made clear that the herbicides would not be "donated" and that it would require formal certifications from the governments in question that Agent Orange was considered legal and "safe for use in the manner intended." Other proposed uses for international development, such as "applying Orange for aquatic weed control in Africa," also failed to develop.[32]

Neither the EPA nor the military was going to allow the use of Agent Orange at home or abroad without modification of the supply to meet existing safety and environmental regulations. Reprocessing the herbicide would require the participation of the chemical companies responsible for the presence of considerable levels of impurities in the first place. They understood better than anyone the processes, risks, and, above all, the costs involved. As early as 1968 Charles Minarik of the Chemical Corps wrote in a memo to the commander of the San Antonio Air Materiel Area, operated by the USAF, that excess supplies of herbicides, if necessary, could be used for a variety of domestic and international applications, but that "discussions with representatives of industry disclose that it is not feasible to sell

Orange to the manufacturer even at greatly reduced prices" because the "cost of conversion to more acceptable formulations" would be prohibitive.[33] In March 1971, before the redrumming operations of Pacer IVY had been completed, Richard Patterson, the government relations manager for Dow Chemical, confirmed Minarik's view in a presentation he made to the Armed Forces Pest Control Board entitled "Modification and/or Destruction of Defoliants." "We [Dow] have been asked, officially and unofficially, whether we would get involved in the three remaining alternatives, i.e., purchase of the inventory, destruction of it on a contractual basis, or recommend a method of destruction," he began, before quickly noting that Dow was "not interested in either of the first two alternatives. . . . The economics of modification or conversion are poor, i.e., this could only be done at a loss; a large portion (probably something more than 40%) of the Orange inventory is unusable because of a dioxin level higher than one part per million; the problem of getting shipping permits and the present condition of the drums would preclude shipment out of the Gulfport facility, and the condition of the drums would present a considerable environmental hazard even if they could be shipped to our manufacturing plant in Midland, Michigan."[34]

At the time of Patterson's presentation, Agent Orange produced by Dow represented just under half of the more than eight hundred thousand gallons in dock at Gulfport, a supply his company believed was too contaminated with dioxin to be worth reprocessing. Given the opportunity to consult on other options, however, Dow was more than willing once again to offer its services to the U.S. government. Patterson outlined Dow's long history of design, building, and maintenance of chemical disposal facilities, including the type of specially designed industrial incinerators that could be used to achieve the high-temperature incineration of Agent Orange. Representatives of the company, he urged, could be available "on short notice" to consult and offer bids on such services. If the DOD ended up incinerating Agents Blue and White as well, Patterson added in closing, Dow hoped to have access to the process to study its effects.[35]

Over the next several years the disposal of Agent Orange would come to signify a great deal more than environmental safety. While Agent Orange had, by the end of the 1960s, become a symbol of the violence inflicted on Southeast Asia by the United States, by the mid-1970s it would serve as a stubborn reminder of the mess that war left behind. As the United States extricated itself from Vietnam, the dilemma of disposal offered a poignant, if painful, reminder of the failures that war represented. Like the painful, toxic memories of the war itself, the barrels, the herbicides, and the dangers they represented could not be easily washed away, so the military did the next best thing: it hauled the remaining stockpile of Agent Orange to a remote outpost, burned it, and made it disappear.[36]

Incinerating Agent Orange

Given the unwillingness of the chemical companies to purchase or reformulate the herbicides, the misgivings voiced by other localities about the alternative disposal methods, and the delicate politics and diplomacy involved in foreign sales of Agent Orange, the USAF decided in 1972 to proceed with incineration of the remaining stock, filing for permits with the EPA and setting out to write an initial EIS. Incineration of dioxin-laded chemicals had been tested in various commercial facilities, including those operated by Dow and highlighted in Patterson's presentation. According to the EIS:

> The data accumulated, together with theoretical considerations and applied thermochemistry, clearly indicate that the production of incomplete combustion products can be minimized to insignificant levels. Incineration will convert the Orange herbicide to its combustion products of carbon dioxide, hydrogen chloride, and water which will be released to the atmosphere. In addition, a relatively small amount of elemental carbon and carbon monoxide will be generated in the incineration process and discharged to the atmosphere. With proper concern for the environment in which such incineration will take place, incineration is an environmentally safe method of disposal of Orange herbicide.[37]

The data presented in the EIS, gathered from stocks of Orange at Johnston and Gulfport, suggested that the total amount of dioxin present in the supply to be incinerated amounted to approximately fifty pounds. With test incinerations resulting in efficiencies of more than 99 percent, the USAF believed that less then 0.05 pounds of dioxin would be released along with the water and other exhaust from the combustion process.[38]

Incineration was the most proven, reliable, and politically viable option for disposal, but given the common concerns raised throughout the process, the question of where the incineration would take place remained. Rejecting earlier proposals for Texas and Illinois, the EPA had urged the USAF to consider incineration in a remote location, a suggestion the project leaders took to heart. In its EIS the USAF described the ideal incineration site as being "as remote as possible from both residential and industrial populations centers," where "women of childbearing age" would have a low probability of coming into contact with the Orange," preferably with a "prevailing wind of nearly constant direction and velocity." Finally, the report noted, the site "should be completely under the control of the Federal Government to minimize the local political controversial effects on state or other government units."³⁹ One ideal site presented itself: an area of the South Pacific near the tiny, isolated group of islands known as Johnston Atoll, where the bulk of the herbicides were already in storage.

Located about eight hundred miles southeast of Honolulu, Johnston had been under the control of the United States since the late nineteenth century, after its annexation of Hawaii. Throughout the 1800s Johnston was a major site of guano cultivation, gathered for fertilizer from the droppings of the many seabirds that nested on the atoll. In 1926 Calvin Coolidge issued Executive Order 4467, declaring Johnston a wildlife refuge to protect the natural bird sanctuary, and during the 1930s Franklin Roosevelt militarized the islands, transferring control of Johnston as well as Wake Island to the United States Navy. As tensions between Japan and the United States mounted, Roosevelt further consolidated these areas, along with several others, as Naval Defensive Sea Areas in 1941. The atoll was originally composed of two natural islands, Johnston and Sand, but the presence of the United States in the area led to coral-dredging operations that eventually resulted in the creation of two additional, man-made islands named Akau (North) and Hakina (East). Johnston and Sand Islands have also been expanded by the military, from their original sizes of 60 and 13 acres to 625 and 22 acres, respectively.⁴⁰

When the USAF assumed control over operations on Johnston after the Second World War, the group of islands became home to a variety of military operations, most of which centered around chemical and nuclear weapons. During the late 1950s and early 1960s Johnston became a primary location for support of atmospheric nuclear testing, and after the United States returned control of Okinawa to Japan in 1970, the military moved large quantities of chemical munitions from there to Johnston. During the most active years of nuclear testing on the atoll, which included Operation Dominic in 1962, Johnston experienced significant levels of radioactive exposure. In June 1962 a Thor missile carrying a nuclear device failed and was destroyed over Johnston, causing "a substantial amount of debris" to fall on and around the area. Navy teams spent the next several weeks recovering and removing radioactive debris from the water around Johnston and Sand.[41] In July another missile failure led the radio team on the ground to destroy the missile shortly after takeoff. No nuclear explosion resulted, but the launching pad area and the lagoon on the northwest corner of the island were heavily contaminated with radioactive debris, particularly plutonium. Agents with the Defense Threat Reduction Agency (DTRA) began work immediately to begin cleanup and monitoring operations, establishing a twenty-four-acre Radioactive Control Area on the island and shipping a large volume of material off the island. The DTRA filled the resulting landfill in the area with "240 tons of contaminated metal debris, 200 cubic meters of concrete debris, and 45,000 cubic meters of contaminated coral soil." Only about 1 percent of the weapons grade plutonium was placed in the landfill, the vast majority of it, according to USAF records, remaining in the lagoon.[42]

While the extensive long-term environmental monitoring of Johnston since the 1970s has shown the atoll to be within acceptable federal limits for air, water, and soil contamination, these same reports attest that Johnston is "an unincorporated, unorganized territory of the United States" and not subject to the same controls as states and other territories. The guidelines of the Clean Air Act, for example, do not apply to Johnston, and the Nuclear Regulatory Commission has no jurisdiction over the island. The military did consult with and submit multiple reports to the EPA, including the eight-hundred-page-plus final report on the at-sea incineration of Agent Orange, but the numerous references in the reports to the liminal jurisdictional and statutory status of Johnston demonstrate that much of the monitoring was left up to the military. It is precisely this unclear status, along with the remarkably isolated geographic location of the atoll, which made it the ideal location for Operation Pacer HO.

In its final incineration plan, the military decided against building an expensive incineration plant on Johnston proper. Despite the seemingly ideal location, the cost of building a new facility was prohibitive, the price of a new incinerator alone estimated at more than $3 million. Instead, the USAF contracted with the owners of *M/T Vulcanus*, a Dutch cargo ship retrofitted in 1972 to become a specialized chemical cargo ship, complete with two incinerators located in the stern.[43] The *Vulcanus* would then transport the materials to a designated "burn zone" due west of Johnston. Like Johnston, the burn area was essentially an extrajurisdictional area, an even more remote portion of the Pacific Ocean. Nevertheless, the USAF ran into one of its first major regulatory problems during the operation when confronted with the Marine Protection, Research and Sanctuaries Act of 1972 (MPRSA). The MPRSA, signed into law while the USAF was drafting its initial EIS, was designed to prohibit "dumping of all types of materials into ocean waters and to prevent or strictly limit the dumping into ocean waters of any material which would adversely affect human health, welfare, or amenities, or the marine environment, ecological systems, or economic potentialities."[44] When the USAF submitted its initial draft of the EIS, the EPA ruled that the MPRSA did not apply because the law was unclear on the issue of air pollution at sea, which would be the primary contaminant produced by Pacer HO. By the time the air force had prepared its final EIS in 1974, however, the EPA had reversed its previous ruling and now required the USAF to go through the lengthy process of obtaining an at-sea incineration permit. This new process, which included several public hearings, ironically forced the air force to revisit the idea of reprocessing the herbicide for domestic use, which it did over the next two years, adding time and cost to the operation but resulting in the same conclusion: at-sea incineration was the safest, most politically viable way to destroy Agent Orange. In 1977 the EPA issued the permit, allowing the incineration plan to proceed.[45]

The USAF ran into another major hurdle when presented with a number of new regulations on the destruction and disposal of chemical and biological weapons passed in the early 1970s. In particular, amendments to legislation covering the regulation of chemical and biological weapons (CBW) made it illegal to dispose of any chemical or biological warfare agent, "unless such agent has been detoxified or made harmless to man and his environment or unless immediate disposal is clearly necessary, in an emergency, to safeguard human life."[46] The counsel's office of the USAF suggested in its recommendation that Agent Orange and other herbicides should not be considered chemical or biological agents because they were not intended as antipersonnel measures, although it did note that this was still an open question, given the growing knowledge about the effects of herbicides on "man and his environment." The debate over the legislation, the counsel's office pointed out, focused largely on agents such as VX and nerve gas, not on herbicides, but the real question was what constituted disposal. The aim of the act, according to counsel, was to prohibit the dumping of CBW agents, especially in the ocean. The USAF, the memo went on, would not be physically dumping the herbicide in the ocean, so that provision would also not apply to Pacer HO. Despite the public health and environmental concerns running throughout the memo, the ultimate justification for the recommendation to proceed was that Agent Orange, "at its present concentrations," is "*not toxic or harmful to man and his environment,* as those terms are used in the act."[47]

The explicit stance of the USAF remained that Agent Orange and its associated dioxin, even at levels well over one ppm, were not dangerous to human or environmental health so long as they were properly handled. This did little to assuage the growing consternation of the public, which became increasingly aware of Agent Orange over the course of the operation. Although the burn zone itself fit the ideal site criteria, the plan still required storage, transport, and shipping of the materials to and from Gulfport and Johnston, projects that involved several hundred airmen, scientists, and contractors. Just as they had throughout Pacer IVY, cries rang out from Mississippi to Johnston Atoll about how the presence of the chemicals would affect the populations and environments involved.

Given how widely herbicides had been used in the United States in the 1950s and 1960s, why was the public outcry now so acute? The timeline of events is critical and underscores the rapidity of change surrounding environmentalist thinking in the 1970s. At the time of the suspensions on Agent Orange missions and domestic use of 2,4,5-T in 1970, the science and politics of herbicides and dioxins were still very new to the American public and their elected representatives. Over the next several years, however, knowledge and awareness of herbicides and their dioxin contaminants spread quickly. By the time of Operation Pacer HO in the late 1970s the environmental movement had helped to solidify public concerns about herbicides, industrial pollution, and government regulations. The environmental legislation signed into law by Nixon in the early 1970s also raised awareness about air and water pollution while carving out a specific regulatory role for the federal government.

Congress had also gotten into the act, holding an ever-increasing number of government hearings on 2,4,5-T and Agent Orange (see chapters 4 and 5), as Vietnam veterans began to raise questions about their health. As the wave of environmentalism continued to spread in the 1970s, the specter of the Vietnam War added the sinister air of ecocide to public perceptions of herbicides, helping to trip alarms about the presence of Agent Orange as various disposal scenarios become known in communities across the country and around the world.

Many southern Mississippi residents were worried about what they viewed as another environmental disaster waiting to happen at the NCBC. This was far from an idle concern for a population that in many places was just beginning to recover from the devastation of Hurricane Camille. As the storage piles at the facility built up over time and became more visible from the shoreline, residents and local officials made known their anxieties over possible contamination of coastal waters and local aquatic life, both of which would have a major impact on the local fishing industry. The executive director of the Mississippi Air and Water Pollution Control Commission requested that the herbicide "be removed immediately and without regard to the final disposition of the material. . . . It is felt this is absolutely essential because of the proximity of the material to recreational and shellfish waters, as well as large densely populated areas, and further because of the history of hurricanes and tornadoes in that particular section of the country. It is our feeling there are many other areas in the continental United States which would provide a much safer depository for this material."[48] The Chamber of Commerce in Gulfport joined the anti-herbicide movement, passing a resolution decrying the negative effects that "human and environmental damage" from the herbicide storage at NCBC would have on local business. Several letters from residents to local papers and to the commander of NCBC fretted about local health issues, most of which revolved around dioxin, about which residents admitted they knew very little. As the mayor of Gulfport expressed it in his letter to NCBC, "Like most people, I know very little about dioxin, but from what I can understand, it is one of the most deadly chemicals known to man."[49]

Sen. Daniel Inouye of Hawaii was similarly troubled by the impact of the Johnston operations on his home state and its environment. In a letter to Secretary Seamans, Inouye requested clarification of the plans for the herbicide stocks in storage at Johnston, which, he noted, was located only eight hundred miles from Honolulu. "It is my understanding," the senator wrote, that the cost of the storage was currently "$11,000 a month; every day approximately 15 drums start to leak; 'herbicide orange' has been found in the local water supply; and a significant amount of the coral under the island has been killed off by these leaks." The USAF responded that the senator's figures on the cost of operations and the level of leaking drums were correct, but that it was unaware of any damage caused to the coral or any dangers related to consumption of the drinking water. Although 2,4-D and 2,4,5-T were showing up in samples of drinking water, the USAF did not "consider the levels to represent a toxic threat." Nevertheless, Seamans explained, the EPA was called in to consult and agreed that the levels did not pose a human health risk.[50]

Inouye and others had expressed their doubts as early as 1973, at least a year before the final decision on incineration had been made, well before the USAF filed its final EIS, and more than four years before Pacer HO began in earnest, on April 29, 1977, when the military invited local and national media to a "kickoff" briefing at NCBC Gulfport.[51] Beginning with that briefing, the USAF continually stressed the safety precautions built into the program and the degree to which the personnel conducting operations at Gulfport and Johnston were protecting themselves and the surrounding environment from the potential dangers of dioxin contamination.

Throughout Operation Pacer HO the USAF seemed to demonstrate a fairly high regard for national environmental policy and, to a degree, for the natural environment itself. This attitude is striking: the same military establishment that in many instances continued to insist on the safety of the chemicals and that, at this very time, continued to deny that reports of health problems among Vietnamese citizens and even among its own troops were the result of exposure to the dioxin present in Agent Orange, was now often going to extraordinary lengths to ensure the safety and protection of the landscape, seascape, wildlife, and people in the path of Operation Pacer HO. This was no accident; it was a result both of the shifting environmental awareness that had taken root in American society and of the laws and regulations engendered by that awareness. Regardless of what the Pentagon, the JCS, and USAF commanders felt about Agent Orange, they were now compelled—culturally, politically, and legally—to provide not only basic safeguards but levels of redundant safety precautions that soldiers and civilians could never have dreamed of during the Vietnam War.

The facility in Gulfport where the barrels would be emptied, rinsed, and crushed included a brand new high-volume ventilation system with charcoal filters that turned over the air supply nearly once per minute.[52] Over the next month more than fifteen thousand drums went through this process at NCBC, after which the herbicide was transferred to rail-based tankers and then pumped on to the *Vulcanus*. Young describes the working conditions of the personnel at both Gulfport and Johnston: "All of the workers were provided daily changes of freshly laundered work clothes, and the men working within the de-drum facility were provided protective clothing including cartridge respirators, face shields, rubber aprons, and rubber gloves. With only a few exceptions, the men rotated through all jobs involved in the dedrumming and transfer operations. All personnel were given detailed pre-operational and post-operational physical exams."[53]

The evidence from photos and motion pictures shot at Gulfport substantiate these claims. The images of Pacer HO personnel wearing aprons,

Figure 7. U.S. troops spraying base perimeter, Can Tho, 1967. (Courtesy National Archives, College Park, Maryland)

American soldiers during the war rarely used appropriate safety precautions when handling Agent Orange and other herbicides. This was particularly true when they sprayed base perimeters by hand, as these soldiers stationed in Can Tho in 1967 are doing.

gloves, protective eyewear, and ventilation masks stand in stark contrast to the images from Pacer IVY, when Vietnamese women working at Da Nang Air Base wore no protective gear whatsoever while working with discarded drums and to images of shirtless U.S. soldiers spraying herbicides around their base (fig. 7). Other instructions reinforced the basic safety precautions to be followed by Pacer HO personnel:

> 1. Clean clothes every day. 2. Wash before eating. 3. Rinse off any significant herbicide in contact with skin immediately and also change any saturated clothing. 4. Immediately notify Sgt. Hatch of any skin irritations or nauseous feelings. 5. Smoke in designated area near wash area after washing hands. 6. Do not discuss project with anyone off base.[54]

While similar precautions were taken at Johnston, what is most notable about the shift in environmental regulation and awareness in the documentation on Pacer HO is the concern for the natural environment surrounding these operations.

In multiple versions of the EIS produced by the USAF and in the voluminous data and notes compiled by its staff, hundreds of pages are devoted to the migratory, nesting, and breeding patterns of multiple bird species; several hundred more reveal the impact of herbicide storage on local coral reefs. Dozens of pages explain the intricate, multistep process for scrubbing the tanks of the *Vulcanus*. Separate volumes and appendices describe the long-term, multiyear procedures for environmental monitoring at both Gulfport and Johnston. Even the bulky report itself notes that it was printed on recycled paper! Hundreds of thousands of man-hours over nearly a decade were committed to demonstrating the potential environmental impact of the operation. Indeed, so much attention was paid to the disposal and incineration of Agent Orange that the National Archives contain nearly as many records specifically pertaining to Operation Pacer HO as they do to the entire history of herbicide use during the Vietnam War.[55]

Among those records are eight rolls of 35mm film shot at Gulfport and Johnston during Pacer HO. Made ostensibly to demonstrate the extent of safety precautions followed during the operations, the title card at the beginning of the sequence is actually entitled "Exposed With Pride," a mantra embraced by the majority of personnel working on Operation Pacer HO in an attempt to counteract the growing claims of Vietnam veterans that Agent Orange exposure was to blame for a variety of health issues. The men working on the draining, rinsing, and crushing of the barrels in the scenes that follow, however, are wearing multiple layers of protective gear, precisely to ensure that they were not exposed in the same manner in which troops and civilians during the war might have been (figs. 8a and b).

The film goes on to trace the incineration process from start to finish, including hours of footage of the machinery and personnel involved, close shots of demonstration slides describing the "flow rates" of the herbicide as it moved through the system, and numerous shots of Pacer HO workers meticulously documenting their activities, jotting down notes and manipulating the knobs of the complex industrial apparatuses used to process the waste. The footage includes striking shots of the vast landscape of weathered barrels populating the beaches of Johnston before their destruction. In scenes that look as though they might have been shot by Hollywood cinematographers, endless rows of barrels fill the screen, an indication of what 2.3 million gallons of herbicides left behind. In the final shot of one roll, just after the Johnston personnel bid farewell to Agent Orange, holding a sign reading, "Good Bye Herbie!!," the camera traces the herbicide out to sea, as the sun sets rather beautifully over the *Vulcanus* (figs. 9 and 10).[56]

Figures 8a and 8b. Safety precautions, Operation Pacer HO, 1977. (Courtesy National Archives, College Park, Maryland)

Unlike American soldiers during the war or Vietnamese civilians participating in redrumming operations in Operation Pacer IVY, those who worked on the disposal of Agent Orange during Operation Pacer HO wore protective gear and closely followed safety regulations.

Figure 9. Documenting Operation Pacer HO, 1977. (Courtesy National Archives, College Park, Maryland)

During Operation Pacer HO, the USAF was required to navigate the newly created environmental regulatory apparatus in the United States. As a result, the efforts to dispose of Agent Orange by incineration became arguably the most heavily documented military operation of the entire Vietnam War.

Not everyone on Johnston was put at ease by the precautions. USAF Staff Sgt. Jerry Firth was extremely apprehensive about the dangers of working at Johnston. In a letter to Sen. William Proxmire of Wisconsin dated August 27, 1973, Firth said he and other men "were literally scared to death to drink any water" on the base. "We are drinking soda pop, milk and the like. We are also concerned about food mixed with the water and food washed with the water." Firth's wife, he noted, had seen a local news program about the presence of dioxin in Agent Orange and was now gravely concerned about the welfare of the men stationed at Johnston.[57] Despite his claim that he spoke for "a majority of the people stationed here," his letter is one of only two pieces of clear, tangible evidence expressing worries about health among the troops at Johnston, and the only one from the time of the operation. The other would not surface until 1999, when Rep. Bob Stump of Arizona, chair of the House Veterans' Affairs Committee, forwarded to the USAF the suspicions of his constituent Neil Hamilton about possible dioxin exposure

Figure 10. The *M/T Vulcanus* incineration ship, 1977. (Courtesy National Archives, College Park, Maryland)

Although the burn zone for Operation Pacer HO was chosen because of its remoteness, incineration had been proven an effective way to eliminate the dioxin contaminant in Agent Orange. As it did throughout most of the operation, the USAF demonstrated a relatively high regard for environmental safety when selecting, monitoring, and cleaning the *M/T Vulcanus*, the Dutch incineration ship responsible for destroying the last remaining supply of Agent Orange.

while he was stationed on Johnston.[58] These examples stand in stark contrast to the mantra "Exposed With Pride." As was true of much of the history of Agent Orange, however, the ways in which operations were described on paper often conflicted with what happened on the ground. A closer look at the operations in Johnston and Gulfport alike calls into question how scrupulously the safety guidelines involving drum disposal, spillage, and waste were followed.

One of the major issues facing the USAF was what to do with the discarded drums, which totaled more than forty thousand by the end of the operation. USAF engineers working on Pacer IVY raised concerns about barrel disposal long before they knew the fate of the Agent Orange supply. "Exclusive of the method finally chosen for the disposition of herbicide Orange," a USAF memo from 1973 noted, "the problem of drum disposal will remain." While the most logical, cost-effective solution was to clean and recycle the drums, "certain political and economic factors discourage this option."⁵⁹ As was the case with Agent Orange itself, which few wanted in their backyards regardless of whether it was reprocessed, reused, stored, or incinerated, few communities were willing to accept the storage, reprocessing, or burial of the drums. Throughout the process the USAF was receiving mixed messages from regulatory agencies about the potential dangers and best methods of dealing with the drums. In response to its revised EIS of 1972, for instance, Assistant Secretary of Health, Education, and Welfare Merlin Duval criticized the excessive focus on drum disposal, stating that they could easily be cleaned and reused rather than destroyed or buried since "their contents were never that toxic." Duval also criticized the USAF for suggesting that "herbicide orange must be considered a very hazardous chemical, which it actually is not." Only four days earlier, the USAF had received a formal reply from the EPA that addressed the multiple dangers of Agent Orange to the areas around the proposed incineration sites and discussed at length, as did later comments by the Department of Interior, the challenges of drum disposal.⁶⁰ Regardless of their assertions about the perils of dioxin and Agent Orange, all of these comments placed a high priority on finding a way to recycle, rather than bury, the remaining drums, further demonstrating the reach of environmentalist thinking within the federal government.

In the end, the drums left over at the end of Pacer HO were largely recycled as scrap metal, but only after extensive operations involving draining, rinsing, "weathering" (that is, allowing the barrels to dry naturally in the outside storage area on Johnston), cleaning, and, finally, crushing. Aside from the contentious matter of burying dioxin and general waste in a landfill, the major advantage of recycling was that while being reprocessed for steel production the barrels would be subjected to even higher temperatures over a longer period (2900°F over six hours) than the herbicides incinerated on the *Vulcanus* (2000°F for just a few seconds), thereby ensuring the destruction of the dioxin residues. As was characteristic of its actions in Pacer IVY and Pacer HO as a whole, the USAF seemed to undertake the tasks involved grudgingly, noting in its description of the barrel processing at Johnston that the extra precautions were taken to mollify worries stemming from the "controversy surrounding herbicide Orange." The final rinse and weathering, the EIS claimed, was of questionable necessity but was nevertheless "accomplished in keeping with the overall intent of minimizing the potential for adverse environmental impact."⁶¹

In the documentary record of Pacer HO, alongside images of protective

Figure 11. "Chemicle Waste Disposal Area": The other side of Pacer HO. (Courtesy National Archives, College Park, Maryland)

Although most personnel involved with Operation Pacer HO embraced the mantra of being "Exposed With Pride" to Agent Orange, others raised concerns about exposure to dioxin. Evidence from the period calls into question how closely new environmental regulations were followed during parts of the operation. Spelling issues aside, official USAF photos reinforce the fact that Agent Orange left "bad stuff" buried in a number of sites around the world.

gear and safety precautions, other images tell of the larger environmental fallout. The endless rows of stacked barrels at Gulfport and Johnston produced large puddles of herbicide drainage and spillage from which dioxin seeped into the soil. Other images show the burial of waste products in small landfills in the ground. In one revealing sequence of images, USAF Lt. Col. C. E. Thalken, who coauthored a number of internal reports and scientific publications on the persistence of TCDD in the soil and was involved in the long-term monitoring of Gulfport and Eglin Air Force Base, inserted handwritten notes on pictures documenting the site of a disposal area on site at the NCBC. A sign at the site reads, "Keep Out: Chemicle Waste Disposal Area," to which Thalken initially responded with a question mark and the reply "chemical." On a close-up of the sign Thalken pasted a note that reads, "This is really bad stuff buried here. *Chemical* is spelled wrong" (fig. 11).[62]

Spelling issues aside, there was indeed bad stuff buried there, and elsewhere too. Along with some of the unidentified waste disposal shown in the images, most of the waste from Gulfport, including all of the protective gear worn by personnel, was buried in a landfill in nearby Bay St. Louis, Mississippi.[63] Of greater concern was the dioxin residue in the soil at and around the NCBC itself. The incineration required long-term monitoring studies to be conducted at both Gulfport and Johnston, studies which remained in place for years and, in the case of Johnston, decades. Early monitoring reports from Gulfport show that the USAF was well aware of soil contamination in the storage areas, but the focus of the reports tended to be on the drainage ditch that ran through the compound and could have transported the dioxin off the base. Some early tests at Gulfport showed TCDD levels as high as two ppb in crayfish, mosquitoes, and fish from the drainage ditch, between seven thousand and nine thousand feet downstream from the storage area. At the time of the studies, the federal government had placed the "permissible limit" of TCDD at one ppb. These initial reports showed that between two and four acres of the fifteen-acre storage site at Gulfport were contaminated with dioxin at meaningful levels, in some cases more than two hundred ppb, but the consensus of the authors was that the dioxin was dissipating over time and that the drainage ditch was not moving dangerous levels of dioxin into the environment. The recommendations of the reports were to continue monitoring but also to leave the area undisturbed to allow for the natural biodegradation of Agent Orange and its associated dioxin.[64]

A few years later, when soil samples from the storage area were still showing TCDD levels in the hundreds of ppb, the USAF decided to undertake a major decontamination effort through the reprocessing of the heaviest areas of contaminated soil by incineration and, eventually, reburial in a contained, underground cement landfill. The final stages of this operation concluded in 2001, but as of 2010 the EPA and the State of Mississippi had not certified the site as environmentally safe.[65]

Similar long-term studies were conducted on Johnston Atoll, which experienced a number of the same problems as those at Gulfport, Da Nang, Bien Hoa, and other major storage sites. Because Johnston was home to the largest number of drums for the longest period of time and because of the effects of salty water and air on the inventory, the attention to barrel condition and redrumming was even greater there. The EIS for Pacer HO claimed that the sandy, coral soil on and around Johnston would absorb the herbicide and TCDD and limit its passage into the surrounding area, including the water. Given the levels of herbicide in and around the storage and transfer areas, however, there was a legitimate basis for concern. One memo from 1978 noted that as many as thirty thousand gallons of Agent Orange may have been spilled over the six years of storage and redrumming operations on the island. Still, throughout the long-term monitoring process, conducted by USAF staff as well as independent contractors and scientists, the water, air, wildlife, and soil on Johnston, outside of the storage area itself, demonstrated insignificant levels of TCDD that consistently dissipated over time. The overall impact of Pacer HO on Johnston and its surroundings appears to have been, as the initial reports claimed, negligible. By 2004 even the heavily contaminated storage area was showing levels of less than one ppb TCDD, allowing the USAF to proceed with its plan to terminate the mission on the atoll pending a final environmental review.[66]

The USAF handled Pacer HO as it would have any military operation: efficiently, effectively, and with attention to detail. Yet the attention to monitoring and the concern for environmental consequences resulting from the operations were built into that mission. While Pacer IVY and Pacer HO should be considered overall successes from both military and environmental standpoints, the clear sense from the available documentation is that the majority of USAF actors involved saw this attention as unwelcome and superfluous. Air force personnel changed their clothes regularly, rinsed and weathered the barrels, and monitored the health of the coral reef and the migration patterns of birds because they *had* to. The point here is not that the military itself had an official change of heart about the potential hazards of dioxin during the process of destroying it. Regardless of the feelings of the USAF, a new regulatory apparatus was in place by the end of the war in Vietnam, one that forced the military to deal with the herbicides and the potential dangers, real or imagined, in a wholly different manner. As the example of Gulfport shows, even with stringent regulations and careful oversight in place, it was difficult, if not impossible, to avoid environmental contamination by TCDD in dealing with a large supply of Agent Orange. But similarly, as the example of Johnston shows, when the safety regulations and procedures were built into the process from the outset and adhered to closely, the impact on the natural environment and the personnel involved could potentially be minimized.

"Ignorance Is Bliss"

No such regulations and precautions were present during the 1960s and 1970s in the area around Times Beach, Missouri. Incorporated in 1925, the town began as a summer retreat for residents of St. Louis and for many years consisted largely of resorts along the Meramec River. Over time some vacationers turned their cottages into permanent residences, while others built mobile homes along the river. As traffic to the areas grew during the 1960s, dust from the dirt roads in and around Times Beach became a growing nuisance, leading to the contract with Russell Bliss for dust suppression via waste oil spray. After Bliss sprayed his dioxin-infused waste oil and the initial investigations revealed that the soil contained dangerous levels of dioxin, the CDC, which was leading the investigation, recommended that the contaminated soil be removed and either buried or incinerated. The matter eventually died in the infighting between various state and federal bureaucracies; much of the contaminated soil was left in place, particular hot spots were cleaned, and the lawsuits from cases against Bliss were settled out of court.

But the story of Bliss and Times Beach did not end there; two events conspired to keep it going. First, as part of an ongoing inquiry into health concerns in Missouri, a former employee at the Verona plant where Agent Orange had once been produced and which many in the area believed to be the source of the dioxin problems revealed to EPA investigators that nearly one hundred drums of chemical waste, later found to contain over two thousand ppm dioxin, had been illegally buried at several sites around Missouri. Second, while the EPA investigation and assessment continued into the fall of 1982, the Meramec River flooded Times Beach. The EPA tests at Times Beach, completed on December 3, confirmed the persistence of heavy amounts of dioxin in the soil. The Meramec overflowed its banks the very next day, leading to widespread worries that the dioxin had now spread across the entire area, making Times Beach virtually uninhabitable. Although the actual hazards posed by staying in the area would be debated for several more years, the federal government and the State of Missouri eventually bought out the entire community, leaving a potentially toxic ghost town.[67]

The manner in which residents of Times Beach came into contact with dioxin were strikingly different from that in which villagers and soldiers in Vietnam and USAF personnel at Johnston encountered Agent Orange. Vietnamese farmers were well aware that they, their crops, and their animals were being sprayed with chemicals, although they did not know the precise nature or known effects of exposure; the residents of Times Beach, by contrast, did not know they had been exposed, and although the scientific understanding of Agent Orange and dioxin had advanced quite a bit in the two decades since the beginning of Operation Ranch Hand, the mysteries of these chemicals continued to outweigh any certainties, particularly when it came to popular perceptions. The Vietnamese were caught in the midst of a war zone, where the spray was but one weapon, one potential danger among many others; people in Times Beach were exposed because of seemingly routine procedures of chemical companies operating in the area. Despite these and other differences, there were powerful similarities in the reactions of community members to chemical exposure in southern Vietnam and in rural Missouri. In their use of *Vietnam* as a signifier which helped them negotiate the meaning of chemical exposure; their concern about the unknowns and uncertainties surrounding these chemical agents; and their internal struggles over whom they should blame for their situation, the experience of the residents of Times Beach suggests a number of ways in which Agent Orange connects the many communities dealing with the toxic legacies of the Vietnam War. As the dioxin-laden landscape of Times Beach slowly moved toward its eventual fate—the abandonment of the town and the incineration of the contaminated soil—the uncertainties surrounding the effects of dioxin, as they had throughout Pacer IVY and Pacer HO, shaped and were shaped by a contested, fluid environmental regulatory apparatus that could detect, analyze, and ultimately dispose of increasingly discrete levels of dioxin but could do little to determine how the chemical would affect human health.

The residents of Times Beach, like most people in the world, had little, if any, knowledge about dioxin when they first encountered it in the early 1970s. Struggling to make sense of the deaths of animals at her farm, Judy Piatt literally followed Bliss's path in trying to determine the various sources of the waste oil he used. Among them was the Northeastern Pharmaceutical and Chemical Company (NEPACCO) plant, where the dioxin-infused still-bottoms originated. When a local veterinarian working on the case with the CDC told Piatt that dioxin was the culprit in the mysterious deaths of several of her horses, she replied, "Dioxin? What's that?" He informed her, "It's the stuff that's in Agent Orange." "I'd heard of that," Piatt later noted. "I knew they sprayed that in Vietnam."[68] Bliss's later defense also spoke to the evolving knowledge and regulation of chemical waste: "They are crucifying me for something I did twelve years ago that wasn't against the law. Back in the seventies nobody knew what dioxin was, nobody knew what PCBs were. You could take whatever you wanted and spread it anywhere you wanted and nobody thought anything about it."[69]

By the time the Missouri dioxin episode began to unravel fully in the early 1980s, the war in Vietnam had been over for several years, and the attention being given to Agent Orange and dioxin was growing because of veterans' claims that they had been exposed to the chemical and were suffering, along with their family members, serious health problems as a result. While the VA and the federal government continued to deny that most veterans were exposed and that exposure was in fact harmful in the manner veterans were suggesting, the issue of Agent Orange became the subject of litigation when a class action suit of veterans from the United States, New Zealand, and Australia launched what would become, to that point, the largest tort liability case in history.[70] Commenting on another case involving railroad workers in Missouri who had been exposed to dioxin after a railway accident, Gary Spivey, who was leading the Agent Orange study being conducted by the VA, argued that "the fear which is generated by current publicity is very likely to be the most serious consequence of Agent Orange."[71]

When several groups connected the health concerns of American veterans with the dioxin exposure in Missouri, residents of Times Beach and veterans expressed their feelings through the metaphors of war, battle, and Vietnam. When the town was permanently evacuated after the flood, one Times Beach resident responded to the EPA's order to abandon his home by placing a U.S. flag in his yard and declaring to a reporter from the *Washington Post*, "When you win a battle, you stick up a flag. This is a war to stay on the land you worked and paid for. It's a free country. I'm staying until they make me leave."[72] Another article in that week's paper quoted a Vietnam veteran as saying sarcastically, "Since the government had never admitted that dioxin was hazardous to the lives of men and women who served in Vietnam, could it be that dioxin is poisonous only when used in the United States?"[73] Even Bliss saw the issue through the lens of the war: "I could have had *dioxin* written on the side of my truck and nobody would have said anything, because no one knew what it was," Bliss later told a documentary film crew. "The Government said *it* didn't know when it sprayed it on our boys in Vietnam."[74]

Absent specific questions about whether or not dioxin was more dangerous than bombs or bullets, the residents of Times Beach quickly reverted to the tropes of warfare to frame the dilemmas posed by possible chemical exposure. These statements imply not only the use of Vietnam as a signifier in attempts to understand the risks of chemical exposure in Missouri, but also a strong similarity in the responses to the uncertainties of Agent Orange in Times Beach and Vietnam. As was the case in Vietnam, the uncertainties faced by Times Beach residents were a function of the extent to which dioxin pervaded their lives and their immediate environment. Confusing and conflicting statements from various authorities further complicated this realization.

After the further damage caused by the flooding, residents were forced to choose among possible risks. "We are afraid of the chemicals, but we are worried about the rats and snakes first," a local resident, Donna Mansker, told the *New York Times* on Christmas day in 1982. "We must put our fears in order. Surviving is the bigger problem." After the government revealed that the dioxin had been contaminating the soil in the area for nearly a decade, many residents adopted a fatalistic attitude about the risks. Many felt that if indeed they had been exposed since the early 1970s, the damage was already done. Faron Rowen, a local bartender, explained this position: "Why, if it's so dangerous, did the government wait ten years to warn us? I figure anyone's who's been contaminated done been contaminated. Why put your tail between your legs and run now after you've been living with it for so long?" Or, as Mansker put it, discussing the vegetable garden in her exposed backyard with a "nervous laugh," "We ate lots of dioxin." Another local resident ridiculed the invisible but seemingly ubiquitous nature of the dioxin threat, joking with his neighbor that "he thought he had seen 'a dioxin' run under" his front porch. Even the mayor of Times Beach, Sid Hammer, seemed to downplay the threat, echoing the claims made by many veterans of Ranch Hand and Pacer HO who claimed to have been "Exposed With Pride": "I've eaten it, inhaled it, and rolled around in it, and I'm in good shape. Only if they said, 'Sid next week you'll have cancer,' I'll start to worry."[75]

For others, that same invisible, elusive threat caused natural and consumer objects that had long been taken for granted in Times Beach to induce fear. "I used to cut wood for my stove right over there at City Park," Ernest Hance, Jr., told a *Times* reporter. Several articles noted the fear that vacuum cleaners and the dust inside of them were striking in the hearts of local homemakers after the EPA reported that samples from vacuum bags had registered unsafe levels of dioxin. Many residents felt anxious about the elusive nature of the threat. "You can't see it, taste it, or feel it," the owner of a local trailer park said "It's fear of the unknown." An epidemiologist at the CDC confirmed this sentiment: "Under no circumstances would I want to raise a family in Times Beach. There are too many unknowns at this point."[76]

Missouri residents were angry and confused by the conflicting information emanating from various agencies and authorities about potential harm from chemical exposure. By the time the episode at Times Beach occurred, dioxin was well established as a carcinogen and teratogen (causing birth defects in fetuses) in laboratory animals. The experience even of large animals like horses and livestock in the areas sprayed in Missouri reinforced these claims. Despite the allegations of veterans and many environmentalists and anecdotal evidence from Vietnam, links between dioxin and medical issues in humans were far from definitive. The tremendous attention given to veterans in the early 1980s certainly raised anxieties that dioxin was a carcinogen for humans as well, but even in heavily contaminated areas like Times Beach the timing and dose of the dioxin would have caused a great variation in individual responses. Still, the complex issues of determining risk, exposure, and harm in Times Beach were often oversimplified in public debates. Media coverage of Times Beach did little to help clear up debates over the science of dioxin, even when the reporting was accurate.

While many articles demonstrated that the effects were debatable—for example, two headlines in the *New York Times* read, "Dioxin's Peril to Humans: Proof is Elusive" and "Concern over Unclear Threat of Dioxin"— they often quoted wildly differing statements from scientists and government officials. Some articles stated without any qualification that dioxin was both a carcinogen and teratogen, but for the most part the press reflected the debates within the scientific community about the threats posed by the toxin. The *St. Louis Post-Dispatch* admitted in November of 1983 that in view of the conflicting information coming from various groups it could do little *other* than print contradictory headlines.[77] Even the American Medical Association weighed in on the matter, passing a resolution at its annual meeting (introduced by the Missouri delegation) stating that dioxin had been the victim of "a witch-hunt by the news media," which had overstated the health hazards it posed, Adding to the confusion were reports and studies showing dioxin to be a naturally occurring contaminant. Dow Chemical, under fire for producing Agent Orange and napalm during the Vietnam War and subjected to growing criticism after the commencement of the veterans' lawsuit, published a study in 1980 showing that dioxin was produced practically every time any substance was burned. Indeed, after dioxin was identified in Agent Orange, studies by industry groups and environmental groups alike reevaluated the presence of various forms of dioxin in pesticides, herbicides, and a wide variety of other consumer products, leading to claims and counterclaims that dioxin was therefore more of a threat because of its seeming ubiquity and, conversely, that its ubiquity demonstrated that it was not nearly as toxic as critics argued.[78]

Some of most confusing and misleading statements about dioxin, ironically, came not from the chemical industry or the media, but from the EPA. "It gets me that some scientists are saying that dioxin is the most deadly chemical known to man," said Rita Lavelle, a high-ranking EPA official dismissed in February 1983 in part for her mishandling of the Times Beach case. "That's not true. It depends on the concentration. In the right concentration, table salt is just as deadly."[79] Although Lavelle's statement was technically accurate and stated in the context of explaining how difficult it was to determine the extent of contamination and exposure at Times Beach, it was statements like this that sent the most problematic mixed messages to the public. Ultimately, the government buyout of the entire town of Times Beach did for dioxin what the veterans' lawsuit would do for Agent Orange: it bypassed the uncertainty among scientists and seemed to confirm the worst fears about the chemicals by suggesting that dioxin was dangerous and even deadly regardless of dose, timing, and the manner of exposure.

As was the case in villages throughout central and southern Vietnam during the war, questions about the uncertainties of the risks related to chemical exposure and about conflicting information from authorities readily translated into questions of who was to blame for the situation. Unpersuaded or unsympathetic to his claims that he was unaware of the poisons in his waste oil product, many people blamed Bliss. But why, he asked, would he have buried and sprayed thousands of pounds of the waste on his own farm if he had known it was toxic?[80] A popular bumper sticker at the time showed a tanker truck with the word *Dioxin* written on the side next to the phrase "Ignorance is Bliss." The *Economist* ran an editorial under the same title in February 1983. The more common target of criticism, however, was the U.S. government. Some residents of Times Beach used language that was virtually indistinguishable from that used by American veterans claiming to have problems from exposure to Agent Orange during their service in Vietnam. "We feel betrayed by the United States government," said Mansker. Others simply wanted to know the facts: "I am not trying to get anything out of this," said Ernest Hance. "I just want to know what's what." At a public meeting held in Hillsboro, Missouri, a shouting match ensued when federal officials tried to explain their evaluation process to local citizens. The government, said an EPA spokesperson, owed the public "nothing less than to give you as much information as possible" and was "as frustrated as you in not being able to do so. We have to make decisions based on the facts and we simply don't have the facts." One member of the audience reacted by shouting, "Bull!"[81]

In the aftermath of the Vietnam War and the Watergate scandal, many Americans came to lose faith and trust in their government. Yet in the area of industrial pollution and chemical spills, many citizens continued to rely on state-based experts in matters of public health and chemical exposure. In Times Beach, this was true even after the revelations of the disasters at Love Canal and Three Mile Island. Such reliance can likely be attributed to the complexity of the issues, the details of which eluded most citizens, but it is also one of the contradictions of American life in the 1970s: even after most Americans lost faith in their government after the debacles of Vietnam and Watergate, they often had little choice but to rely on government in situations like Times Beach. But reliance is not the same as trust. Rural residents of Missouri may have had little recourse outside of government in the case of Times Beach, but unsatisfactory results from the state quickly provoked their disappointment, their anger, and their voicing of distrust.[82]

The central problem was that neither the government, nor the chemical industry, nor independent scientists had easy answers. As difficult as it was for Missouri residents to understand at the time, there are no easy answers to dioxin. Although Lavelle mischaracterized the risk by equating dioxin with table salt, the reality was that determining the levels of the toxin present in the affected communities and its potential harm to local residents even before the flood was a painstaking process likely to result in information that was less than definitive. Many citizens of Times Beach, however, found it easier to regard the government's response as just another way of letting them down. The government remained much more part of the problem than part of a solution.

The frustrations over this uncertainty intensified when the matter of cleanup and disposal arose. Should the whole town be abandoned? or only the hot spots? Should the sites be cleaned up? if so, how? by incineration? by removal and burial? Even the more immediate concern of the level of danger associated with remaining in the area, however, revealed the difficulties in reaching agreement over risk determination. Bill Keffler, the EPA scientist in charge of dioxin testing, summed up the issue for the *Washington Post*: "Is it in the houses? Is it in the garbage? Did it get into the water supply? We've found it on the shoulders of roads, but how much is in the roadbed? How deep did it go down? Is it in the ditches? Will it be cheaper to buy out the town or clean it up?" Many residents were unwilling to move unless the government could prove that the levels of dioxin around their homes were dangerous to humans. Yet the type of certainty for which citizens clamored remained elusive, as the demand for answers continued to far outweigh supply.[83]

During the debate, residents became aware of the similar dilemmas posed by the Love Canal and Operation Pacer HO, but that did little to resolve their immediate concerns. "[The Air Force] got rid of Agent Orange by burning it," one EPA official related, "but such a huge amount of dirt is not feasible to burn. You can move it, but you can't get rid of it. Our approach is not to destroy it, but to assure public exposure can be limited to healthy levels."[84] Once the EPA and the CDC confirmed the presence of dangerous levels of dioxin in dozens of sites by early 1983, the federal government decided to proceed with a full buyout of the town. The EPA announced its program in February, allotting $33 million to purchase all eight hundred homes in Times Beach. Over the next several months, as the families relocated, many spent as much energy disputing the health issues about dioxin as they did the appraisal prices offered for their properties. "People have the perception that the government is trying to get in and out of this thing as fast and as cheaply as possible," James Cisco told reporters in October, a sentiment he believed applied to both home valuations and health studies. Residents were divided as to whether the government was understating or overstating the dioxin threat, but they were nearly unanimous in their distrust of EPA recommendations.[85]

By June 1983 only about a dozen self-described "river rat" families remained in the town, which by that point had no water or sewer systems, no functioning government, and no police force. Armed security guards kept watch at the town border to keep out would-be visitors. The holdouts had to carry their trash more than a mile to dispose of it. The post office refused to deliver mail to Times Beach. Adam Johnson, a former chemical plant worker who remained in Times Beach with his wife and two children, told the *New York Times* he was not worried about the dangers: "I've talked with chemical engineers, doctors, pediatricians, even my own veterinarian, and they've all told me the same thing—that the only thing dioxin has been shown to do to humans is to give them a rash. The only rash we get around here is poison ivy." "The kids are better off here," his wife, Willow, added. "This is our home." One week after that interview, the ongoing EPA investigation turned up a site, sprayed by Bliss more than ten years earlier and just a few miles from the Johnsons' home, with concentrations of up to eighteen hundred ppb dioxin in the soil.[86]

The initial cleanup plan called for excavation of the contaminated soil and burial in a concrete-lined protective landfill. The EPA estimated the total cost of the effort, including the home buyouts and relocations, at around $33 million. By early 1984 the costs had exceeded $41 million, and fewer than ten of the more than thirty contaminated sites had been excavated. As the government continued its multiple lawsuits to recoup some cleanup costs from the chemical companies, Bliss, and other defendants, it became more and more obvious that soil removal and burial might not be sufficient. Multiple studies showed that while the dioxin was not migrating into other areas or into the ground water supply, it was entering the food chain through animals. As it had in Operation Pacer HO, the EPA decided that the best way to destroy the greatest volume of dioxin was through incineration. Few volunteered to let the incinerator be built near their cities or even to allow the contaminated soil to be shipped through their state by rail or truck, even just across the river to the familiar site of the Monsanto/Sauget incinerator.[87]

The government had no remote area like Johnston Atoll at its disposal for this process, but Times Beach, a virtual ghost town, was as close to remote as possible and would not require crossing of state or county lines. By 1986, although the buyout process continued, the last residents of Times Beach had been relocated. Over the next several years the EPA implemented a multistep plan to remove and incinerate the contaminated soil, while improving the levee system along the Meramec to limit damage from future floods. At public hearings about the incineration plan residents from the surrounding area made objections over noise pollution, residual chemical contamination, and worries that the incinerator would become the processing site for toxic materials from other states in the region.[88] In February 1990 several hundred former Times Beach residents, many wearing gas masks, gathered on a highway overpass just outside the town to protest the decision to build the incinerator. Otis "Barney" Nelson, the mayor of nearby Eureka, which was downwind, worried that the supposedly temporary incinerator would never go away: "They have state after state with soil contaminated with dioxin. And they have those states all looking for a place to send it. You don't have to be that intelligent of a person to guess that the EPA doesn't want to go from state to state and fight this all over again." Nelson and other members of the newly formed Times Beach Environmental Task Force suggested shipping the waste to Sauget. The case became a classic example of NIMBY environmental politics, with cities like Eureka and Fenton threatening to sue the Syntex Corporation, the current owner of the IPC/NEPACCO plant, and the EPA should the incinerator plan move forward. Adding to the controversy, several stories in the press quoted current and former EPA officials to the effect that the dioxin threat was overstated and the forced evacuation of Times Beach unnecessary.[89]

A federal court ruled in 1991 that the incineration must proceed as planned in Times Beach.[90] By the end of that year what remained of the town was leveled, collected, and designated for the burn pile. Between 1996 and 1997 the incineration facility treated more than 250,000 tons of contaminated soil from twenty-seven Missouri sites at a cost of around $110 million, more than three times the EPA's original estimate.[91] In yet another parallel to Pacer HO, later studies showed that the incineration had been extremely effective in destroying the dioxin and that exposure to the emissions did not result in increased TCDD levels among residents in the surrounding area. In September 1999 the four-hundred-acre site where Times Beach had once stood was reincarnated as Route 66 State Park. Two years later the EPA conducted its last evaluations and labeled the area safe and the cleanup complete.[92]

New Plymouth, Dioxin, and the Burden of Proof

Thousands of miles from Times Beach, in New Plymouth, New Zealand, local constituents likewise debated, for several decades, the dangers of herbicides and their dioxins. They negotiated shifting frameworks for thinking about the relationships between chemicals, the body, nature, and the state, based on a rapid expansion of global knowledge about the effects of chemicals on humans and the natural environment. Like the residents of Times Beach, those of New Plymouth often reached their understandings through the lenses of Vietnam and Agent Orange. But local and federal governments in New Zealand were more likely than their counterparts in the United States to dismiss citizens' worries about dioxin, relying upon the framework of scientific certainty to make their case.

Tension between developing large-scale agriculture and maintaining a healthy natural environment had always existed in New Zealand. As the geographer Bruce Wildblood-Crawford explained in his doctoral dissertation of 2008, the country's "grasslands revolution" of the twentieth century relied on the use of pesticides and herbicides like 2,4-D and 2,4,5-T to transform millions of acres of land into viable grazing lands for sheep and cattle, especially after the Second World War.[93] Wildblood-Crawford shows convincingly that the revolution was fueled not only by chemical herbicides but also by the intertwined discourses of progress through technology and a so-called war on nature.⁹⁴ These conversations were also present in American culture during the same period, articulated in the development of herbicides themselves and in their various applications, from agriculture to the military.⁹⁵

A safety film produced by New Zealand's governmental Film Unit in 1960 makes the analogy clear. "War on the Farm" explicitly used the metaphor of war to discuss both the promise and peril of the growing use of pesticides and herbicides. "War on the Farm" opens with a comparison of modern herbicides and pesticides to the chemical agents developed and deployed during the First World War, warning that although they are safe when used properly, modern chemical agents are still very dangerous, even lethal, if safety guidelines are not followed. As an example, the film shows a shirtless farmer carelessly filling his tractor-pulled sprayer with the pesticide malathion and spilling it over the tank, the ground, and himself. The farmer then licks a rolled cigarette and sets about his task, spraying his orchard while virtually showering himself in pesticide. "Luckily" the film continues, there are a number of growers who have been "cooperating with the Departments of Health and Agriculture" by following the safety guidelines found on all herbicide and pesticide sprayers. The film concludes by reminding users to destroy empty containers, showing the farmer's child eating out of an empty carton of pesticide lying on the ground by his shed. The farmer, oddly undisturbed when his wife finds the child, picks her up, and carries her into the house, is scolded by the narrator: "Remember: You're at war with insects. You don't want to die like a fly. Don't be a casualty."⁹⁶ In New Zealand, no entity was as crucial to the development of this discourse as Ivon Watkins Dow (IWD) (fig. 12).

From the 1950s through the 1980s the IWD chemical plant in New Plymouth had been the sole producer of 2,4,5-T herbicide in New Zealand. While it did not produce herbicide directly in support of the war in Vietnam, the firm dramatically increased its sales during the late 1960s to coincide with a spike in rising demand fueled by the Pentagon's thirst for Agent Orange. After New Zealand brought its troops home in 1972–73, however, its citizens confronted the many troubled legacies of that war, including new knowledge about the effects of herbicides on human and environmental health. Much of this knowledge was based on other dioxin incidents around the world, including an explosion in 1976 at a chemical factory in Seveso, Italy, concerns among the residents of Alsea, Oregon, that women exposed to herbicides in the community were experiencing higher rates of failed pregnancies during the mid-1970s, and the mixed messages circulated by American regulatory agencies.

Figure 12. Ivon Watkins Dow plant, New Plymouth, New Zealand, 2011. (Photo by author)

Communities around the world were forced to deal with the potential threat of dioxin contamination in the decades after the Vietnam War. Among those most directly affected was New Plymouth, New Zealand, home to the Ivon Watkins Dow chemical plant that manufactured 2,4,5-T, but not Agent Orange.

Both the Seveso and Alsea incidents have been written about extensively. The key points as they affected the decisions of policymakers around the world in the 1970s have to do with what the ensuing data suggested about the effects of dioxin on human health. On July 10, 1976, the Seveso explosion released half a pound of dioxin into the air, with concentrations of around five hundred ppb. Chloracne, the one universal sign of exposure to high levels of dioxin, appeared on much of the population in the immediate area and dissipated over time. Some long-term studies have since postulated elevations in various health conditions, but as policymakers in the United States, New Zealand, and elsewhere made their decisions about 2,4,5-T in the late 1970s through the mid-1980s, there was no evidence of other health deficits resulting from the explosion.[97] In Alsea, which was located near federal forestry areas where 2,4,5-T and other herbicides had been in use, a local teacher named Bonnie Hill drew attention to what she and others believed was a high rate of miscarriages among the town's population. As the EPA investigated the claims, which were based on data that were far from scientifically viable, a number of scholarly studies weighed in, finding little evidence to support the assertions based on the small sample. A second study, commonly known as Alsea II, was commissioned by the EPA and completed by researchers at Colorado State. It found a meaningful increase in the rate of spontaneous abortions in Alsea compared with other populations. This study would later be largely discredited for failing to account for a number of potential variables, including other likely causes of the condition in question. Still, the Alsea studies became the basis for the EPA's decision, made in February 1979, to suspend 2,4,5-T permanently. Over the next several years the EPA held hearings on the cancellation of registration for 2,4,5-T, which the chemical companies initially fought before eventually suspending their production.[98]

When the EPA announced its decision, anti-herbicide activists in New Zealand flooded members of parliament and the minister of health with letters calling on the government to follow suit. Leo and Suzanne Neal, residents of Auckland, said the government's failure to follow the EPA's lead was highly irresponsible, only to be told by Minister of Health George Gair that he found no reason to pursue the ban in New Zealand.[99] Local papers covered the news from the United States, describing the seeming disjuncture between the growing public outcry, particularly from rural areas where 2,4,5-T was in heavy use, and the measured response of the government.[100] Far more than their American counterparts, however, New Zealand residents wrote explicitly about the burden of proof in such cases. In hundreds of letters written in 1979 alone, residents rehearsed the argument that the chemical companies and the people attesting to the safety of the chemicals must prove them safe. Barbara Faithful, writing to the health minister in April was incensed that the minister had recently described 2,4,5-T as being "quite harmless" to humans: "I suggest that the people of this country are virtually being held ransom by wealthy firms who challenge us, the people, to prove the chemical harmful," Faithful wrote. "I would respectfully suggest, Sir, that the onus should, rather, be with the chemical firm to *prove their chemical safe* and that it should be automatically removed from use until or unless this firm is able to do so." Stephanie McKee, a resident of Wellington, echoed these sentiments, writing to Minister Gair that "the demand is growing that the onus should be on the manufacturers and sellers of the herbicide to prove that it is safe, rather than on the population at large to prove that it is unsafe." Unswayed by these arguments, Gair and various members of parliament took the opposing view that since 2,4,5-T could not be "proven safe" there was no reason to discontinue its use. For his part, Gair came to believe that the media were fanning the flames of the story, producing what he regularly termed an emotive response, which he saw as being linked to the contentious domestic views of New Zealand's involvement in the Vietnam War.[101]

The government went to some lengths to discredit the decision of the EPA, which it saw as fundamentally flawed. The ministry sought counsel from a number of physicians, all of whom seemed to regard the decision as seriously flawed. One report noted "serious errors of judgment" in the EPA's handling of the data on animal experiments, the Alsea study, and follow-up studies on Seveso. The ministry was also provided with a thorough response to the decision by Dow Chemical, on company letterhead, with a note reminding the health minister that the reports were "courtesy of IWD."[102] Michael Bates, the senior toxicologist at the Department of Health, reviewed a number of studies, including Alsea, in preparation for submitting a report to the minister and found no conclusive evidence that 2,4,5-T was a health hazard. In a memorandum issued to "all Medical officers of health," Bates wrote that although the "evidence indicates that there is no clear threat to the public," there is usually "some outcry" around the issue every year. "In view of the emotional aura surrounding 2,4,5-T, we believe it is prudent to minimise public exposure as much as possible."[103] Particularly revealing is a letter from the director of the Department of Health in Nelson, on New Zealand's south island, to Bates asking him for his views on 2,4,5-T and noting his displeasure "with the blasted Americans suddenly banning it as dangerous after so conclusively proving a year or two ago, with their Vietnam investigations, etc., that it was quite safe. This latest news is going to be grist for the mills of all the environmentalists, and anti-spray, anti-chemical people we have in such abundance in this district."[104]

It was not only decisions in the United States that influenced the thinking of health officials in New Zealand. The Ministry of Health kept close watch on developments in Australia, Canada, and England as it made its recommendations on 2,4,5-T, and those nations were likewise in contact with New Zealand. In March 1979 the New Zealand embassy in Ottawa cabled Wellington that Canada was not yet planning to ban 2,4,5-T, preferring to wait until its Ministry of Health could study more "raw data." The same month the Australian government reiterated its position of allowing the use of 2,4,5-T, joining their colleagues in New Zealand in criticizing the Alsea study and the EPA decision. In its final report in May 1979 the Weed Control Authority in Australia quoted the New Zealand Health Ministry's findings, noting, "Quite frankly, we agree with the New Zealand investigators who stated, 'it must be faced that there is no way in which *any* substance, including common foodstuffs, can ever be proved absolutely safe. The best that can be achieved for any substance is a high degree of 'assurance' of safety based upon a rational and experienced scientific judgment of the available evidence."[105] At the end of the 1970s Australia, New Zealand, Canada, and the United Kingdom allowed 2,4,5-T to continue to be used without restriction, declaring it safe for humans, animals, and the land. To help make the point, an Australian farmer and veterinarian staged a public demonstration witnessed by many, including several newspapers. At the annual meeting of the New South Wales branch of the conservative National County Party, Michael Cobb drank a small glass of 2,4,5-T. "The level of dioxin I drank wouldn't give a grasshopper a headache," he told the press, "but the gimmick was the best way to get the message across."[106]

Yet the issue of 2,4,5-T and dioxin did not go away in New Zealand, as committed citizens and activists focused their attention on the community of New Plymouth and the IWD plant. In 1980 the Ministry of Health announced the release of a new study on workers at the plant, showing "no signs of adverse health."[107] That same year the annual meeting of the Royal Society of New Zealand featured a series of presentations focusing on the "Toxic Hazards of 2,4,5-T." The synopsis of the sessions read, "As is common in many such controversies, arguments are often based on emotional grounds, rather than on scientific evidence." On the basis of the available evidence, however, "there are few grounds for regarding 2,4,5-T itself as having any important health hazards. The contaminant dioxin is extremely poisonous and has a potential for causing damage to the developing fetus, mutagenicity and carcinogenicity. In the concentration currently present in the herbicide distributed in New Zealand, the possible hazard represented by the contaminant dioxin appears to have been negligible."[108] Many in New Plymouth remained unconvinced, especially when it came to the issue of incineration.

During the 1980s, as IWD continued to produce 2,4,5-T in New Plymouth, the local group Residents Against Dioxin (RAD) pressed for a parliamentary investigation into the safety of the herbicide. Founded in 1985, RAD had engaged in a number of popular epidemiology investigations of its own. Requesting government records through the Official Information Act, the group used environmental monitoring reports from the plant to develop airflow charts for the surrounding area while compiling an extensive collection of scientific studies and anecdotal information from the community. RAD was chiefly worried about the waste being released from the IWD incinerator during controlled burns, although the Ministry of Health regularly monitored the plant, particularly during scheduled incinerations, and found levels to be within government safety parameters.[109] Other citizens around the country developed their own studies, which they then sent along to RAD. Sheila Cregg of Wainuiomata, just outside Wellington, developed a detailed map of areas in her community sprayed with 2,4-D and 2,4,5-T, adding red dots indicating houses of families she knew to be suffering from illnesses they believed to be caused by exposure.[110] The activists were given a boost by a "Seveso-like event" at the IWD plant on April 15, 1986.

At 3:10 a.m. on that day, a disc in the No. 2 Autoclave of the plant ruptured, releasing hazardous gases, including several hundred milligrams of TCDD, into the air.[111] Despite the efforts of groups like RAD and Greenpeace, opinions about the manufacture and use of 2,4,5-T in New Zealand remained sharply divided even after the incident at the plant. In 1986 the New Zealand Environmental Council released a report entitled "The Use of 2,4,5-T in New Zealand," based largely on information gathered by members of RAD. The council met with representatives of RAD, managers from the IWD plant, and a variety of other groups but found opinion on the herbicide far more divided than the RAD materials suggested. The head of the Playcentre Federation, representing child-care centers around the country, wrote annually to the Ministry of Health between 1983 and 1988, citing a variety of international studies on the risks of 2,4,5,-T and supporting RAD's call for a ban on its production and use.[112]

A variety of other individuals and groups continued to oppose such restrictions for a number of reasons. Alan Parsons, a pediatrician in New Plymouth, blamed the media for the increased anxieties about the herbicide. In a letter to the official Committee of Inquiry convened to examine the impact of the incident at the plant, Parsons explained that he was dismayed to see one of his patients, whom he had recently treated for "a straightforward viral infection," profiled on the local news as having conditions caused by the IWD plant. "It seems that despite the growing evidence and the clear observations of intelligent people," he wrote, "that the media fail to understand the power that they exert over the population."[113] A far more common defense was the pragmatic and economic utility of 2,4,5-T. The Federated Farmers of New Zealand also took issue with media coverage but focused their defense on the ongoing usefulness of 2,4,5-T as a "cost effective approach to weed and brush control." Workers at IWD and other chemical plants opposed further regulations for fear of cutbacks in the workforce as a result of shrinking profits. The Food and Chemical Workers Union indicated that they were pleased to cooperate with the investigation if only to show it was "entirely satisfied that none of its members are exposed to health risks, and, secondly, to ensure that our members have a reasonable degree of job security that is not prejudiced unnecessarily by unsubstantiated and/or emotional and/or unreasonable claims made about the manufacture of, or application of this or that product which they produce." Both the farmers and workers were regularly supported by news articles calling chemical herbicides and pesticides, for example, "essential tools for New Zealand agriculture."[114]

As time went on, however, opponents of the herbicide made some inroads by arguing that the continued use of herbicides could have a negative impact on the economy, limiting agricultural exports to countries that had banned 2,4,5-T and calling into question the Pure New Zealand marketing campaign of the tourist industry that played on the nation's natural resources and environmental quality. Peter Whitehouse of Greenpeace New Zealand wrote to the minister of overseas trade and marketing in June 1986 to state he had received "a number of strong requests from Germany, Japan, and the United States" to supply information on levels of 2,4,5-T in exports that "would enable trade boycotts to be carried out against New Zealand."[115] Dozens of letters to the minister of health drew attention to the fact that New Zealand was now the only country in the world producing 2,4,5-T, which hardly fit the profile of the Pure New Zealand campaign. Many of these letters made a point of connecting 2,4,5-T to Agent Orange and the Vietnam War. Kayla Mackenzie-Komp described the herbicide as "the substance made notorious as Agent Orange during the Vietnam War."[116]

The final report of the Committee of Inquiry, commonly referred to as the Brinkman report after its lead author, G. L. Brinkman, the dean of the Otago Medical School, went even further than many of the letters, claiming in its introduction that preoccupation over 2,4-D and 2,4,5-T had arisen over the past fifteen years "as a consequence of the controversy surrounding the use of Agent Orange in Vietnam."[117] Although the report did not offer overwhelming evidence of the role of the war in shaping attitudes about the herbicides, the overall tenor of the report is that the suspicions were largely based in socially constructed attitudes that were ultimately unsupported by scientific evidence. It describes the perceptions of illness and risk among New Plymouth residents, noting, "There is also what appears to be diminishing faith among some lay people in expert scientific findings indicating that fears of this kind are without foundation." The committee compared these perceptions to the experiences of farmers and commercial chemical workers, who "have used the chemicals for years, and are often soaked to the skin when spraying" and yet did not report adverse health conditions for themselves or their families.[118] Ultimately, the report's conclusions rejected the claims of New Plymouth residents: "We can find no substantiated evidence that the manufacture of these pesticides has had any ill effect on the health of the residents of New Plymouth. There is certainly no indication from overseas experience that acute exposure to 2,4,5-T or dioxin has any long-term deleterious effect on health. The risk of prolonged chronic exposure is still under investigation but the fact that proof of any adverse effect is so difficult to obtain, even after 30 years of use, indicates any such effect must be subtle rather than catastrophic."[119] The committee also concluded that to ban 2,4,5-T would place "an unnecessary financial burden" on farmers and chemical workers. While other countries had banned it, the Brinkman report made its recommendation on the basis of the "special case" of New Zealand, observing that "the dioxin level in 2,4,5-T made by IWD is now the lowest ever achieved in any commercial manufacturing process" and that "brush weeds are a particular New Zealand problem."[120] While ultimately recommending that dioxin levels in 2,4,5-T made at IWD be further limited to one ppb after 1987 and calling for ongoing research in the area, the upshot of the report was clear and very much in line with the New Zealand government's approach to dioxin issues since the 1960s: it would continue to allow the manufacture and use of 2,4,5-T within the safety guidelines established by the Ministry of Health.

Not surprisingly, the Brinkman report did not bring an end to the nation's concerns about 2,4,5-T, Agent Orange, and dioxin. Neither, in fact, did IWD's announcement in 1987 that it had found an even more effective and cost-effective replacement herbicide called Grozon.[121] Despite voluntarily phasing out production of 2,4,5-T in 1987, the IWD plant and the incinerator continued to unnerve activists and residents of New Plymouth. In late 1988 Parliament was considering a ban on the herbicide, a prohibition that would have prevented farmers and other users from exhausting the remaining stockpiles of 2,4,5-T, which were expected to be depleted by 1991. In 1990 IWD voluntarily relinquished the last 2,4,5-T registration, and by the end of the year 2,4,5-T was seemingly gone for good.[122] The legacies of Agent Orange, however, endured.

Conclusion

On the cover of its special retrospective on the Missouri dioxin episode, "Dioxin: Quandary for the '80s," the *St. Louis Post-Dispatch* featured two pictures: the first showed a group of local residents peering anxiously through the locked doors of a local hotel, awaiting the announcement of the government's proposed buyout of their community. Their faces register a mixture of concern, frustration, anger, and uncertainty. The second photo shows an abandoned garage in Times Beach, spray-painted with the message, "Gone and Forgotten by the U.S. Government." These images stand in stark contrast to the smiling faces of Operation Pacer HO personnel as they waved goodbye to "Herbie," the men who continued to claim the pride of exposure while safely encased in their protective gear. The juxtaposition of these images offers poignant summations of the great uncertainties surrounding debates over Agent Orange in the late 1970s and the 1980s, the growing divide between those who believed that dioxin was no more dangerous than table salt and those who believed that this invisible, elusive threat was responsible for a growing list of human and environmental health disorders. As the threat of dioxin and other chemicals became a more common feature of everyday life around the world, this divide was no longer easily broken down into categories such as military and political or soldier and civilian.

Over the course of the late 1970s, the 1980s, and the 1990s a growing number of global citizens, soldiers and civilians alike, confronted the same types of questions and apprehensions faced by the residents of Times Beach and New Plymouth. Vietnamese villagers, war veterans from Vietnam, the United States, Korea, Australia, and New Zealand, and residents of polluted areas near manufacturers of chemicals and military test sites all came to deal with confusion, frustration, and uncertainty as they encountered the global legacies of Agent Orange as well as the global reach of the chemical–industrial complex it sprang from. In facing the elusive enemy of uncertainty, however, these and other groups would come to realize that uncertainty itself was in many ways a product, like Agent Orange, of the chemical–industrial complex and of the legal, political, and economic forces that supported it.

CHAPTER FOUR
THE POLITICS OF UNCERTAINTY
Science, Policy, and the State

In March 1965, a crucial month in the escalation of the Vietnam War, the United States implemented a series of actions and policies that would forever link the fate of the two nations and shape the future of millions of people around the world. On March 2 the United States formally launched Operation Rolling Thunder, a sustained bombing campaign against the revolutionary forces of Vietnam. Over the next three years the campaign would drop a greater tonnage of bombs on northern Vietnam than were dropped throughout the entire Pacific theater during the Second World War. On March 8 two Marine battalions totaling around 3,500 men arrived at Da Nang, the first U.S. ground combat forces to be deployed in support of the American war to preserve the South Vietnamese regime. Within three years their numbers would total nearly 550,000. While these two decisions—a massive escalation of the air war and the introduction of ground combat forces—would come to define the terms of battle for the United States for years to come, another decision was made that month that would have similarly significant long-term consequences for those involved on all sides of the war. Yet this decision was not made at the White House, the DOS, or the Pentagon. Nor was it made by a government official, elected or otherwise. Nor was it made even in Washington, D.C. This fateful decision was made in Midland, Michigan, home of the Dow Chemical Company, one of the largest manufacturers of Agent Orange.

On the evening of March 23 a number of small private planes landed on the runway at Midland's small airfield. They were carrying representatives of the four other largest manufacturers of Agent Orange: Diamond Alkali, Hercules, Hooker Chemical, and Monsanto. The representatives—plant managers, toxicologists, and medical directors among them—were shuttled by limousine to the Midland Country Club. The country club had hosted executives from chemical companies before, but this meeting was different. It came about at the urging of Dow, which summoned its chief competitors to Midland in the hope of finding a solution to a pressing problem they all shared: the toxicity and health hazards of 2,4,5-T.

In the invitation he sent to his colleagues, Vernon K. Rowe, director of the Biochemical Research Laboratory at Dow and, at the time, president of the International Society of Toxicology, explained the purpose of the gathering: "To discuss the toxicological problems caused by the presence of certain toxic impurities in certain samples of 2,4,5-trichlorophenol and related materials. As I told all of you with whom I have talked, we have been doing analytical and toxicological research on this problem and wish to share our findings to date with all the producers of 2,4,5-trichlorophenol for the sole purpose of lessening any hazards to health that might be attributed to this and related products."[1] Rowe concluded by noting that it was his and Dow's hope "that through this meeting, we will acquire a better understanding of the problem and that each company will then proceed independently as it sees fit to institute such self-imposed controls on its production as are necessary to insure the safety of its products."[2] Another Dow employee put it somewhat differently in a memo to Rowe three weeks before the meeting: the crux of the agenda was for Rowe to "sell the group on the idea that industry must police itself or the government will." "If they accept this philosophy," the memo concluded, the companies would need to decide "if they should go and who should go to the appropriate federal government agencies."[3]

It remains unclear exactly what took place at the meeting and what, if any, collective decisions were made about dealing with the dioxin levels. Records indicate that Rowe divulged Dow's experience with dozens of employees who had experienced chloracne outbreaks as a result of dioxin contamination. The internal Dow summary of the meeting says only that the viewing of laboratory animals subjected to heavy amounts of dioxin "appeared to have considerable impact" on those in attendance.[4] Given, however, that the chemical companies never went to "the appropriate federal government agencies," let alone the Pentagon, to inform them about the presence of "toxic impurities" in their product, and given the wide discrepancies in dioxin levels in the Agent Orange produced by the various manufacturers, apparently the only decision made at the Midland summit was to continue business as usual. Tests performed on the inventory during Operations Pacer IVY and Pacer HO would show that the herbicide produced by firms at the meeting, including Monsanto, the single largest producer of Agent Orange during the war, had average levels of TCDD well above the level of one ppm sought at the time. Dow, however, continued to work on the problem, producing herbicide with comparatively low levels of dioxin. Still, the company did nothing to alert authorities to the potential dangers they knew at the time—prior to the widespread introduction of Agent Orange into the Vietnam War—were present in their product.[5]

Two months after the Midland summit Rowe sent a memo to the bioproducts manager of Dow Canada urging him to invite "big customers" to come to Midland and see for themselves what he had shared with the other chemical companies. "We are not trying to hide our problems under a heap of sand," Rowe noted. "The primary objective is to avoid situations which will cause the regulatory agencies to become restrictive." Rowe closed with a reminder to his colleague to be judicious with the information contained in the memo and accompanying reports: "Under no circumstances may this letter be reproduced, shown, or sent to anyone outside Dow."[6] Rowe's desire for transparency had its limits. Presumably, sharing the issue with the U.S. government, already Dow's largest customer for herbicides, would trigger the very regulatory oversight it wanted to avoid.

In 1970 Julius Johnson, then director of research at Dow, would testify before the Senate Energy Committee that Dow indeed had known of the presence of dioxin in 2,4,5-T in 1964. With Rowe seated next to him, Johnson described at length that Dow had invited its colleagues from other chemical firms to Midland and had notified the Michigan Department of Health, the University of Michigan, and "other health oriented individuals in private medicine and industry" about its procedures for manufacturing clean herbicides, that is, with no detectable traces of dioxin. Dow was not asked by Congress to substantiate its efforts to contact the State of Michigan. The company has never been able to show evidence that it gave any notification either to the U.S. Department of Agriculture (USDA) or to the Pentagon, its two largest herbicide clients in 1965.[7]

The chemical war coincided not only with the explosion of environmentalist thinking in the United States and elsewhere, but also with the exponential growth in the ability to measure and detect trace amounts of molecules like dioxin. In the spring of 1965 technical knowledge about dioxin and its effects on humans was limited to a relatively small group of people around the world, nearly all of whom were scientists and several of whom attended the summit at Dow's headquarters. By the end of Operation Ranch Hand in 1971, largely because of the policies of herbicidal warfare pursued by the United States in Southeast Asia, that knowledge had spread considerably. As Operations Pacer HO and Pacer IVY and the responses to dioxin in Times Beach and New Plymouth showed, by the late 1970s technical experts were able to detect discrete amounts of dioxin in water, soil, and wildlife. The precision of measurement, however, did not translate easily into determining the toxin's effects on humans, linking dioxin exposure to particular illnesses and conditions, or to prescribing treatments for remediation. This disjuncture between scientific expertise at the molecular level and an inability to predict or determine accurately what the effects of exposure to those molecules would be led to a series of political and legal battles over the legacies of Agent Orange from the late 1970s through the 1990s, many of which remain unresolved to this day.

Questions of exposure, risk, and consequences were debated ad nauseam in courtrooms and congressional hearings in the United States, Australia, New Zealand, and elsewhere in the late seventies and eighties, veterans, scientists, and policymakers repeatedly squaring off at the powerful confluence of experiential reality, scientific knowledge, and state authority. Veterans stridently pressed their case on exposure despite evidentiary gaps in the historical record, and they grew frustrated both with lawmakers, who demanded proof of exposure and causal links to specific illness, and with scientists, who could not provide such proof. As veterans and other citizens affected by Agent Orange sought specific, knowable facts about what Agent Orange might have done to them and their loved ones, they situated their own experiences and beliefs as a powerful counternarrative to those set forth by science and the state.

But narratives are rarely equal, and knowledge is never neutral. In her study of "biological citizenship" and the Chernobyl nuclear disaster, Adriana Petryna writes, "The processes of making scientific knowledge are inextricable from the forms of power those processes legitimate and even provide solutions for."[8] The same is true of debates over the effects of Agent Orange, which were shaped by "multiple, intersecting, and competing epistemologies."[9] While veterans and their advocates made powerful cases through their personal narratives, they were immersed in legal and political structures that continued to privilege objectively measurable effects over experiential bodily trauma. In their challenges to these structures, Vietnam veterans around the world have relied on memory, personal testimony, and the power of their experience to contest both scientific knowledge and state authority. As I argue, however, they have often done so in problematic ways, relying on the language of conspiracy and coverup to offset historical and evidentiary gaps. Whether in the form of so-called popular epidemiology or populist protest, one must consider the experiential narratives of the veterans alongside what the equally imperfect and always incomplete documentary record, both scientific and historical, reveals about the limits of Agent Orange exposure and the likelihood that it is (or is not) to blame for a variety of health conditions among veterans, civilians, and their offspring.

By tracing these battles across geographic and temporal boundaries, I seek to follow the call of environmental and public health scholars to situate uncertainty itself as "an historical artifact, produced between particular ways of apprehending the world or by clashes between different versions of the world."[10] The participants in the debates over Agent Orange exposure and causality, whether veterans, scientists, or agents of the state, articulated competing ideas about chemicals, nature, and the body. As several works in this literature have demonstrated with regard to other chemicals and other constituencies, much of the debate over Agent Orange was constructed by all sides within a model that relied on conceptions of risk, exposure, and clear distinctions between individual human bodies and the surrounding environment that had been rendered increasingly problematic, if not obsolete, by a world in which humans, chemicals, and environments have become largely inseparable.[11] Just as courts rewarded uncertainty in Agent Orange litigation while policymakers punished it when dispersing benefits, the chemical industry's production of uncertainty in Agent Orange cases relied upon the epistemological and ontological framework of separable, individuated chemicals, bodies, and environments. Too often, veterans and others who encountered Agent Orange worked within this framework even when attempting to counter the assumptions of the opposition. They were seeking answers to a number of questions that could not, and likely never will be, answered.

Framing Uncertainty

From a policy perspective, scientific uncertainty poses a number of challenges to lawmakers. The U.S. Senate first began to explore the issue in April 1970 at a Commerce subcommittee hearing entitled "Effects of 2,4,5-T on

Man and Environment." After two full days of testimony by experts from universities, government agencies, and industry, testimony during which the committee discussed, dissected, and placed into the Congressional Record hundreds of pages of authoritative reports and scientific studies, the Senate came to the only conclusion plausibly available at the time: there was little, if any, certainty in assessing the effects of 2,4,5-T on "man and environment." The chairman of the subcommittee, Sen. Phillip Hart, in fact, focused his opening statement, in somewhat hyperbolic terms, on this dilemma:

> I suggest that what is at stake at these hearings is virtually impossible to evaluate at this moment, in light of the uncertainty about this frequently used herbicide. The questions which have been raised about the hazards of 2,4,5-T and related chemicals may in the end appear to be much ado about very little indeed. On the other hand, they may ultimately be regarded as portending the most horrible tragedy ever known to mankind. What does emerge from this uncertainty is that we must take steps to eliminate it. In view of the potential disaster that could befall us—or conceivably has insidiously already befallen us—absolutely no delay is tolerable in the search for answers to the questions posed.[12]

As public health historians have observed, the type of uncertainty Hart describes is not easily eliminated. In the case of 2,4,5-T and Agent Orange it would continue well into the next century, as increased knowledge about the effects of dioxin complicated rather than simplified determinations of exposure and causality. But Hart also suggested a need to return to a policy of preventive regulation based on presumptive and potential consequences rather than waiting until such implications could be verified objectively through laboratory and epidemiological studies. Given the inferences made through the demonstrable effects of various herbicides, particularly 2,4,5-T, on animals, a number of lawyers, scientists, and senators at these hearings discussed shifting the burden of proof on the dangers of herbicides back onto the chemical companies. In an exchange with the legal scholar Harrison Wellford, Hart pressed the matter of recent studies showing 2,4-D and 2,4,5-T to be teratogenic (causing birth defects) in humans:

> HART: I have been told, however, that the tests that have been conducted are merely preliminary and that although they suggest certain conclusions they cannot confirm them. Is that correct? In other words, as of now, we can't say that we know that the currently produced 2,4,5-T is teratogenic, can we?

WELLFORD: Certainly, as far as the effects on human beings, you are entirely right. There is no clear evidence—we haven't been able to find evidence through epidemiological surveys for the reason that I described, that these herbicides are definitely a danger to human beings that come into contact with them. It is also true that there has not been time for the tests, which raise the suspicion to be checked and rechecked by many other scientists. Again, it is a question of the burden of proof.

HART: You would turn it around and say—well yes, you can't say they are harmful, but you can't say they aren't?

WELLFORD: I would say the more recent tests have expanded the burden of proof upon those who wish to prove that they are not harmful.[13]

Yet within the U.S. government, such skeptics as Ned Bayley, the director of science at the USDA, articulated his belief that the regulation of hazards should be a last resort, and, as such, the burden of proof should be on those wishing to regulate or restrict public access to the product in question. Using a comparison that would be echoed through the seventies and eighties, including by EPA officials during the Times Beach dioxin disaster, Bayley argued that "one can never be absolutely sure that a hazard does not exist, even if we are talking about table salt. In fact, we know table salt is hazardous if taken improperly, and we don't even register it."[14]

At the conclusion of the hearings the preponderance of the experts called to testify concurred (notwithstanding strong dissents from Dow Chemical) with the decision to suspend registration of 2,4,5-T and recommended that the burden of proof in reversing the decision should indeed be on those, such as Dow and Bayley, who claimed there was no direct evidence of harm to humans. At no time in the hearings did statements about human health concerns focus on U.S. veterans of the war in Vietnam; and only Arthur Galston explicitly drew attention to the potential consequences for the people of Vietnam. That the war was ongoing and still very controversial partly explains this lack, but up to this point few, if any, Vietnam veterans had publicly raised questions about their exposure to Agent Orange.

While government investigations went on largely ignoring the fate of veterans, two events brought the issue of Agent Orange to the attention of veterans, Congress, and the American public: the efforts of Maude DeVictor, a VA counselor in Chicago, and a lawsuit filed in 1978 by the Vietnam veteran Paul Reutershan that morphed into the largest class action suit to that point in U.S. history. DeVictor had been working with a number of veterans in the Chicago area who were suffering from a variety of mental and physical health issues. After speaking at length with the wife of the veteran Charles Evans, who was dying of cancer he believed was caused by exposure to herbicides during the war, DeVictor began looking into the use of chemicals in Vietnam.

DeVictor's search for answers was recounted in the made-for-television movie *Unnatural Causes,* starring Alfre Woodard as DeVictor and John Ritter as a composite veteran character afflicted with terminal cancer. In the film, which she regularly described during a brief publicity tour at the time of its release as being accurate and documented, DeVictor learns about the spraying of chemicals in Vietnam from her conversations with Evans's widow, who recounts the story of her husband seeing the smog around Los Angeles and saying that it reminded him of the spray during the war. "He said sometimes it was so thick you couldn't see your hands in front of your face," she reports. Initially encouraged to pursue the issue by her immediate supervisor, DeVictor eventually reaches someone in the USAF who suggests she speak with a "plant physiologist." The scene shows DeVictor talking on the phone with the scientist and jotting down notes, including the words: "Operation Ranch Hand, Agent Blue, Agent White, Agent Orange, 2,4-D, 2-4-5-T, and 'tetrachlorodibenzoparadioxin.'" The scene ends with DeVictor saying, "So what I'm looking for is Agent Orange."[15]

In real life the voice on the other end of the phone belonged to Maj. Alvin Young of the USAF, who had recently completed his work in support of Operation Pacer HO, would soon become a key member of the White House Agent Orange Working Group (AOWG), and does indeed have a doctoral degree in plant physiology and toxicology. Years later DeVictor remembered the conversation:

> He said it's 2,3,7,8 tetrachlorodibenzo-paradioxin. He spelled it and he gave me the everyday routine name, Agent Orange. And he also stated that there were other chemicals used in Vietnam. Agent Orange, Agent White, Agent Blue, Agent Purple, all named for the color of fog they produce. We decided on Agent Orange because it would kill you just as bad as the others, but in terms of the staff handling it, loading it, whatever, it was less toxic and less dangerous to handle for the American troops. But it would kill you just as good as Agent Blue and White. Agent Purple was a little bit too much, a little bit too wild and so they quickly had to stop using Agent Purple. So Agent Orange was the poison of choice. He told me that it had been used on certain regions of Vietnam and, let me see, he told me that it was a teratogen, which

is fetus deforming. He told me it was a carcinogen, which means cancer producing, and that it was 150,000 times more toxic than—now I don't know my chemistry, I don't know if this is inorganic or organic, but it's more toxic than arsenic. I don't want to say the wrong thing because some chemistry professor would say there's an error in the article, so bullshit to the whole thing."[16]

There are, indeed, many errors in DeVictor's account. Young, who remembers the conversation with DeVictor somewhat differently, says she called him to obtain more information about unspecified chemicals used by the United States in Vietnam. Young later claimed that DeVictor's initial questions revolved around a helicopter pilot, not a grunt, and that the chemical her claimant had been spraying was Agent Blue, not Agent Orange. Years later Young recalled telling DeVictor that there was no evidence of human health risks from exposure to Blue or White but went on to add, "'The only information that we are unsure of, in terms of toxicological and environmental fate and human exposure, is the dioxin associated with the 2,4-D and 2,4,5-T in the Agent Orange formulation.' That was our unknown. We did not have a great deal of knowledge about the impact on the human."[17] The discrepancies in these accounts and DeVictor's confusion regarding some of the specific details about the herbicides are not simply the stuff of historical memory; they became an important piece of evidence in the growing case against the VA's handling of Agent Orange cases in the early 1980s. But even setting aside those details momentarily, the basic uncertainty described by Young offered DeVictor an opportunity to press her claims and begin making her own case that the veterans with whom she was working were suffering as a result of exposure to Agent Orange. According to DeVictor, her further investigations into the matter were stifled by the VA, which had no interest in pursuing veterans' claims. DeVictor then took her story to Bill Kurtis, the anchor at WBBM-2, the CBS television affiliate in Chicago. The resulting documentary, *Agent Orange: Vietnam's Deadly Fog*, completely changed the political landscape of veterans' allegations about their exposure to Agent Orange.[18]

The documentary, which aired on March 23, 1978, opens by introducing shots of several Vietnam veterans. The men have never met, viewers are told, but they have a common bond of mysterious, unexplained illnesses: "The government of the United States is now investigating claims that they are the victims of our own chemical warfare in Vietnam. Call it Agent Orange." After describing the story of the veteran James Simmons, who was suffering a variety of health problems, from numbness in his fingertips to a decreased sexual drive, DeVictor appears on-screen and explains that all of the men in the "more than forty" cases she had collected, including Evans, whose wife repeated her husband's remark that the defoliation spray was "thicker than Los Angeles smog," had been exposed to Agent Orange. Kurtis goes on to narrate a brief history of herbicide use in Vietnam, culminating in the at-sea incineration during Pacer HO because "it was considered too deadly to be used again as a herbicide." He summarizes the study by the National Academy of Sciences (NAS), which he says was unable to determine the effects of dioxin on humans despite the troubling anecdotal evidence from Vietnamese civilians. Kurtis then interviews a number of doctors and scientists, focusing on the reports of medical concerns coming from Vietnam during the war, then transitioning back to his focus on U.S. veterans through the issue of birth defects among veterans' children.[19]

Against the backdrop of a cut from grainy black-and-white footage shot at a hospital in Hanoi and showing the truncated hand of a Vietnamese child to the color image of a similarly deformed hand, Kurtis narrates, "Few people dreamed the photographs showing birth deformities in 1968 would bear a chilling similarity ten years later to birth defects among the children of U.S. Vietnam veterans." The hand in the contemporary shot belonged to Richard Ross, the son of Milton Ross, who served in special forces operating in the central highlands in and around areas that were repeatedly defoliated. Richard was born in 1971 with congenital deformities in his legs and fingers.

Kurtis later recalled the powerful impact that seeing Richard's fingertips "dangling grotesquely from the rest of his fingers," had on him:

> After seeing Richard's birth defects—whether they would be attributed to Agent Orange or not—we knew that the veterans were sincere, that their complaints were not imagined. Yes, Milton Ross and his family had serious problems. Yes, he had been exposed to Agent Orange. Suddenly the story had grave implications. Many similar interviews in the course of our interviews would confirm the belief in the veterans' plight, but none would have more impact than the sight of the boy's fingertips. The picture needed no narration. In an instant it communicated a message with an emotion that couldn't be forgotten. That is television's strength. It also carries a responsibility. Because certain pictures are so strong, the reporter must ensure they are properly qualified. In this case, we would use the videotape of Richard's fingertips not as absolute proof of a connection with Agent Orange but to demonstrate the kinds of problems the veterans were talking about and to question whether they could have been caused by Agent Orange.[20]

The power of the image is indeed hard to deny; so too are the fundamental contradictions in Kurtis's account: the image, while offering no proof of being connected in any way to Agent Orange, was evidence that veterans were "sincere" and that their complaints were "not imagined"; the image "needed no narration" because of its emotion but also had to be "properly qualified" because of its "grave implications." For neither the first nor for the last time a horrifying image of a grossly deformed child suggested a powerful connection to Agent Orange without "proper qualifi[cation]." In *Agent Orange* Kurtis also interviews Robin Agena, a woman who worked as a secretary in an army logistics office in Vietnam and went on to have two spontaneous abortions and several other complications during four failed pregnancies. Neither Agena nor Kurtis explains how she came to be exposed, but at that point she was the first female veteran to file a claim with the VA related to Agent Orange.

"There is no proof," Kurtis continues in the film, that dioxin caused the conditions described by Ross, Agena, and others interviewed in the piece, but scientists questioned by Kurtis at the time felt they had a number of clues hinting at what even low levels of dioxin exposure administered over time were having on laboratory animals and thus potentially on humans. Summarizing the results of multiple studies linking small amounts of the toxin ingested by rats and monkeys to increased rates of cancer and birth defects, Kurtis describes the dramatic effects of the experiments against a backdrop of severely deformed animals. Noting that the levels ingested by the animals were as low as five ppt, Kurtis attempts to qualify the implied link between animal studies and human effects. "But ten years ago, in Vietnam," he goes on, "10,000,000 gallons of Agent Orange were spread over 10 percent of Vietnam's land area, with levels of dioxin 100,000 times greater than the levels that produced toxic effects in Dr. Allen's monkeys."

Agent Orange: Vietnam's Deadly Fog explores other incidents of human exposure to dioxin, including industrial accidents in Europe, the Alsea sprayings, and the Times Beach episode. Judy Piatt, the owner of one of the Missouri sites contaminated with dioxin (see chapter 3), tells Kurtis about the symptoms she and her children experienced, which were very similar to those experienced by the U.S. veterans who appear in the piece. But Kurtis does little to contextualize these exposure scenarios and ignores the higher levels of dioxin involved in the incidents at Seveso and Times Beach compared with those likely experienced by Vietnam veterans. Other problems with the documentary included, most glaringly, Kurtis's claim that 2,4,5-T was still used commercially in the United States, albeit at perhaps as much as one-fiftieth the level used in Vietnam. Kurtis mentions ongoing domestic use several times, against the image of shelves full of "Ortho Weed-B-Gone," a commercially available herbicide. Various products containing 2,4,5-T, Kurtis observes, were still "used extensively, to control vegetation in national forests, around corn, and rice crops." In fact, the EPA suspension of 2,4,5-T was still in effect at the time the documentary aired, and Weed-B-Gone, which is still available today, contains 2,4-D, not 2,4,5-T. Some activists have argued that 2,4-D, still used in a variety of commercial herbicides like the popular Round-Up, may be an endocrine disruptor, but it did not and does not, like 2,4,5-T, contain dioxin.

As he explores these domestic applications, Kurtis remains focused on exploring the possible link between Agent Orange exposure and veterans' health claims. "Putting aside the Vietnamese health studies," Kurtis notes, "critics emphasize that not one human case in Vietnam has been linked to dioxin." The critic whom Kurtis chose to represent that view was Major Young. Young had run tests on beach mice at Eglin Air Force Base, Kurtis points out, and while he found no significant consequences for the mice, others challenged his data and interpretations, just as Young had challenged the work of Ton That Tung, a North Vietnamese surgeon, and other "North Vietnamese scientists."[21]

At the time, DeVictor told *The Progressive* that Young changed his story for the cameras and was now less certain about the connections between 2,4,5-T and the health issues described by veterans than he had been during their conversation. Young maintains that his part of the Kurtis piece was selectively edited, but the charge that he changed his tune on the dangers of Agent Orange is dubious, to say the least. He has long held (to a fault, according to his critics) that there is little evidence to support the majority of health claims made by alleged victims of Agent Orange, whether American veterans or Vietnamese civilians.[22]

In fact, Young is a central figure in the politics of uncertainty surrounding Agent Orange. Aside from his involvement in the early testing, developing, and monitoring of military herbicides, his essay "Agent Orange: At the Crossroads of Science and Social Concern," originally written as his thesis for the Air Command and Staff College and later reprinted in a number of publications, was the first major piece to adopt the idea that diseases believed to be related to Agent Orange exposure might actually be in some way social diseases, constructed in part around the intersecting matrices of memory, media, trauma and legitimate disease concerns. In his essay Young did not assert that Agent Orange was *not* responsible for veterans' conditions; instead, he argued rather forcefully that the science was simply not, at that point, able to show that most of these conditions were related to dioxin or to Agent Orange exposure. Looking back on the language used in the essay, one finds not the impassioned language of an activist or an ideologue, but the detached rhetoric of a scientist: "In other words, there are no epidemiologic data associated with any long-term health effects in humans other than chloracne, but, as noted by epidemiologist W. H. Wolfe, *neither is there strong evidence to validate the absence of such effects.*"[23] Young proposed that Agent Orange had become a way for some veterans to make sense of their experiences, their trauma, and perhaps their other medical conditions: "If Agent Orange is not responsible, then some other factor associated with the war may be responsible or, perhaps, the symptoms are afflictions of aging and attendant psycho-social aberrations."[24] The crucial point Young made with regard to the politics of uncertainty surrounding the effects of Agent Orange was this: "When these perceptions are manifested as fear of the unknown, such as the risk associated with a poisonous chemical environment, the public does not always react to that fear in proportion to the seriousness of the threatened harm." Young went on to agree with others, such as the epidemiologist who claimed in an article in 1980 that many of the issues described by veterans after the WBBM documentary aired "may have nothing to do at all with Agent Orange as a scientific fact, but [are] grounded in other problems affecting the Vietnam veteran population and [have] been launched into celebrity by a self-generating series of press and television stories."[25] Young's primary example of such stories was Kurtis's documentary. The notion of Agent Orange as a social disease would resurface a few years later in Australian investigations.

Such sentiments did little to endear Young to his fellow veterans and their supporters. But while his position on the effects of Agent Orange was clear and somewhat controversial, his critics are largely unaware—or choose to ignore—the fact that he advocated for expanding, rather than limiting, veterans' benefits, both as a member of the AOWG and afterward. In 1981 then-Colonel Young wrote a memo to the VA arguing that even though he was convinced that all but a tiny fraction of the health conditions were caused by something other than Agent Orange, the VA should not contest claims of exposure and whenever possible should avoid placing the burden of proof on veterans. Essentially, Young was saying, given that this will ultimately be a political rather than a scientific decision, let us not compound the problem by doing a disservice to the veterans. Despite the lingering resentment felt by many Vietnam War veterans and their advocates toward Young and others (including many Ranch Hand veterans, most of whom shared his views), Young's early advocacy for not contesting exposure claims may have helped accelerate, rather than slow down, the eventual process of allowing hundreds of thousands of veterans to file for benefits without providing definitive proof of exposure.[26] At the time *Vietnam's Deadly Fog* aired, however, Young was derided by DeVictor and other critics as an example of the military's refusal to acknowledge the dangers of Agent Orange.

In the concluding segment, Kurtis acknowledges that there is no "absolute proof" that Agent Orange was the culprit. "But the symptoms are suspiciously similar," he notes, "to what scientists describe as dioxin poisoning," an assertion that is then backed up by a number of scientists, including Barry Commoner and Matthew Messelson. Kurtis closes by noting that the VA was currently denying disability claims based on Agent Orange exposure, but that "after researching this report, and listening to the recommendations of the leading dioxin scientists in the country, we feel there is a need for immediate testing of all Vietnam Veterans who handled Agent Orange or went into sprayed areas. Not only for the sake of those who have told us of their symptoms, but for the countless others, whose lives and whose children's lives could be blighted by the dioxin poison in Agent Orange."[27]

Kurtis produced two additional documentaries on the subject: *Agent Orange: The Human Harvest,* in March 1979, and *Agent Orange: The View from Vietnam,* in April 1980. Both were thoughtful explorations of the environmental and human impact of dioxin on Vietnam, but neither had the impact of *Vietnam's Deadly Fog.* According to the sociologist Wilbur Scott, immediately after the airing of the documentary in March 1978 hundreds of veterans called the VA in Chicago to inquire about the possible effects of and benefits related to Agent Orange exposure. Kurtis later held a special screening for members of Congress and other government agencies. Within a few weeks Congress requested that the Government Accountability Office (GAO) explore the possibility that sizable numbers of U.S. troops had been exposed. As Scott puts it, "In one dramatic swoop, Agent Orange went from the private rumblings of a handful of veterans to the center of national attention."[28]

Still, by October 1978 only five hundred veterans had filed for disability claims. That was a major increase from the three hundred filed for the period from 1962 though the end of 1977, but it was a far cry from the thousands filed by the end of 1979.[29] After the initial impetus provided by DeVictor and Kurtis, the key factor in increasing the number of claims and further raising awareness was the lawsuit brought against the chemical companies in 1978 by Paul Reutershan.[30]

The lawsuit has been covered extensively by other studies, but several of its key points bear on the issues I raise here.[31] The legal scholar Peter Schuck has argued that the case was made possible by a number of historical factors, including the treatment of veterans upon their return home, the climate created in part by DeVictor and *Vietnam's Deadly Fog,* and changes in the legal apparatus that made it easier for plaintiffs in liability litigation to address the issues of causality and damages. Even in view of those changes, however, the lawyers representing the veterans knew from the outset that proving causation was going to be extremely difficult, if not impossible. As one member of the legal team later told Schuck, "Our clients could not help us prove our case. After all, they had not been hit by a truck, victimized by a doctor, or injured by a drug or other consumer product. They didn't know what had happened to them or when. They only knew something had gone wrong. Our causation case, therefore, had to be proved almost entirely from documents in the defendants' files."[32] Originally, this meant files from the government and the chemical companies, both initially named as defendants. The court threw out the case against the government because of the doctrine of sovereign immunity, which shielded the federal government from service-related torts brought by former soldiers.[33] The chemical companies were left to fend for themselves in a climate in which veterans were winning the public relations battle and under a legal system that left them potentially vulnerable to large awards by sympathetic juries. Luckily for Dow, Monsanto, and the other defendants, they still had scientific uncertainty on their side.

The original judge in the case, George Pratt, ruled in 1982 that the chemical manufacturers could not use the "government contractor defense," which held that they were simply providing a product to the DOD in line with contractual obligations. The ambiguities, to use Schuck's words, in Pratt's ruling over questions such as "what it meant to 'know' something as elusive, subjective and indeterminate as whether a chemical agent was 'safe'; [and] what level of official must have that knowledge in order for it to be imputed to the government would come back to haunt Judge Pratt and the parties" because they were ultimately impossible to answer definitively.[34] On the question of what the government knew and when it knew it, the case offered no clear resolution. According to Schuck, testimony would suggest that Monsanto had informed the army in 1952 about the dangers of 2,4,5-T if applied improperly, but the evidence for this claim came out only in the so-called fairness hearings on compensation, after the case had been decided. No other direct evidence that the government knew about the risks to human health posed by the 2,4,5-T in Agent Orange was offered during the case.³⁵ The chemical companies' documents did reveal matters like the Midland summit meeting in 1965, which the veterans' lawyers initially thought was the smoking gun, but even then the question of responsibility was not distributed evenly among the companies. Dow and Monsanto had extensive knowledge of the dangers of dioxin, but others, including Riverdale and Hoffman-Taff, were released from the case because Pratt found them to be ignorant of the hazards involved.³⁶ Decisions such as this led to increased tensions among the defendants; Dow continued to assert that it had, at the summit and afterward, attempted to get its competitors to act more responsibly in limiting dioxin levels in 2,4,5-T and began raising the possibility of a separate settlement. Infighting was also rampant among the veterans' legal team, which became more and more divided over tactics and strategy. Neither side was overly optimistic about its chances should the case go to trial.

When Pratt abruptly, and much to the consternation of all parties involved, recused himself from the case, Judge Jack Weinstein succeeded in negotiating an out-of-court settlement of $180 million, announced on May 7, 1984, only hours before jury selection was to begin. Through considerable arm-twisting and thinly veiled threats, Weinstein convinced the veterans' lawyers they did not have the type of scientific evidence that would hold up in court, particularly against the well-funded, expert-heavy chemical companies; he likewise persuaded the companies that the politics of uncertainty on which they would rely would find little sympathy among a working-class jury from Brooklyn or Long Island.³⁷ As in Congress and a variety of medical studies carried out in the 1980s, in this case it was impossible to separate the uncertainties of science from the realities of politics.

Wrangling over such matters as determining exposure and awarding compensation would go on for years, while uncertainty reigned and causality remained elusive. Point nine of the fourteen-point settlement negotiated by Weinstein stated that the "defendants denied all liability," meaning they admitted neither that Agent Orange was to blame for the veterans' health issues nor that it was dangerous in general.³⁸ Immediately after the settlement was announced, the two sides kept on litigating the issue of causality in the press. Victor Yannacone, the lawyer who first filed the veterans' case, told reporters that regardless of the companies' denials, "a quarter of a billion dollars is one hell of an admission." A spokesman for Dow immediately shot back that the company maintained that "overwhelming scientific data proves that exposure to a small amount of dioxin is harmless and that no veterans were exposed during the war to levels high enough to cause any damage."³⁹ The chemical companies were relatively united in their belief that they had escaped with minimal financial damage; the stock of all publicly traded companies involved in the suit went up the day of the announced settlement. Veterans, however, were seriously divided over whether or not the settlement was just. Throughout the fairness hearings over the next year the settlement was more often called a sellout than a victory. Schuck writes that "the great majority of witnesses" at the hearings opposed the settlement.[40] In the absence of an admission of liability and of the element of causation by either the companies or the government, the veterans sensed that the settlement funds, which would amount to a maximum of around twelve thousand dollars per member of the class, might be the only compensation they would receive. "If we prove causation," the veteran John Kopystenski told the *Washington Post,* "the Veterans Administration will have to pay disability benefits to us and that will be worth more in the long run to veterans and their families than $180 million."[41] Kopystenski was right.

The settlement funds, which began to be dispersed in 1989, were exhausted by 1994. According to the VA, of the more than one hundred thousand claims made to the fund, just over half were awarded; the average payment was thirty-eight hundred dollars.[42] The ongoing legal fight would only get more difficult for veterans. Since the original settlement, every single Agent Orange case brought in American courts against the U.S. government or the chemical companies has been dismissed owing either to a lack of scientific evidence or, in the global arena, to jurisdictional issues and the complexities of international law, especially in cases related to war.[43] The fight with the VA would be just as long and complicated, particularly once the Congress got involved.

The Burden of Proof

By the end of the 1970s there was no avoiding the implications of 2,4,5-T for U.S. veterans of the war. Given the publicity generated by DeVictor's investigations and the veterans' lawsuit, the issue of dioxin-related conditions stemming from exposure became the focal point of the veterans' lobby and the veterans' affairs committees in Congress. Between the end of Operation Ranch Hand and the filing of the veterans' lawsuit, both the NAS and the USAF had completed their initial investigations into the effects of Agent Orange exposure. Neither of the two studies showed any direct health effects on Ranch Hand veterans, but questions of certainty, evidence, and the burden of proof remained at the center of debates about the possibility and effects of exposure. Their persistence was owing not only to the nature of scientific inquiry and, in particular, of epidemiology, but also to the efforts of the chemical manufacturers, the DOD, and the VA to use the politics of uncertainty to marginalize veterans' claims.

At a hearing in 1978 on medical benefits, representatives of the USAF and the VA continued to deny that large numbers of veterans were exposed to a significant amount of dioxin and to discount the link between exposure to Agent Orange and specific claims by veterans; they also hedged against future claims should that link become less uncertain. Paul Haber, the assistant chief medical director of the VA, noted that the evidence about the extensive domestic use of 2,4,5-T and 2,4-D was itself going to further complicate the matter. "If later proof is produced that human health is significantly impaired by dioxin," he testified, "the VA's task will be to distinguish harm which veterans might have encountered through the use of herbicides during the war from harm which may have come to them through nonmilitary domestic exposures to chemicals. We do not anticipate that this will be easy."[44]

It was not. By the time Max Cleland, a Vietnam veteran and the director of the VA during the administration of Jimmy Carter, testified on Capitol Hill in 1980 at a hearing on Agent Orange exposure, more than 10,000 veterans had gone to the VA with conditions they believed were caused by the chemical; 1,233 of those veterans filed claims for service-related disability resulting from exposure. Of those, the VA processed 21 who it believed had service-related conditions, and of those 21 only 2, both of whom had chloracne, were determined to have conditions caused by exposure to Agent Orange.[45] From the VA's perspective, as Cleland explained, the key determinant in approving a claim was showing not that a given condition was caused definitively by Agent Orange but that it was service related. For the two soldiers who developed chloracne during their service this was not difficult. For the overwhelming majority of others, however, whose claims were based on everything from vague descriptions of tingling, irritability, and difficulty sleeping to acute cases of testicular cancer, making such a decision was nearly impossible, and that fact inevitably steered questions back to the issues of evidence and scientific certainty.

In response, for instance, to a committee member's observation that the government agencies were "long on planning and devising protocols for studies, but up to this time are very short on conclusions," the witnesses repeated their conviction that the conclusions sought by veterans and members of Congress would likely not be forthcoming, even after exhaustive epidemiological studies. In restating this point, Joan Bernstein, the legal counsel to the Department of Health, Education and Welfare, noted that if the science was not yielding the certainty desired by the parties, other options were available: "There are solutions other than scientific solutions. There are legislative solutions. There are administrative or policy-type solutions in which one simply makes the judgment that there is enough association even though it may not constitute causation in a legal sense. The Congress could decide, as has been suggested in a couple of legislative proposals, simply to act on that amount of association of those people who were there and decide to compensate them at a certain level."[46] Ultimately, the decision on compensation would indeed be based largely on politics rather than on science, but that did not stop Congress, the VA, other government agencies, and veterans from continuing their quest for scientific certainty.

At one point Rep. Tom Daschle of South Dakota, who would become a leading advocate of veterans' claims, even asked Cleland himself to identify his exposure scenario through the VA questionnaires. Using an actual VA form, Daschle asked Cleland, "When and where did these exposures occur, and what was the length of the exposure?" to which Cleland quickly replied, "Landing Zone Nancy, north of Hue, early 1968, for a period of about two weeks." Pressed by Daschle on other details, Cleland and his staff pointed out that veterans' responses were not the determining factor in moving forward with service-related claims. The forms and questions, the VA staff clarified, were designed "to get the veteran's statements about what he thinks his exposure was. Now that helps us, then, in the epidemiological survey, but obviously in many instances the veteran himself may not know the answers." At this point, the VA policy was still to "accept for compensation purposes any veteran's claim that he was exposed to Agent Orange unless we have positive evidence to the contrary. We will accept the veteran's word that he was, in fact, exposed to Agent Orange."[47]

As veterans' groups advanced their claims in the late 1970s and early 1980s, however, they said the VA was not taking their word and was instead regularly, repeatedly, even systematically denying it. As Bobby Muller, the president and cofounder of Vietnam Veterans of America (VVA), stated in his testimony before the Veterans' Affairs Committee, his group was "concerned about the continuing inability, or unwillingness, of local VA hospitals to respond fully and compassionately to the needs of Vietnam veterans exposed to Agent Orange."[48] Muller's assertions about the relationship between science and policy, however, highlighted the difficulties in moving the issue forward: "This turn toward science has important consequences. For a scientific determination is not a policy question. We cannot determine, as a matter of policy, that TCDD causes cancer in rats. Only science can. Accordingly, waiting on science means we can wait to make policy. But there is science and then there is science: there are different kinds of scientific evidence. Saying science decides is fine as long as we have determined what we mean by science."[49] Muller argued further that "science is not on the leading edge of the health question at all," before going on to explore at length the methodological problems associated with determining Agent Orange exposure. As would be the case throughout the next two decades, veterans and their backers were trapped by the dilemmas and discourses of scientific certainty. On the one hand, they sought to use medical discourses of disease, exposure, and associated conditions to push for acceptance of conditions as being related to Agent Orange and for a corresponding expansion of benefits; but when those discourses were mobilized by the VA and other agencies to throw doubt upon veterans' claims, groups like the VVA sought to marginalize the language of science and certainty as insufficient to address the needs of veterans "fully and compassionately."

By the summer of 1980 the tide seemed to be turning against the VA. At least seven states had taken steps to address Agent Orange–related issues, including the establishment of Agent Orange Commissions in California, Maryland, New Jersey, and New York. As the lawsuit crawled along toward a possible trial, several states, hoping to recoup costs associated with veterans' health care and with the cleanup of dioxin-contaminated sites, began exploring the feasibility of suing the chemical companies. Even the National Council of the Churches of Christ weighed in that year, passing a resolution condemning the use of herbicides and calling on the "scientific community, along with government agencies and corporations involved, to exercise their moral responsibility as well as their scientific acumen in conducting these studies in a factual, balanced, and objective manner." And at yet another congressional oversight hearing in July the VA was continually placed on the defensive by veterans, scientists, and members of Congress.[50]

A surprising amount of time and energy at these hearings was devoted to covering the same issues discussed in the same forum just months earlier. The tone of the hearings in July, however, seemed to reflect the growing frustration with the VA being expressed by veterans, who were gaining greater public support. The deference shown to medical experts and bureaucratic challenges in February gave way to confrontational assertions that the VA was hiding behind uncertainty. Rep. David Bonior of Michigan chastised the agency for being slow to respond to veterans and to congressional admonishments to improve its services. "Is the VA pursuit of certainty grounded in a responsible quest for additional science or is it a deliberate measure designed to produce delay?," Bonior wondered, before answering his own question: "When confronted with their record, the VA response has been to point towards the gaps in our scientific information. Gaps are there. Gaps there certainly will be. But the VA cannot continue to hide behind a façade of scientific prudence. For the question no longer remains whether or not there is some uncertainty. The question now is, how much science will the VA require, how much certainty will they hold out for until they make a humane decision to resolve this pressing problem?"[51] Like Muller and the VVA, Bonior and others in Congress juxtaposed *science,* a term itself increasingly referenced in terms of quantity (*how much* science; *how many* studies) rather than as a process, with *humanity* and *compassion.* In doing so they pushed the VA, both implicitly and explicitly, to adopt procedures based not on science but on morality and, ultimately, on politics.

Yet Congress refused to let science go by the wayside and simply move to pass legislation to extend benefits to any and all presumptively exposed veterans. Instead, Congress and veterans focused their frustration on the VA, holding that it was incapable of managing the health studies. In his opening statement Daschle noted that the purpose of the hearings was "to establish what we now know and what we do not know" about Agent Orange, but he went on to imply that the real purpose was to downgrade drastically the role played by the VA. "Up to this point," Daschle began, "the VA's actions on this issue have been insufficient and inherently biased to preclude them from doing an accurate and thorough study. For this reason, I strongly recommend to the committee that they should not conduct this [epidemiological study of veterans' health], but rather that the Center for Disease Control of the National Academy of Sciences is far more credible and therefore acceptable to immediately take over this responsibility."[52]

Daschle and Bonior had several allies in this battle, none more crucial than Jeanne and Steven Stellman, scientists who would play a critical role in future scientific, medical, and historical investigations of Agent Orange. The Stellmans had long held that military records were a sufficient basis on which to determine likely exposure, a position they consistently promoted in the 1980s and well into the twenty-first century. Testifying at the Oversight Hearings in 1980, the Stellmans castigated the VA for its handling of health studies and its failure to inform veterans effectively. They also met uncertainty head-on, surprising Daschle and others by declaring that scientists knew more about the causes and effects of dioxin exposure than about the causes of multiple sclerosis.[53] The talent and reputation of the Stellmans lent their statements a great deal of weight with policymakers. As independent scientists having no links to the military or to industry the Stellmans, over the next decade, helped shift the politics of uncertainty more in favor of the claims of veterans, arguing that there was sufficient evidence to support presumptive benefits for a variety of health conditions. Their efforts did not go unchallenged: other witnesses at the hearings subscribed to a "wait-and-see approach," one that would not further inflame the "emotionalism" of the Agent Orange dispute.[54]

The Stellmans were at the vanguard of this renewed attack on the VA, but exhibit A was an "unsigned memorandum" from 1977 that Bonior and Daschle thought revealed a VA conspiracy to conceal information about the health implications of Agent Orange and Agent Blue. The seemingly mysterious memo, the congressmen believed, "crystallized" their doubts about the ability of the VA to deal with the issue of Agent Orange "forthrightly."[55] Before their scathing letter to Director Cleland even reached the VA, Daschle and Bonior held a press conference on the steps of the capitol, excoriating Cleland and his staff and distributing the unsigned memo to the media. They reiterated their conviction that the VA was dragging its feet on Agent Orange–related disability claims for veterans and concealing knowledge about the human health effects of multiple herbicides used in Vietnam. The congressmen's statements, Cleland was careful to point out in his reply, also implied that he had lied in his testimony to Congress.[56]

The memo in question was entitled "Re: Defoliation Operations from 1965–1969 during Vietnam War" and contained a number of statements that the author of the memo indicated had come from Alvin Young. Among the claims seized on by Daschle and Bonior, the following three were potentially the most damning:

- There were two basic types of defoliants used in Vietnam, Agent Orange, and Agent Blue. Both agents contained chemical 24D and 245-T. They are mutagens and teratogenics. This means that they intercept the genetic DNA message, process it to an unborn fetus; thereby resulting in deformed children being born (similar to the thalidomide situation). Therefore the veteran would have no ill effects from exposure, but we would produce deformed children due to this breakage in his genetic chain.

- The military is observing these proceedings with extreme concern, in the event that favorable decisions in the case of such charges should open the way to possible litigation.

- Agent Blue has been clinically shown to be a human carcinogen.[57]

Each of these statements, as it turns out, was demonstrably false. They were also unlikely to have come from Young, who, as Daschle and Bonior admitted, was "probably one of the world's leaders in knowledge of plant herbicides." Young as well as anyone knew the technical names, vernacular names, and technical specifications of every herbicide used in Southeast Asia. He certainly knew that there were more than two, and he would never have claimed that Agent Blue had been "clinically shown to be a human carcinogen." It strains credibility to believe that Daschle and Bonior took this discrepancy, presented only as an unsigned memo, as evidence of a cover-up rather than as some form of miscommunication, but they nevertheless pressed Cleland to provide a quick and thorough explanation of the questions raised by the memo.[58] Cleland was able to do just that, solving most of the mysteries surrounding the document: the unsigned memo was none other than a typewritten summary of the conversation DeVictor had with Young on October 12, 1977.

The unsigned memo was taken nearly verbatim from DeVictor's handwritten notes about the conversation, which included all but one of the statements mentioned above. The only discrepancy—one that Daschle and Bonior took pains to point out in their follow-up message to Cleland—was that DeVictor's notes do not state explicitly that "both agents contained chemical 24D and 245-T." DeVictor's notes, which are included in the hearing transcripts, lead one to infer, not unreasonably, that the two were accidentally conflated in the transcription of the notes on that section into the typewritten memo. Whether it was confusion or conflation, whoever transcribed the notes knew little about either the herbicide program in general or the specific herbicides in question. DeVictor maintained that she did not write the memo. Cleland was unable to determine who wrote the memo, which had made its way at some undetermined point from the Chicago Regional VA Center to the National VA Department of Medicine and Surgery. Most likely DeVictor's notes were transcribed by office staff at the direction of one of DeVictor's supervisors and packaged along with the corresponding notes and articles to be sent to VA headquarters in Washington.

Both Cleland and Young provided a line-by-line refutation of the entire memo, pointing out the many errors of fact and misspellings of technical terms as well as correcting the record of what Young had actually told DeVictor, the specifics of which are confirmed by Young's own handwritten notes, his formal written response to DeVictor, the memo he prepared for his supervising officer immediately after the conversation, and two interviews conducted years later. The answer to the supposed mystery of the unsigned memo, in short, was fairly straightforward: DeVictor, who was finding out for the first time about the herbicide program and had never before encountered the term *Agent Orange,* let alone trichlorophenoxyacetic acid or tetrachlorodibenzo-para-dioxin, confused or conflated technical information about the herbicides. As the handwritten information taken by DeVictor moved its way up the chain of command, it resulted in a perverse form of the children's game telephone, in which the initial information transmitted bears little, if any, resemblance to the final version. While this perversion of fact had potentially serious consequences for VA policy and the many ongoing investigations about Agent Orange, the explanation should have been fairly obvious. Yet months later the House subcommittee was still debating the origins and consequences of the memo, and critics persisted in using it to construct an air of conspiracy surrounding the VA.[59]

Daschle and Bonior remained unsatisfied. In tersely worded correspondence to Cleland they seized on the discrepancies between DeVictor's notes and the memo on the subject of Agent Blue to imply again that Cleland and the VA were withholding important information, a charge they also levied in a letter to the secretary of the USAF.[60] In his reply to the congressmen, Joseph Zengerle, assistant secretary to the air force for manpower, reserve affairs, and installations, maintained that the memo writer was likely not "capable of assessing complex, technical health information," and he steered the discussion back to the language of uncertainty. Responding to Bonior's and Daschle's observation that the VA had ignored five studies demonstrating links between industrial workers exposed to dioxin and various forms of sarcoma and stomach cancer, Zengerle wrote, "It is rare that such research will produce unequivocal results; rather the results more likely will produce inferences subject to differing interpretations. And of course it is necessary for a determination of their reasonableness that anyone drawing conclusions from such a study be prepared to justify on the basis of the weight of evidence then available, the detailed basis for his position, and why other hypotheses were rejected. This is especially important in the area of health-effects research, where it is virtually impossible to prove that a given exposure is absolutely safe."[61] Had such assertions of uncertainty by the military or the VA come a few years earlier, around the time of DeVictor's and Young's conversation, they might have been accepted by members of the Veterans' Affairs Committee. But after two years of media exposure, numerous congressional hearings, the beginning of the veterans' lawsuit, and growing frustration with the VA, the discourse of uncertainty was losing its grounding in the realm of scientific expertise and was being effectively mobilized by veterans, their advocates, and their allies on Capitol Hill to argue that the VA was hiding behind a cloak of indeterminacy for legal and economic reasons. And while the VA and the USAF were emphasizing studies that showed the effects of exposure to be unproven, Daschle, Bonior, and the veterans' groups were being no less selective in emphasizing studies which confirmed their beliefs about the malignant effects of dioxin. Both of these points of view rely on the idea of uncertainty, and both were shaping and being shaped by the larger context in which they were being voiced, negotiated, and mobilized. After the summer of 1980 the politics of uncertainty helped to discredit the VA, the chemical companies, and the Pentagon and to help veterans and their advocates take the offensive in their battle for compassion and justice. The pendulum would later swing back, but in the meantime the VA had been effectively marginalized.

A GAO investigation in October 1982 found that VA personnel were not well informed about how to identify and deal with Agent Orange claims, that the VA was keeping poor records on the almost ninety thousand veterans in the Agent Orange Registry, and, most tellingly for veterans and their advocates, that 93 percent of the fourteen thousand plus Agent Orange claims filed had been denied, overwhelmingly because it could not be shown that the condition in question was service related.[62] By the end of 1982, the handwriting was on the wall: the VA was not to be trusted, by Congress or the public, let alone by the veterans it was designed to serve.

These initial debates over Agent Orange exposure took place as the status and image of Vietnam veterans were undergoing a profound shift in American society. As Patrick Hagopian demonstrates in his book *The Vietnam War in American Memory* (2009), the late 1970s and early 1980s were a critical moment in the cultural construction of the Vietnam veteran, a moment which "played a central role in shaping the remembrance of the war." Witnessing public debates over post-traumatic stress disorder, homelessness, and drug use among veterans set against a backdrop of troubling images of veterans in popular culture in films such as *The Deer Hunter, Coming Home,* and *Taxi Driver,* Americans in the late seventies were likely to regard veterans as victims, a feeling, Hagopian writes, that demonstrated "more pity than respect."[63] This feeling would shift dramatically during the eighties as both the image of the veteran and the cultural memory of the war underwent a major transformation. At the time, however, the victimization of the veteran was a powerful force in American society, and the VA rightly became a focal point of the anger felt among veterans and their advocates. Agent Orange fit easily into the narrative of victimization and served as a powerful symbol of the U.S. government's ongoing failure to support and care for its veterans.[64]

The Politics of Experience

In response to congressional legislation and to the recommendations of the GAO, in 1979 the Carter administration established an interagency committee designed to coordinate the federal government's responses to veterans' concerns about Agent Orange. Under President Ronald Reagan, this committee would be rechristened the White House AOWG.[65] For the next decade, the AOWG would wrangle among itself and with federal agencies and researchers over research protocols, exposure models, and the effects of Agent Orange on veterans. GAO investigations and congressional testimony had made clear that ground troops had indeed been exposed to Agent Orange, but the science panel of the AOWG could not agree on what that meant when it came to devising studies about the actual effects of exposure on veterans or determining whether individual soldiers were in fact exposed. AOWG records reveal a group of experts and government officials that is unable to reach any satisfactory resolution of the dilemma that lay at the heart of all major studies undertaken during this period: how to determine who was exposed to Agent Orange in Vietnam and under what conditions.

In 1982 debate over such questions escalated rapidly when Congress, having reached a breaking point in its frustration with the VA, transferred the veterans' health study to the CDC. Michael Gough, a member of the AOWG who worked in the White House Office of Technology Assessment (OTA) at the time, would later write that CDC had a proposed protocol for the study prepared for submission to the AOWG just months later, before the official transfer was even complete. The initial CDC plan called for three separate studies: the Agent Orange Study, to examine the link between Agent Orange exposure and health effects on veterans; the Vietnam "Experience" study, to examine whether service in Vietnam was associated with adverse health effects among veterans; and a third study to determine whether exposure to Agent Orange *or* service in Vietnam resulted in higher rates of cancer among veterans.[66] The proposed model seemed to answer many of the protocol questions raised by the VA, Congress, and the AOWG over the previous few years and to offer a larger, more diverse set of subjects than the Ranch Hand study had. The central question of determining exposure, however, would not go away.

The Ranch Handers were the only large group of veterans known to have been exposed to Agent Orange. Critics of the Ranch Hand study, however, pointed out a number of legitimate concerns with using Ranch Hand veterans as a case study. As Gough describes it, the argument against the study was "not so much that [ground troops] were more exposed to Agent Orange as that the conditions of exposure accentuated dioxin's effects":

> Ranch Hands slept in clean beds, ate in permanent mess halls, and bathed and changed into clean uniforms after flying missions and servicing aircraft. Infantrymen, as in all wars, were dirty; they stayed in the field for days, sleeping where it was possible, supplementing their rations with what fresh fruits and vegetables they could find, and taking water from available supplies. Infantrymen's exposure to Agent Orange depended on the extent to which it was present in the jungle, on soil, on food, and in the water. Once an infantryman got Agent Orange on his skin, it was likely to stay there for days until he returned to a base camp for a shower and clean clothes. On the other side of the coin, there are often questions about whether infantrymen were exposed at all.[67]

Given the number of pilots and officers in Ranch Hand, their overall levels of health, education, and economic resources tended to be higher on average than a similarly large cohort of ground forces. Thus when the USAF released the initial results of the Ranch Hand study in 1983, which showed that, far from being adversely affected, the health of Ranch Handers was actually better than that of comparison groups and their rates of cancer, heart disease, and overall mortality rates lower, many veterans and their advocates were unconvinced.[68] While Gough's distinctions between Ranch Handers and grunts held up generally, they did little to further the case for individual exposure. As the AOWG soon demonstrated, the number and complexity of the variables in the everyday experience of ground combat troops in Vietnam made it nearly impossible to determine with any accuracy whether or not an individual soldier was exposed to Agent Orange.

In constructing an exposure model to study the effects of Agent Orange on ground forces the CDC considered three options: it could ask veterans to self-identify, to voluntarily state, without offering any evidence or documentation, whether or not they believe they were exposed to Agent Orange; it could take blood and tissue samples to test for elevated dioxin levels; or it could use military records of herbicide missions and troop movements to try to determine the location of specific individuals in relation to areas where Agent Orange had been sprayed—the hits method, as it was labeled by the CDC.[69] Finding the first option too unscientific and the second both costly and scientifically unproven, the CDC opted for the hits approach. The first several years of the study were often overtaken by debates over whether exposure could be determined through military records alone and by such questions as whether battalions or companies should be the baseline unit of analysis.[70]

An early draft of a proposal for an exposure model from the AOWG subcommittee on exposure gave an indication of how complex such models could be and how much uncertainty was built into them even before the attempt to reconstruct a given soldier's presence in the area. Drafted by Young, the model noted no fewer than thirty-five variables related to the nature of the chemical and target, how much time had elapsed after application of the herbicide, and distance from the application site.[71] These issues, which remained at the forefront of the AOWG's deliberations, led to a series of clashes in the mid-1980s.

The most heated debate came over the interim reports of the CDC Exposure Assessment Team, sent to Congress and the AOWG in February and November 1985. The February report employed the hits method to develop exposure scores for various individuals in the study, the research for which had been carried out largely by the retired army colonel Richard Christian, who was working for the army's Environmental Support Group (ESG). Scott, who later interviewed Christian, described the daunting project in his book *Vietnam Veterans since the War*: "Christian's mission was to assemble the necessary records and to copy the information so that company level operations—a unit size of about 200 men—could be tracked for spray zones of various widths and time frames. Christian set about his task with zeal. Working with a staff of up to fifty-five employees, Christian in 1984 gathered 40,000 linear feet of records occupying more than 40,000 boxes and weighing more than 800 tons. The painstaking process of extracting and computerizing the data began."[72] Christian's task was both Herculean and Sisyphean. Not surprisingly, he became frustrated, or "miffed," to use Scott's words, when, after he had worked for more than a year, the OTA, AOWG, and even some people at the CDC found his research to be inadequate as a basis for an exposure index.

The initial interim report argued that the use of companies as the baseline unit for determining individuals' locations was not possible based on the available data. Instead, the study would employ battalions—units of up to one thousand troops potentially spread out over a much larger area. In the early rounds of analysis, data were showing a high degree of uncertainty about individual locations because those locations were not correlating with company locations. A pretest using documentation from the company level found an average of 38.5 percent uncertainty for determining an individual soldier's location relative to herbicide missions.[73] Even though the researchers switched to the battalion level, however, and applied complex algorithms devised by the ESG and CDC to determine exposure, the interim report was not highly optimistic about the model developed from Christian's data:

> There are two main reasons why accurate estimation of the "true," biologically effective exposure to Agent Orange or dioxin among Vietnam veterans is impossible. First, individual veterans cannot be precisely located in relation to Agent Orange applications. The military records were not collected for the purposes of epidemiological research and allow only assigning veterans the "average" location of their battalions. Second, even if we knew the exact location of veterans in relation to every application of Agent Orange in Vietnam, we could not accurately estimate the actual exposure to Agent Orange or dioxin. Not enough is known of the ecological, toxicologic, or physiologic properties of Agent Orange to know what constitutes a certain "exposure level."[74]

The best that the revised model could offer was the "likelihood" or "opportunity" for exposure. What Christian and his team found, however, surprised them: even when they used revised models for likely exposure, the incidences of exposure were remarkably low. The initial studies resulted in startling discoveries by the ESG, which had anticipated much higher levels of exposure. For Christian, though, these results were attributable to the gaps in the historical documentary record. Aside from not being able to pinpoint specific individuals in precise locations, Christian noted that the records for spray missions were incomplete for perimeter base sprayings, during which many soldiers could have been repeatedly exposed at close range over time: "In the 9th Infantry Division's area alone there were about 365 such locations that might or might not have been sprayed. Yet, the records only show a total of 478 perimeter sprays for all of Vietnam during the entire war, although there were 11 plus U.S. Divisions, 2 Korean Divisions, 1 Thai and 1 Australian Division in Vietnam."[75] The science panel and the AOWG exposure subcommittee were not pleased with the interim report and engaged in lengthy correspondence with the CDC and Christian's ESG team about revising the protocol. Internal documents from the AOWG, however, suggest that many in the group felt that the CDC was simply not capable of carrying out this type of study. For example, handwritten notes by Young on drafts of a revised interim report that the CDC would deliver in November indicate a variety of dismissive responses from the AOWG science subcommittee. In response to the CDC's explanation that it "originally thought that company level morning reports would provide additional information but for various reasons discussed below this is not the case," Young wrote, "Can't wait to read this!" Where the CDC explained that its team was ill-equipped to deal with the unfamiliar names of many Vietnamese locations, Young jotted in the margin, "*True*! CDC is not qualified!" And, finally, where the report notes the addition to the CDC team of Shelby Stanton, a Vietnam veteran and amateur military historian, are the phrases "Self-Proclaimed Expert," and "Ha! Ha! Ha! No criteria. No justification. Just a blanket guess!"[76] Young, who had long been skeptical about the health effects of Agent Orange, was unconvinced that the CDC had the right framework for this study. The comments on the draft are dismissive and condescending, but the larger point Young made in his full comments and in his work on the science panel subcommittee of the AOWG was that it was not CDC's fault; there was simply no way to determine if, when, and how ground forces were exposed to Agent Orange. Young would later be criticized for having been instrumental in killing the CDC study so that the government would not have to make payments to veterans, but this is only partially true. He actively worked to have the study canceled but in doing so made it possible for proposed settlements for veterans to be part of the cancellation.

The November report again acknowledged that even if the records were sufficient to determine the relative location of the troops in question and their proximity to a herbicide operation, the terrain, jungle cover, and weather conditions would play a major factor in determining if the spray ever reached the troops. Equally impossible was determining the levels of dioxin (or dose) in the Agent Orange used on a particular mission. Even when a mean level was used, the report noted, "estimates of dose would depend on knowledge of bioavailability, absorption rates, clothing worn, as well as behavioral factors such as amount of time spent in contact with vegetation, soil, and grasses, and consumption of local food and water. Data are insufficient to determine dose."[77] These multiple variables were widely accepted by the AOWG as essential to determining risk: probabilities of exposure could determine the likelihood that soldiers might have been exposed, but the level of that risk or possible correlations to health effects could be ascertained only through a sense of the dose involved in a given exposure scenario. Again, the CDC's own reports seemed to acknowledge the fundamental limitations of such an investigation. Science could simply not answer the types of questions being asked. The determination of exposure would have to be decided politically.

According to Scott, Young called Christian into his office on Christmas

Eve in 1985 to tell him that his work was going to be reviewed by John Murray, a retired army general. According to Christian, tensions ran rather high between him and Young, who indicated that Christian's work was suspect, but Murray and Christian would come to express admiration for each other. They were both veterans and career military men. They seemed to agree on the limitations of the documentary record and the corresponding limitations the gaps in that record placed on CDC's investigations.[78] Christian would later describe Murray's review as a peer review, but it was clearly designed to be the final stake in the heart of the CDC Agent Orange epidemiology study.[79]

General Murray's report, under the seemingly innocuous title "Report to the White House Agent Orange Working Group Science Subpanel on Exposure Assessment," is a remarkable document. Over 350 pages long, including attachments and addenda, the report begins with a cover letter addressed to Young that reads as follows:

> Attached is my report. People don't read such reports. For the many who don't, I hope, if given the opportunity, you will stress with heavy Richter scale reverberation, that:
>
> - Vietnam was not designed as an epidemiological laboratory. As a result, the data does not support a scientific cause and effect relationship between Agent Orange and Veterans' ailments alleged to it.
> - The combat records vividly disclose the need for reconsideration of the Executive Order that deprives the military from the first use of herbicides and the instant, ready, first use of riot control agents to save lives of Americans in combat, and routing the enemy.
> - The Department of Justice has denied the military services from producing the records, the expert interpretation of them, and full disclosure of the data available for the benefit of the Veterans entitled to individual awards from the chemical companies in a settlement without fault, before trial.
> - Dropping the study does not mean dropping concern for the Veterans' hurts; nor does it mean compensation that will add to the Country's budgetary ills in order to palliate those of the Veterans.
> - The Veterans can be compensated by a salatia from cutting out the current and projected costs of interminably continuing the epidemiological study, or its ill-advised options.
>
> Sincerely,
> John E. Murray, Major General, JUSA (Ret)[80]

While there was deep disagreement over Agent Orange, its possible relation to veterans' health, and the allocation of benefits for those health conditions, it is a rare document indeed that renounced the ban on the weaponization of herbicides, called for an end to ongoing epidemiological studies, and at the same time called for extended compensation to veterans. Murray's rambling foreword to the report begins, "I'd rather play around with the Apocalypse than deal with epidemiologists. At least the end-of-the-world comes to a conclusion. But that's tongue in cheek and probably why, in this case, bureaucracy got a bum rap." Murray went on to deride the "litigious society" that had brought American society to the "threshold of the ridiculous," arguing that Agent Orange was a "transcendent topic, since it approaches tripping over that threshold into an abyss of nonsense." His colloquial style notwithstanding, Murray's report came to a very clear determination, one that was shared by many who had been involved in the ongoing health studies, including Christian: "What did exist was the continued, stubborn insistence of the Vietnam War to produce anything but frustration, and the War's refusal to retroactively serve as a nice scientific showcase to produce findings that have otherwise—even under laboratory or more benign conditions—avoided scientists from drawing reasonable conclusions except for chloracne, other than: 'We don't know.'"[81] The conclusion of his report was consistent and clear: "In view of the limitation of records, restriction of the criteria, the unlimited expression of doubts, and the scientific inability because of these doubts to arrive at conclusions the continuance of this Agent Orange Study is an exercise in futility."[82]

Even though in Murray's view there was an absence of evidence of either widespread exposure among ground troops or definitive harm resulting from it, Murray recommended that the government award affected veterans a solatium, or "a legal compensation for loss or grievance without admission of fault." He noted that the solatia could be modeled on the recent court settlement between veterans and the chemical companies, an ironic compromise given Murray's low opinion of the lawsuit.[83] The question of benefits and future studies would be up to others, but Murray's report served its immediate purpose as final confirmation that the veterans' epidemiology study was dead in the water.

Testifying in front of the House Subcommittee on Hospitals and Veterans later that summer, Christian, after detailing the "exhaustive work" undertaken by him and his staff over the past three years, confirmed major gaps in the documentary record required to locate specific individuals: "Of riveting importance," he noted, "is the fact that ESG can identify a combat company's locations on a given day. However, records do not permit the location of individual sub-elements or individual troops within these sub-elements at each location. ESG is not qualified to answer the scientific problems this creates. This is an issue the scientists must address. ESG can only report what is in the records." Christian concluded by reiterating that "the records do not support continuance of the Agent Orange epidemiological survey."[84] The Agent Orange Study would languish for several more years, as the CDC tried in vain to develop a better exposure model. The Murray report, despite playing a key role in the eventual termination of the CDC epidemiology study, was never released to the public.[85]

For years military, medical, and scientific experts had attempted to pinpoint the location of specific soldiers on certain days, correlating that information with specific spray missions to determine even the statistical probability that particular units, areas, and, most important, bodies were exposed to Agent Orange. The VA, CDC, and AOWG were all working within a traditional epidemiological framework, a "risk-based" paradigm, in which sovereign bodies, distinct and separable from the surrounding environment, come into contact with chemical a resulting in condition b. Such a framework ruled out any possibility that, as Murray put it, the laboratory of the Vietnam War was going to produce "anything except frustration." Murray, Young, and the AOWG were undoubtedly right that from military records alone certainty about which troops were exposed to what levels of Agent Orange was unattainable. But the larger point here is that regardless of the ultimate exposure probability models, the VA study and the CDC study were effectively called into question and ultimately canceled because the AOWG, populated with more than a few skeptics, was able to use the politics of uncertainty to place the burden of proof largely back on individual soldiers.

Returning to the ideas generated more recently by modern science studies, environmental studies scholars, and public health historians, one should understand the uncertainty constructed in and around the Agent Orange health studies less as scientific fact than as a historical construct shaped by social and political dynamics and "produced by particular ways of apprehending the world or by clashes between different visions of the world."[86] As an alternative to the risk-based, epidemiological paradigm, for instance, the biologist Joe Thornton has proposed an ecological paradigm for understanding the relationship between chemicals, bodies, and the environment. Such a model "recognizes the limits of science: toxicology, epidemiology, and ecology provide important clues about nature but can never completely predict or diagnose the impacts of individual chemicals on natural systems."[87] Adding to this model the concept that bodies and chemicals are inescapably part of the same entity, rather than ontologically distinct categories, forces one, according to the historian Nancy Langston, to "envision the body as permeable to the environment, just as earlier generations of physicians did."[88] Even within such a structure, however, the mobile nature of troop movements and the often limited duration of tours may have further complicated the exposure scenarios for many Vietnam veterans.

A parallel problem is that veterans and their advocates were locked in a largely medical, epidemiological paradigm. Instead of arguing explicitly for a more ecologically oriented exposure model, veterans often tried to show they had been hit. While urging a more expansive definition of exposure, veterans and their supporters unwittingly played into the larger logic of the scientific uncertainty, striving for more exacting exposure models based on military records that could locate and mark specific bodies as "exposed." Under the benefits regime and legal apparatus then in place, this model would inevitably put the burden of proof squarely on veterans and their families. Benefits for Agent Orange claims expanded over time, the VA steadily adding conditions presumed to be service related, but the decisions continued to be based on politics, not science. A risk-based medical and epidemiological model stayed in place, but the VA assumed increasing latitude and discretion in determining which conditions should be covered.[89]

Dioxin Down Under

The United States was not the only nation investigating the effects of herbicide exposure on its veterans. In the early 1980s Australians and New Zealanders, too, debated Agent Orange in the public arena, through the press, through government investigations, and, in Australia, through the Evatt Royal Commission on the Use and Effects of Chemical Agents on Australian Personnel in Vietnam, more commonly known as the Royal Commission. Royal Commissions, a form of government investigation long established in British law and extended to many of its imperial outposts, are roughly equivalent to American counterparts like the Warren Commission and the 9/11 Commission. Prior to the Evatt Royal Commission, however, the Australian and New Zealand postwar encounter with Agent Orange ran a course remarkably similar to that experienced in the United States. The story of Agent Orange in both countries broke after the WBBM documentary aired in Chicago but was complicated by local outcry over the domestic spraying of herbicides containing 2,4,5-T and 2,4-D over Yarram, Australia, an area in the southernmost region of Victoria province. In the months after the documentary aired, conflicting reports by the media and government agencies created confusion about the effects of the chemicals while also revealing the extent of their application in Australia.[90] The lack of clarity about the issue carried over into the investigations, including the Royal Commission's, into the health of Vietnam War veterans of both nations.

During the late seventies and early eighties the Vietnam Veterans Association of Australia (VVAA) clashed repeatedly with the Liberal government, particularly the Department of Veterans Affairs (DVA), over disability compensation ("repatriation" in the Australian terminology) and the need for larger investigations. These studies faced the same limitations as those experienced in the United States, namely, the spottiness of records, above all those on ground uses of herbicides, and the inability to locate precisely an individual soldier at a given moment when the HERBS tapes show he might have been exposed to a Ranch Hand mission. As in the United States, veterans and their advocates regularly condemned the government, charging it with conspiracy and cover-up. In Australia, however, the DVA and subsequent government investigations took a different tack than their American counterparts in rejecting the claims of veterans. Rather than simply rely on the uncertainties inherent in the scientific and epidemiological studies under way, the Australian government proposed two alternative hypotheses to explain the health concerns of Vietnam veterans. First, they argued that veterans had likely been exposed to pesticides, particularly malathion; and, second, that many veterans were likely suffering from psychological conditions inflicted by the trauma of war, some of which might be manifesting themselves physiologically. Young and others had made such claims in the United States, but the government there never went so far as to propose social or psychological conditions as an official, alternative explanation for veterans' health concerns.

An Australian government report in 1980 comparing the herbicides and pesticides used in Vietnam with those used domestically went to some lengths to argue that considerable numbers of Australian troops were not exposed regularly to Agent Orange. It did, however, readily admit that those troops were regularly and repeatedly exposed to pesticides.[91] In 1982 another government study, "Report on the Use of Herbicides, Insecticides, and Other Chemicals by the Australian Army in South Vietnam," continued this line of inquiry, again making the case that although possible Agent Orange exposure was sporadic and limited among Australian troops, exposure to pesticides was likely widespread.⁹² The VVAA countered this holding with anecdotal evidence from veterans who described the perimeter base sprayings and the regular helicopter missions employing Agent Orange, neither of which was well documented in Australian and American military records. It also developed a form of popular epidemiology, conducting its own studies of veterans' illnesses and their links to Agent Orange exposure.⁹³ But the VVAA also met head-on the speculation that pesticides, rather than herbicides, were to blame, exploring the increasingly well-documented health effects of pesticides, most notably DDT. The government's counterclaim about the widespread exposure of Australian troops to malathion and DDT may have represented a somewhat persuasive alternative to claims of Agent Orange exposure, but it opened the door to the dangers of pesticide use, which further investigations confirmed had potentially dangerous consequences for human health. The veterans used the concerns about pesticides to their advantage in promoting a slightly more ecological view of health, risk, and exposure: "The total possible effects of the chemicals is [sic] enormous," the VVAA later told the Royal Commission, detailing not only the risks associated with herbicides and pesticides, but also the many chemical solvents and additives used in combination with them: "Having then looked at the herbicides, insecticides, and solvents and contaiminents [sic], one has to consider the possible synagistic [sic] effects of the various chemicals," they concluded. "Some of those effects are known; however the possible combination of one chemical acting with another on a person in Vietnam is enormous and it is not possible to consider all combinations."⁹⁴ In proposing this more holistic, ecological approach, the Australian veterans were years ahead of their American counterparts but still largely unsuccessful in achieving the desired results from their government.

The proposed link between psychological damage and unexplained health problems was even more controversial. This proposal was largely the work of the Senate Standing Committee on Science and the Environment and of Sen. A. J. Messner, who became minister of veterans affairs in 1980. The Senate committee also focused its investigations largely on the effects of pesticides on human health, but took special care to explore the possible link between psychological trauma and health. Messner traveled to the United States in 1982 to meet with representatives of the VA and came back apparently convinced that neuroses stemming from war experiences were to blame for veterans' health issues. In an interview upon returning from the States, Messner said that stress and trauma, not Agent Orange, were to blame for veterans' health problems.⁹⁵ In its first major report on the topic, the Senate committee made strident claims about its theory but offered no more evidence for its case than veterans had for Agent Orange exposure. Once again the VVAA used this counterclaim to its advantage, attempting to link the "total effects" of chemical exposure to a variety of psychological conditions, but again it failed to persuade the DVA or the Senate. Under the seal of authority of a government committee on science, the "achievement of the Senate report," according to Jock McCulloch, the Australian author whose book *The Politics of Agent Orange* came out just prior to the initial work of the commission, "was to transfer the issue of veterans' health from the field of physical medicine to that of psychopathology. This in turn brought the weight of popular prejudice to bear against the veterans and lay the blame for their condition on their own shoulders."[96]

In New Zealand the government was far more reactive than its Australian counterpart, adopting what was regularly described as a wait-and-see approach. Deferring to ongoing research in the United States and Australia, the Departments of Health and Veterans Affairs in New Zealand answered a barrage of letters from veterans' groups by saying it was unlikely New Zealand veterans were exposed to herbicides and that, as one Health Department report put it, "*NO* scientific study has so far proven damage caused by Agent Orange to either Vietnam servicemen or their children."[97] V. R. Johnson, a veteran and longtime activist on Agent Orange issues in New Zealand, wrote to Health Minister Thompson in 1982 that he was unimpressed with the government's response thus far and pressing for the government to initiate its own inquiry. The response from the government remained consistent: Committing government resources before the American and Australian studies were completed would be "a waste of time."[98] Both the government and the veterans in New Zealand closely monitored events in the United States and Australia. The files of the Department of Health are filled with clippings from newspapers of both countries on the lawsuit, the Ranch Hand study, and the responses of both governments to veterans' claims. Yet no inquiry on Agent Orange would be forthcoming for many years. In 1983 the new Labour government in Australia delivered on its promise to veterans: aiming to settle, once and for all, the question of the relationship between Agent Orange and the health of Australian veterans, it set up the Royal Commission. McCulloch recognized immediately that although "the terms of the Commission are generous and appear on the surface to favor veterans," the proceedings were largely overdetermined by the existing power dynamics of the politics of uncertainty, the same confluence of industry and state experts which persisted in rejecting categorically the claims of veterans. For more than a year the commission gathered testimony from multiple constituents at hearings around the country. Scientists, representatives of the chemical manufacturers, and the VVAA all submitted evidence as to whether or not the herbicide was to blame for a variety of health concerns. Like hearings and inquiries in the United States, Korea, and Canada, those of the Royal Commission were decidedly political events. Veterans and their advocates were of the view that the "total possible effects" of exposure were enormous, while the opposition, including Monsanto's Australian subsidiary, rejected that contention.[99]

Although focused on the health of veterans, the commission and its various constituents relied upon the existing international structure of scientific knowledge and expertise to address the problems. The Australian veterans who, along with those from New Zealand, were active members of the class action lawsuit under way in the United States networked with their American colleagues, sharing their experiences of working with their respective governments. The chemical companies, Australian-based subsidiaries of Dow and Monsanto, drew on their parent companies' legal and scientific resources to reinforce the sense of uncertainty surrounding the health effects of Agent Orange exposure and cast doubt on veterans' allegations. The commission and the Australian government itself also drew on international expertise, particularly the resources and experience of the AOWG in Washington, including Young, but also those of European scientists who had studied the effects of dioxin and of Vietnamese scientists, whose studies had been largely dismissed by the AOWG, VA, and CDC. The international networks had been involved in evaluating the incidents and reports from Alsea, Seveso, and Times Beach.[100]

In December 1983 Young visited Australia, serving as an official scientific consultant to the Royal Commission. In a follow-up letter to Phillip Evatt, the chair of the commission, Young wrote that the issues connected to Agent Orange "cross national boundaries." Young discussed with the commission staff the history and use of herbicides in Vietnam, the toxicology and environmental fate of dioxin, and the ongoing veterans' health studies in the United States. The commission asked him to advise it on the design of their morbidity study and, later, to sign off on it.[101] Given Young's active role in the heated debate over the design and implementation of the VA and CDC studies, Australian veterans may have been right to be skeptical of his involvement, but, as Young wrote to the commission, the Australian veterans had a number of collective advantages over their American counterparts. Although they were faced with the same gaps in the records of local spraying and the location of individual soldiers, Australian troops were far less dispersed during their time in service; they were more homogenous in their demographic makeup, reducing potential variables; and the total number of troops (around forty-five thousand) was large enough to develop viable cohorts yet small enough to facilitate the contacting and tracking of subjects in a long-term study. Far from advising against it, Young strongly encouraged the Australians to begin an epidemiological investigation and eventually signed off on its design as "epidemiologically sound and feasible."[102] In 1984 several members of the commission visited the United States to consult with Young and other members of the AOWG. In the end, the commission and the government were unable to agree on the funding and model for the study, and as a result no epidemiological study of Australian veterans was initiated during the tenure of the commission, which focused instead on neuropsychiatric studies of veterans.[103]

The final report of the commission, released in July 1985, attempted to deny that Agent Orange caused veterans' health problems while simultaneously situating their concerns in a larger historical context to posit the explanation that the Agent Orange issue was social and political in nature rather than scientific or medical. "The tragedy of the Agent Orange controversy," the report begins in its remarkable opening statement,

> is that Vietnam veterans were doubly vulnerable. First, they were raised in the era of environmental upsurge, when pollution in general and chemicals in particular were "to blame." The "natural" was better than the "unnatural" and the use of "artificial" or synthetic" was said to be heinous, greedy, and usually capitalistic.
>
> Second, conflict has taken its toll on many of them, as it had on their fathers and their shell-shocked grandfathers. But their forbearers came home heroes, the conquerors of Kaiser-Bill, Hitler, and Tojo: their wounds, visible and invisible, could be worn with pride.
>
> We sent the cream of our youth, with strong value systems and a belief in themselves and those values, to Vietnam. We trained them to kill men. They learnt by necessity to kill women and children. They saw in the co-incidence of contest and village home-life (which is guerilla warfare) and in like fighting like, the futility of the conflict and of their participation in that conflict.
>
> When they returned to Australia, they were ostracized by many and any sense of purpose in their sacrifices evaporated.
>
> Is it any wonder that they felt poisoned?[104]

Among the many memorable sentences in the opening statement and, indeed, in the entire nine-volume report, this last is perhaps the most striking.

Summing up the commission's unique attempt not simply to deny veterans' claims but also to posit an alternative explanation for their afflictions, the question, "Is it any wonder that they felt poisoned?" powerfully acknowledges that the problems of Australia's Vietnam veterans were the collective responsibility of the entire nation—the "we" who sent them to war and asked them to do the unthinkable. Despite the implication of collective responsibility, the report sees their distress as being largely imagined, that is, socially constructed and perhaps psychologically induced by the trauma of war and the generational shift in environmental and chemical awareness, rather than by exposure to toxins. Finally, by using the past tense ("they felt poisoned"), the sentence relegates not only the veterans' experiences but also their questions about their health to the past, problems to be put collectively behind "us." Notwithstanding its unique endeavor to arrive at an alternative explanation, the Royal Commission was no more successful than U.S. courts or the Congress in reaching closure on the Agent Orange issue.

In reconstructing the history of the Agent Orange controversy, as the report labeled it, the Royal Commission strove to prove its case that the issue was a social rather than a medical one by citing the role of media coverage in the rise in veterans' claims. The report traces the impact of DeVictor and Kurtis, following the story from Chicago to Sydney, as the Australian version of *Rolling Stone* and then the Sydney *Sun Herald* picked up the story, which then spread like wildfire across the country with provocative headlines that focused disproportionately on birth defects in veterans' children.[105]

After laying out the social context in which the controversy arose, the commission detailed the legal complexities involved with Australia's repatriation laws and what they meant for veterans' claims. Focusing on risk and causality, the commission relied heavily on the class action suit, the ongoing epidemiological studies in the United States, and the experience of American veterans with the VA to lay out a procedure that, in the end, still left the burden of proof for both exposure and causality on the claimants. The veterans, for their part, were largely represented by the VVAA. Although individuals gave testimony at the exposure hearings and at the informal hearings conducted around the country, the VVAA tasked itself with presenting the collective case of veterans to the commission. The group's report to the commission, however, met with a resounding thud. It was, as the final report described it, "an anti-climax," given the VVAA's role as the primary advocate for the formation of the commission: "The 67 page document submitted over the signature of Senior Counsel for VVAA was a farrago of generality, hyperbole, mis-statement, mis-spelling, and bald assertion. No indication of any statistical or other scientific basis for the generality of the claims appeared. By way of example, no distinction was drawn between acute and chronic, or even between immediate and delayed effects of chemical agents. Indeed, questions of dose, absolutely critical to an intelligent analysis, were simply ignored."[106] While the veterans' attempt at popular epidemiology was filled with a number of evidentiary gaps, the commission's dismissive response points to the competing epistemological frameworks brought by the various sides. Working within a legal, medical, and scientific system that emphasized direct causality supported by statistical evidence and scientific certainty, the commission sided with the chemical companies and with consultants from the AOWG, which believed that the VVAA was incapable of supplying the type of evidence "absolutely critical to intelligent analysis."

When veterans testified in the exposure hearings about what they had seen and experienced during the war, the commission disputed their stories. Raymond Daniel, who served in 1966–67 at Nui Dat, a major Australian base, exhibited photographs and slides he claimed to have taken during the war that purported to show the effects of defoliation in and around the base. Nui Dat was at the center of a controversy during the hearings that entailed the large number of troops who served there and the lack of indisputable evidence in military records as to whether or not Agent Orange had been sprayed by air around the base area and whether it had even been stored at the base. Daniel testified that he sent the film to his wife, who developed it and sent it back to him, but the investigators determined that many of the "before" slides were processed considerably later than the "after" slides showing what he claimed were the effects of defoliation around the base. As Daniel's case fell apart under cross-examination, the commission concluded that none of the photographic evidence he presented "show[s] anything that could be described as the result of defoliation." On the basis of the uncertainty the commission constructed around Daniel's case, it resolved that "no inference can be drawn from his evidence that Agent Orange was sprayed over Nui Dat in August 1966."[107]

Both the transcripts of the hearings and the final report contain dozens of reports by veterans who described their exposure scenarios. Whenever possible the commission investigated the stories in some depth, seemingly proposing to discredit the veterans' case rather than corroborate it. Often such undercutting was done in cross-examination by lawyers for the chemical companies.[108] If neither the commission nor the Monsanto lawyers could cast sufficient doubt on the veterans' version of events, the commission assumed the default position dictated by the current politics of uncertainty: in view of the lack of documentation to confirm exposure, the commission's beliefs about the toxicology and environmental fate of dioxin, and the fact that the commission was steered largely by the guidance of the AOWG advisers and the scientific experts of the chemical companies, there was ample doubt that the veterans had been exposed to sufficiently high levels of dioxins to cause health problems. Where direct exposure seemed likely, the commission regularly found that it resulted from the failure to follow safety provisions for the handling and storage of chemicals; in cases of indirect exposure it relied on scientific experts to demonstrate that the environmental fate of dioxin under normal conditions would not have allowed for significant levels of exposure through food or water; and in other instances the commission simply relied on its assumptions, backed by the weight of the expert testimony, that the dose available in the exposure scenario was unlikely to have caused real harm. A number of gaps in the cases presented by individual veterans and the VVAA did exist, but the commission relied on and trusted the scientific counsel of the chemical companies and AOWG advisers such as Young and privileged that information by continuing to place the burdens of proof and causality on the veterans. In doing so, they emphasized the uncertainty of veterans' claims while ignoring, if not suppressing, the appreciable degree of uncertainty about dioxin and its effects on humans and the environment.

In its conclusions the Royal Commission went beyond previous studies and investigations. While it could not disprove conclusively all veterans' claims of health issues stemming from chemical exposure, it did present an alternative narrative of psychological trauma and social concern that sought to close the book on the Agent Orange controversy. It went so far as to label the issue a hoax. On the charge that Agent Orange caused the health problems of veterans, it rendered a verdict of not guilty and moreover saw this as a triumph for the veterans: "No one lost; this is not a matter for regret but for rejoicing. Veterans have not been poisoned. This is good news and it is the Commission's fervent hope that it will be shouted from the rooftops."[109] That verdict, as explained in final report, was based on what the authors called "intractable uncertainty."[110] The aftermath of the verdict was, unsurprisingly, very politicized as well. Critics charged the commission with creating a tendentious environment around veterans and their supporters and with plagiarizing testimony from Monsanto when, without proper acknowledgment, it adopted wholesale the company's summary of reports on the health effects of 2,4,5-T.[111] In New Zealand the response was equally scathing. After being told for years to wait for the commission to complete its work, the

VVANZ criticized the final report as a sham, renewing its claims for an independent inquiry. V. R. Johnson of VVANZ was still fuming about the commission years later. Writing directly to "Her Majesty, Queen Elizabeth II," in 1990, Johnson accused the commission of being "tainted" by "perjury and fraud."[112] The debate over the commission and its report, however, revolved mostly around the inexorable topic of uncertainty.

In the introduction to *Evatt Revisited*, a volume of essays drawn from a conference held in 1989, the editors write, "Many scientists who appeared before the commission felt that the legal treatment of the scientific evidence failed to bring out the degree of uncertainty in the data on the issues and its interpretation by scientists."[113] The accompanying essays, including contributions by scientists who had been involved to varying degrees with the commission, highlight a number of studies, some of which were not available at the time of the final report, that called into question claims about acceptable levels of dose and overall safety and effects of Agent Orange and its associated dioxin. Others advised that given the admittedly high degree of uncertainty involved on all sides of the debate, the commission could have rendered a verdict of insufficient evidence rather than not guilty.[114] Although many of these criticisms were also made when the final report was issued, there was little immediate impact on the outcome of the commission's findings. In a sense, the commission's findings resulted in the worst of both worlds: by seeking to dismiss and discredit the veterans' claims and by proposing a social-psychological alternative narrative, it succeeded in undermining its own recommendations for granting additional support to veterans. Young would later write to Evatt in support of the report, arguing, by reference to recent work by Gough, that the Agent Orange decisions being made in the United States were also largely political, not scientific. "In reflecting back to the Final Report [of the Royal Commission], it is apparent that you recognized the limits to which Agent Orange was a science problem," Young wrote. "Indeed, I believe you correctly concluded that Agent Orange was but a symptom of a larger problem—our society's inability to deal with the Vietnam Veteran. I hope your Commission's Report has led the way to settling the difficult issues of re-adjustment by both the Vietnam Veteran and the Australian public."[115] The Royal Commission and its final report did no such thing.

Conspiracies and Coverups

In the United States, while the Agent Orange Study languished, the "Postservice Mortality Study among Vietnam Veterans," more commonly known as the Vietnam Experience Study (VES), went on. Released in February 1987, the VES compared the mortality rates of more than nine thousand U.S. Vietnam veterans with those of a similar cohort of veterans who served in Korea, Germany, and inside the continental United States at the time. The CDC selected the participants randomly from the U.S. Army National Personnel Records Center. Trying to place veterans' concerns about their overall health in a wider context, the VES explored "a wide variety of health-influencing factors, such as psychological stresses associated with the war, infectious diseases prevalent in Vietnam, and exposure to the herbicide Agent Orange." The study found that although "total mortality in Vietnam veterans" was 17 percent higher than in the control group, that percentage was heavily correlated with the first five years after discharge from the service and was largely owing to "motor vehicle accidents, suicide, homicide, and accidental poisonings."[116] Going beyond previous approaches in its attention to the causes of death, the VES was the most comprehensive study to date of the health of Vietnam veterans. Like any such study, however, it had limitations, which the full report acknowledged. Despite the large number of participants, the VES was considered by its authors to be a "relatively small sample size" and potentially more subject to irregularities. Notably, the study pointed out that the time elapsed since the end of the war was highly limited: "If the 'Vietnam experience' does place veterans at an increased risk for certain fatal chronic diseases, the time interval between exposure and death may be longer than our current 10–15 years of follow-up."[117]

Indeed, around the time the CDC released the preliminary findings of the VES, other studies began showing different results. In late 1986 the National Cancer Institute (NCI) announced that its joint study with the University of Kansas found that farmers exposed to herbicides containing 2,4-D had much higher rates of cancer than control groups.[118] Even more significant, however, was the study of Vietnam veterans in 1988 conducted by the Stellmans and sponsored by the American Legion, which found higher correlation between Agent Orange exposure and various health conditions. As Scott has noted, the Stellmans used their study to reignite the issue of the HERBS tapes and whether or not they could be used to indicate individual troop exposure. The Stellmans attempted to correlate the self-reporting of herbicide exposure by veterans with the hits model that had been dismissed previously by the AOWG and CDC. Although they were still using battalions as their primary unit of analysis, the Stellmans believed that the combination of the ESG data and the self-reporting of veterans in the study made the exposure likelihood models much more accurate. Provoking the ire of the AOWG, the Stellmans were forced to defend both their study and their reputations in front of a congressional oversight committee that once again called the entire CDC operation into question. A report by the House Committee on Government Operations eventually found that the AOWG had improperly acted to quash, in Scott's words, the CDC studies and raised additional questions about the OMB's role in stressing the financial and legal implications of admitting health consequences from dioxin exposure among veterans and civilians alike.[119] While the report partly justified the charges made against the AOWG, the VA, and the CDC over the past decade, it gave little comfort to veterans and their families, who remained caught in the web of uncertainty spun and then reinforced by these charges, conflicting studies, and ongoing misgivings about sample sizes, exposure likelihood models, and growing but still inconclusive evidence.

The Agent Orange Study was canceled in 1989 after expending $43 million of the $56 million appropriated by Congress. The response from the veterans' community was scathing. The *Veteran,* the official publication of the VVA, accused specific members of the VA advisory board of suppressing evidence that would prove the link between exposure and cancer. The VVA attacked Ted Colston, a physician at Boston University and a VA adviser, for asserting that the evidence from the Kansas/NCI study was limited and proposing the need for additional studies. "The Kansas Study is the smoking gun," Muller argued in the piece. "As a member of the VA committee Dr. Colston has an obligation to recommend compensation when sound scientific evidence is found to support such an action. He's found the evidence—it's time for him, and the rest of the committee to keep its mandate."[120] The *Washington Post* headline on the cancellation of the study called the CDC's efforts either "botched or rigged" and wondered if various groups had exerted political pressure to suspend and later terminate the studies.[121]

Veterans weren't the only ones who believed the VA and the CDC were hiding the truth. In August 1990 the House Committee on Government Operations issued a report entitled "The Agent Orange Cover Up: A Case of Flawed Science and Political Manipulation," which accused the White House of pressuring the CDC to terminate the study and concluded that "the Federal Government has suppressed or minimized findings of ill health effects among Vietnam Veterans that could be linked to Agent Orange exposure."[122] Relying largely on testimony by the staff of the National Institute of Medicine (IOM), the report (which was accompanied by a substantial "dissenting views" rejoinder from several committee members) contended in fact that military records could provide a high level of "probability for likely exposure."[123]

Responding to the committee's report, two veterans' groups filed a lawsuit in federal court alleging that the cancellation of the CDC study was part of a larger government coverup. In its coverage of the suits, the *Veteran* again resorted to the language of conspiracy: "This never-ending legacy of the war in Vietnam has created among many veterans of that war and their families deep feelings of mistrust of the U.S. government for its lack of honesty in studying the effects of the rainbow herbicides, particularly Agent Orange, and its conscious effort to cover up information and rig test results with which it does not agree."[124] Accusations of conspiracies and coverups aside, these stories elucidate the idea of many veterans and their advocates that individual veterans and the VVA, not scientific advisers to the White House and the VA or scientists at the CDC, should determine what constitutes "sound scientific evidence."

More fuel was thrown on the Agent Orange conspiracy fires in 1990, this time by the eminent figure Adm. Elmo Zumwalt Jr., a decorated veteran of the Second World War who by 1968 had risen to become the commander of all navy forces in Vietnam. During the late 1960s Zumwalt pushed for and eventually ordered increased defoliation along waterways patrolled by American boats in southern Vietnam. Among the Swift Boat commanders whose missions were supposed to be made easier by the defoliation missions was the admiral's son, Elmo III. After the war Elmo III and his son, Elmo IV, were afflicted with conditions the family believed to be caused by Agent Orange. In a *New York Times* magazine article in 1986 Elmo III is quoted as saying, "I am a lawyer and I don't think I could prove in court, by the weight of the existing scientific evidence, that Agent Orange is the cause of all the medical problems—nervous disorders, cancer and skin problems—reported by Vietnam veterans, or of their children's severe birth defects. But I am convinced that it is." Elmo III died in 1988 of complications from lymphoma.[125]

Several years after Elmo's death, the VA appointed Admiral Zumwalt a special assistant "to determine whether it is at least as likely as not that there is a statistical association between exposure to Agent Orange and a specific adverse health effect." In the thorough report that resulted from his study Zumwalt recounted the long, disputed history of the Ranch Hand study and the evolution of knowledge about connections between Agent Orange and a variety of health concerns. The primary reason for the attention the report continues to garner is its inclusion of the infamous Clary letter. The Zumwalt report held that, despite the Pentagon's claims that it had no knowledge that Agent Orange and other herbicides were dangerous to human health, "evidence readily suggests that at the time of its use experts knew that Agent Orange was harmful to military personnel."[126] The only seeming confirmation of this claim is a letter written to Daschle, at that point a senator, in 1988 by James Clary, who had worked on the development of spray equipment at Eglin Air Force Base. Clary wrote,

> When we (military scientists) initiated the herbicide program in the 1960's, we were aware of the potential for damage due to dioxin contamination in the herbicide. We were even aware that the "military" formulation had a higher dioxin concentration than the "civilian" version due to the lower cost and speed of manufacture. However, because the material was to be used on the "enemy", none of us were overly concerned. We never considered a scenario in which our own personnel would become contaminated with the herbicide. And, if we had, we would have expected our own government to give assistance to veterans so contaminated.[127]

Clary added that he found the current literature on dioxins and human health to be understating the risks to health. These claims, which Daschle repeated regularly in committee hearings and on the floor of the Senate, were held up by the senator and by longtime critics of the government's approach to Agent Orange as proof of a coverup. Speaking in the Senate on November 21, 1989, Daschle argued that "if [Clary's accusation is] true, then several agencies of the Federal Government have spent decades trying to keep the truth about Agent Orange from the general public. You need only read Dr. Clary's letter to come to that conclusion."[128] No other new evidence addressing the government's knowledge of the dangers of Agent Orange appears in the Zumwalt report, but this one letter, written more than twenty years after the fact, is repeatedly held up as evidence of a coverup. On the website of Friendship Village, a long-term care facility specializing in Agent Orange victims located outside of Hanoi and built by an international group of Vietnam War veterans, the following statement appears: "Scientists involved in Operation Ranch Hand and documents uncovered in the US National Archives, present a different picture. There are strong indications that not only were military officials aware as early as 1967 of the limited effectiveness of chemical defoliation in military strategy, they also knew of the potential long-term health risks of frequent spraying and sought to censor relevant news reports."[129] The sole documentation of this statement is Clary's letter. Other documents, many of them discussed here in earlier chapters, describe the minimal impact of herbicides on enemy forces and possible ecological effects, but none discuss their effects on human health. Journalists too have seized on the letter, citing it as proof that the government knew about the risks of Agent Orange. Even the widely acclaimed *Chicago Tribune* series on Agent Orange in 2009, which claimed to "unearth new documents" showing awareness of the agent's toxicity, uses long-available documents from the chemical companies, including those from the Midland summit, and the Clary letter to back its assertions.[130] The letter has fueled the fires of conspiracy, but to this day it stands as the only evidence that the U.S. government or its military knew prior to the late 1960s that Agent Orange was harmful to human health.

Why, more than half a century after the beginning of Operation Ranch Hand, is the Clary letter the unique evidentiary text? The preferred answer of many veterans and critics of the U.S. government is that a massive, extraordinarily well-orchestrated coverup destroyed other evidence. Leaving aside the fact that no such proof has surfaced in the myriad lawsuits and government investigations over the past several decades, the conspiracy explanation is profoundly ahistorical. It ascribes to government officials a mindset and knowledge base that was almost certainly not available to them at the time. In the early 1960s millions of Americans had grown accustomed to widespread spraying of herbicides and pesticides in their communities, and only a handful of people were questioning their safety. Why should one expect that members of the Kennedy administration or military officers in Southeast Asia would find such chemicals suspect? It is not, in other words, that these officials were incapable of suppressing answers to questions about the dangers of herbicides; it is that they were extraordinarily unlikely to be asking such questions in the first place.

This does not mean the Clary letter is a hoax or a fraud. It, too, is a product of its time. As we have seen, the presumed risk of damage owing to Agent Orange exposure has grown over time in direct proportion to the knowledge of potential danger stemming from dioxin exposure. Just as one would not expect government officials in the early 1960s to have ideas and concepts that were not widely available at that time, one should not expect veterans and others in the 1980s to reflect on their experiences in the war except through the filters they have developed in the ensuing decades. Given what is now known about Agent Orange and its associated dioxin, and given the politics of the war with which it is inextricably linked, it is not surprising that so many veterans from so many countries believe they were subjected to Agent Orange, or that they often blame the governments involved for that exposure. It is not surprising either that individuals like Clary and Admiral

Zumwalt himself felt a sense of guilt or remorse over their role in the chemical war.

That such views are understandable does not make them historically accurate. Those who think the Clary letter points to widespread wrongdoing should be asked the following question: Is it more likely that a massive government coverup managed to destroy all other evidence corroborating Clary's claims and that this campaign has prevented others like Clary from coming forth decades later? or that Clary's memory, like that of so many who served in Vietnam decades ago, has become subject to revision as it was shaped by time, knowledge, and the enduring trauma of that war? In this case, the reasonable response is that the burden of proof should rest with those who continue to advocate explanations based on conspiracies and coverups.

Conclusion

By the dawn of the 1990s, more than a decade removed from the end of the Vietnam War, the politics of uncertainty surrounding Agent Orange had undergone a profound shift. As a result of lawsuits, congressional investigations, and a dizzying array of competing health studies, the idea of scientific uncertainty—which for many years had given skeptics and apologists for Agent Orange sufficient cover to avoid costly litigation, to refuse to compensate veterans and their families more robustly, and to acknowledge the many other widespread economic, legal, medical, and moral implications that would likely have accompanied more conclusive evidence of its dangers—had come to reinforce the assumption that the government, not veterans, bore the burden of proof in determining Agent Orange exposure. While there was no evidence of an organized government coverup, the battles over the science and politics of Agent Orange exposure from the late 1970s to the 1980s had nevertheless paved the way for a major shift in how the disability claims of Vietnam veterans would be dealt with.

The Australian Royal Commission found that scientists' assertions of the uncertainties in epidemiological studies were justification enough for denying veterans' claims, while the VVAA and others believed that in light of that same uncertainty, veterans should be given the benefit of the doubt. The same logic has been applied to legal cases in the United States, where every single case brought against the manufacturers of Agent Orange since the veterans' class action suit was settled in 1984 has been dismissed for lack of convincing scientific proof.[131] In Agent Orange cases the uncertainty and the lack of convincing scientific proof have rewarded governments in Australia, the United States, and elsewhere as well as the chemical manufacturers of Agent Orange, with the benefit of the doubt, while placing the burden of proof on individual veterans and other victims of the chemical war. Despite the contentions of a variety of courts and commissions, the scientific knowledge brought to bear on these cases is not neutral; it is inscribed within particular legal, political, and economic frameworks that have, in nearly every instance, benefited states and corporations at the expense of citizens.

In largely discrediting the medical claims of veterans, however, the various battles over Agent Orange exposure in the United States and Australia ironically brought into high relief the impossibility of separating science, law, and politics when discussing the health of Vietnam veterans. By clarifying that the decision regarding benefits for exposure was inevitably, in part, a political one, the battles allowed the U.S. Congress to press forward with a new paradigm for Agent Orange–related benefits. The resulting legislation, the Agent Orange Act of 1991, passed unanimously by the House and Senate in January 1991, made a number of changes in how the VA would handle claims. First, the act allowed the secretary for veterans' affairs, a cabinet-level political appointee, "to obtain independent scientific review of the available scientific evidence regarding associations between diseases and exposure to dioxin and other chemical compounds in herbicides, and for other purposes." Whenever the secretary determines "on the basis of sound medical and scientific evidence, that a positive association exists between (A) the exposure of humans to an herbicide agent, and (B) the occurrence of a disease in humans, the Secretary shall prescribe regulations providing that a presumption of service connection is warranted for that disease for the purposes of this section." In other words, the secretary had a great deal of discretion in determining what was considered eligible for disability benefits. In addition to the increased discretionary authority, the secretary was to be advised by the NAS, which was now charged with evaluating all available scientific evidence, making recommendations to the VA based on those reviews, and providing "any recommendations it has for additional scientific studies to resolve areas of continuing scientific uncertainty relating to herbicide exposure."[132]

The biggest changes stemming from the Agent Orange Act related to the determination of exposure. The NAS and the secretary of veterans' affairs were charged with identifying which conditions had an "association" sufficient to be eligible for service-related disability benefits. Under the new benefits regime there was no attempt to link that association to the individual exposure of specific veterans. In view of the challenges in trying to ascertain specific exposure scenarios through the seventies and eighties, Congress decided it was best to avoid such efforts in the future. Instead, once the secretary had resolved that an association existed between a disorder and "an herbicide agent," the veteran had to show only that he or she had the condition and that "during active military, naval, or air service, served in the Republic of Vietnam during the Vietnam era and while so serving was exposed to that herbicide agent, "for it to be considered" to have been incurred in or aggravated by such service, notwithstanding that there is no record of evidence of such disease during the period of such service." In practice this meant any U.S. soldier who had a condition listed as having a positive association to Agent Orange exposure and who served in Vietnam during the Vietnam era (that is, the period of direct U.S. military involvement in Southeast Asia), regardless of how long, where, or in what capacity, was eligible for benefits. They were now *presumed* to have been exposed simply by having set foot in South Vietnam.

As we will see, even this gross simplification of the process did not solve the question of exposure. In the 1990s and well into the twenty-first century veterans who served off the coast of Vietnam pressed their claims for inclusion in the new benefits regime, as did those who stretched the limits of the ill-defined Vietnam era. Still, there could be no mistaking the sea change this legislation represented, given the fierce battles fought over the previous decade. The scientific uncertainty surrounding Agent Orange, dioxin, and its impact on human health remained, but the politics of uncertainty that had long been stacked against veterans had shifted significantly in their favor. What that meant for other constituencies, however, remained unclear. Despite the millions of pages of testimony, scientific studies, and other evidence offered in this debate, despite the efforts of the Royal Commission, and despite the Agent Orange Act of 1991, the Agent Orange controversy not only failed to recede in the United States and Australia, but grew in size and scope as governments, veterans, scientists, and an increasingly large and diverse group of citizens around the world came to believe that they, too, were victims of Agent Orange. Nowhere was this truer than at the ground zero of Agent Orange: Vietnam.

CHAPTER FIVE
"ALL THOSE OTHERS SO UNFORTUNATE"
Vietnam and the Global Legacies of the Chemical War

In the absence of tangible forms of documentation and data that might afford the type of certainty, causation, and proof that would satisfy the legal, political, and diplomatic forces ensconced in the politics of uncertainty that continue to play out so unevenly around the world, the bodies of Agent Orange victims themselves have become, in the twenty-first century, contested evidence of the long legacies of herbicidal warfare. In Vietnam the bodies of children and the preserved fetuses of the unborn are regularly put on display in the service of drawing attention to the unsettled legal and economic legacies of Agent Orange. In the United States, Australia, New Zealand, Canada, and elsewhere the bodies of veterans, their children, and other alleged victims are being reconceptualized as repositories of dioxin as a new way to explain often decades-old conditions. On these bodily canvases scattered around the globe are inscribed—by the state, by scientists and advocates, and by historical memory itself—narratives and counternarratives of poison, disease, and injustice. But not all narratives are created equal. Some fit the parameters established by various agencies and governments and result in inclusion, compensation, and perhaps healing. Others are burdened with different expectations and different standards of proof, leaving them excluded from those same benefits.

I trace some of these stories here across geographic and temporal boundaries, exploring the legacies of the chemical war in Vietnam, Korea, New Zealand, Canada, and the United States. In each country a number of factors have framed how communities and actors have responded to the seemingly ongoing threat of Agent Orange. First, each of these cases continues to be shaped in part by the politics of uncertainty; while knowledge of dioxin and of its effects on humans and the natural environment continues to grow—making possible both better diagnoses of and more helpful solutions to a variety of problems—it also continues to raise nearly as many questions as it answers. Second, the reactions to Agent Orange in these diverse locations have been shaped by transnational networks of knowledge and power. In response to the resistance of states and corporations to compensate victims, veterans, their allies, and a growing group of legal, political, and environmental activists have forged bonds across borders, seeking justice and compensation for alleged victims, remediation for cleanup of poisoned landscapes, and accountability by the governments and companies they believe to be responsible for the ongoing effects of dioxin contamination. In the face of scientific uncertainty, these groups continue to pose experiential counternarratives and alternative moral, legal, and epistemological frameworks. Finally, the response to Agent Orange in the global arena has been shaped by the memory of the Vietnam War and the relationship of the affected nations and communities to the prosecution of the war, to the United States more generally, and to the veterans of those countries who served in the war. In some cases the memories have enabled former enemies to forge alliances; in others they reflect both historical and contemporary concerns about the place of the United States in the world and the relationships between various actors and nations.

Hot Spots

Several major studies released in the early twenty-first century transformed the way scientists, historians, veterans, and other constituencies around the world think about, write about, and grapple with Agent Orange. None of the studies had a more powerful global impact than that published by Jeanne and Steven Stellman in 2003 in *Nature*. The outgrowth of a long-term study mandated by the Agent Orange Act of 1991 and supported by the NAS and the VA, the Stellmans' piece offered dramatically revised estimates of the overall volume of herbicides and dioxins spread over Vietnam during the war.[1] Returning to the original USAF flight records from Operation Ranch Hand, the Stellmans identified major gaps in the HERBS tapes and used after-action reports from the National Archives to add a sizable number of previously uncounted missions to the record. The results were staggering. The revised estimates showed that between 1961 and 1971 the USAF sprayed 7,131,907 liters (1,884,050 gallons) more than previously believed. The revised estimate showed the totals from the NAS examination in 1974 to be more than 10 percent off. The revised total brought the estimate of total herbicides sprayed to nearly seventy-three million liters, or more than nineteen million gallons. For Agent Orange the revised numbers were nearly forty-six million liters, or about twelve million gallons.[2]

The total volume in and of itself, while shocking to those who had studied the issue, was not as alarming to the affected communities as what it suggested about the corresponding revisions of total dioxin content contained in the herbicides. The Stellmans admitted that determining anything approaching an average dioxin level in herbicides containing 2,4,5-T was virtually impossible, given the wide variance in the levels from producer to producer, manufacturing runs, and changes in procedures over time. They based their estimates on the data gathered by the USAF for the EIS required during Operation Pacer HO. Although an average value of thirteen ppm was perhaps more realistic based on the data, the Stellmans took a conservative level of three ppm TCDD for Agent Orange, resulting in the conclusion that more than 366 kg of dioxin were sprayed over the course of the war.[3]

Notwithstanding the attention garnered by the study, a number of questions remained unresolved. Just as in the debates over veterans' exposure to Agent Orange in the 1980s, discussion of the Stellmans' findings often centered on the issue of locating specific bodies within the temporal and geographic boundaries cited by the HERBS tapes, on the questionable records of local Vietnamese populations during the war, and on models provided by Geographic Information Systems (GIS). For all of the new data and revised estimates, the *Nature* article essentially consisted of an updated, more sophisticated, more technologically advanced version of the exposure opportunity indices (EOI) that were so contested in the eighties. Using GIS databases, the Stellmans combined the "cleaned and updated" HERBS tapes with military records to "permit assignment of quantitative exposure opportunity indexes (EOIs)" to military units or individuals.

The final and arguably the most controversial finding of the Stellmans' examination was its use of population data to revise radically the estimates of how many Vietnamese civilians had been exposed. The data came from the Hamlet Evaluation System (HES) run by CORDS in the late 1960s. Of the 20,000 plus "unique hamlets" identified in the records, 3,181, according to the Stellmans' estimate, were sprayed directly during herbicide missions, representing between 2.1 and 4.8 million people.[4] The unknown variables in this equation, however, underscore the problems of using military records to make such calculations. CORDS records for the HES were often wildly inconsistent, and the accuracy of the HERBS data about the actual areas exposed during particular missions is questionable. The importance of the overall volume of herbicides and dioxin used was at times lost in the emphasis placed on the large number of Vietnamese civilians who may have been sprayed directly by Ranch Hand missions. If the point of more recent advancements in knowledge about the persistence of dioxin in the soil and food chain was, in part, to diminish the need to show that a specific body was sprayed by a specific chemical at a fixed moment in space and time, debates over direct exposure of civilians have tended to reinforce outdated chemical exposure models rather than support the shift to a more holistic, ecological view. Yet the broad estimate of civilian exposure became the basis for Vietnamese claims about the extent of damage caused by Agent Orange and its associated dioxin.

Throughout the exposure battles of the 1980s the Stellmans argued consistently against Alvin Young and the rest of the AOWG that military records were sufficient for determining likely exposure. The *Nature* article stood in many ways as the culmination of that methodological debate, but longtime critics as well as more agreeable allies were not entirely convinced by the Stellmans' revelations. In response to their updated EOIs, Young continued to maintain that military records were insufficient to determine the precise location of small units and especially of individual soldiers. In two articles published in 2004 in *Environmental Science and Pollution Resolution* Young pushed the argument even further. First, in a piece coauthored with Paul Cecil, a Ranch Hand veteran and historian, he wrote, "Herbicide missions were carefully planned and . . . spraying only occurred when friendly forces were not located in the area." The authors pointed to the detailed, lengthy approval process required of all Ranch Hand missions and to the role played by the forward air controller, whose responsibility it was to check the targeted area prior to missions for weather conditions and the presence of friendly forces. In response to the abundant anecdotal evidence of veterans that they were sprayed directly by herbicides during the war, the authors maintained that antimalarial spray missions were the more likely culprit.[5] In the second article Young and several coauthors (scientists, in this case, not historians) doubled down on the issue of dioxin exposure. Leaving aside the issue of direct spraying, they contended that "the prospect of exposure to TCDD from Agent Orange in ground troops in Vietnam seems unlikely in light of the environmental dissipation of TCDD, little bioavailability, and the properties of herbicides and circumstances of application that occurred. Photochemical degradation of TCDD and limited bioavailability of any residual TCDD present in soil or on vegetation suggest that dioxin concentrations in ground troops who served in Vietnam would have been small and indistinguishable from background levels even if they had been in recently treated areas."[6] In defense of this argument, the authors pointed to the technical specifications of the spray equipment installed in the C-123 Ranch Hand aircraft, which was exhaustively tested in the fifties and sixties to produce the most efficient droplet size and an accurate spray swath. They also refer to the "consistent enforcement" of a variety of procedural mechanisms, including those carried out by the forward air controller. The most important corroborating evidence is the nearly complete absence of elevated TCDD levels in Vietnam veterans. Even Ranch Handers and Chemical Corps veterans, who would have been regularly exposed to Agent Orange and other herbicides containing dioxin, have failed to show elevated levels of TCDD.[7]

While the rebuttals of Young et al. do cast some shadow of doubt on the exposure of Vietnam veterans to direct spraying by herbicides, they are not without their own problems. In defense of their assertions about proper procedures being followed during Ranch Hand operations, the authors point occasionally to MACV documents but more often to sources from Ranch Hand veterans, including Cecil himself, in an article he cowrote.[8] Indeed, much of the debate about the usefulness of military records for determining exposure centers on the accuracy and reliability of the records. I have noted here several examples, such as the discrepancies between safety policies and actual practice for the spraying of herbicides, in which military records offer dubious historical evidence. In the case of the HERBS tapes, both the Stellmans and Young ran afoul of some inconsistencies on these issues.

In their work the Stellmans were still limited by what Ranch Hand veterans like Cecil describe as regular deviations from the flight paths. In their GIS article the Stellmans acknowledge that many of the records, particularly from earlier missions, are "likely flight paths" rather than exact locations.[9] The same can be said of their reliance on military records for the location of U.S. troops and Vietnamese civilians at the time of spray missions. For Young, Cecil, and others who faulted the Stellmans' study, a similar case can be made. They consistently hold that safety guidelines were strictly followed, as were other policies and procedures that reinforce their claims. In the same article, however, they point out the often gaping inconsistencies in military records, arguing that contemporary scholars should not expect to find complete, well-organized records produced during wartime.[10] One could argue that both groups of scholars are attesting to the accuracy and reliability of military records when it suits their purpose and calling other records into question when it does not. Still, a comparison of these studies and their competing assumptions opens a window on the dilemmas posed by Agent Orange and the limits of scientific and historical analysis in resolving them. As is true throughout the history of Agent Orange, in the absence of clarity and definitive answers the debate was framed largely by politics.

Critics could easily point to Young's long-standing defense of Ranch Hand, his connections to Dow and the military, and his strongly held positions about veterans' exposure and the environmental fate of TCDD. Yet even Arnold Schecter, who has devoted his life to the study of dioxins and human health and who has sided overwhelmingly with the Stellmans rather than with Young in the exposure debates, has noted the limitations of the Stellmans' study. According to Schecter, the Stellmans "stopped short of validating their exposure model by actually measuring dioxins in people, or soil, or wildlife, or food." They essentially constructed an elaborate exposure model but did not ultimately confirm that the areas they identified as being highly exposed actually resulted in increased dioxin levels. Schecter has proposed that someone, if not the Stellmans themselves, return to those areas and perform a validation study, ideally by using high-resolution gas chromatography and mass spectrometry, which both Schecter and Young describe as the gold standard for identifying dioxins.[11]

More recent studies have asked the same questions. Notable among them is an article by Michael Ginevan, John Ross, and Deborah Watkins entitled "Assessing Exposure to Allied Ground Troops in the Vietnam War: A Comparison of AgDRIFT and Exposure Opportunity Index Models." The authors take note of the Stellmans' study of 2003, noting that it "does not estimate actual exposure." After leveling several pages of criticism at the Stellman model, the authors conclude that they are "pessimistic about anyone's ability to develop individual-level exposure assessments for Vietnam veterans."[12] While they are correct in pointing out the limitations of the Stellmans' studies, the authors provided plenty of ammunition for their own critics: the acknowledgments indicate that the work "was supported by the Dow Chemical Company and Monsanto Company."[13]

To date, no validation study has compared the dioxin levels of actual people under the path of the spray missions with those predicted by the new EOIs. The cost of such studies, as VAVA has regularly pointed out, can be prohibitive; but just as important is the realization that the more pressing worries in Vietnam center not around the areas sprayed by Ranch Hand missions but around a variety of dioxin hot spots, the vast majority of which appear to be former military storage sites for the rainbow herbicides.

When Ranch Hand began in 1961, the conventional wisdom among scientists was that the environmental fate of herbicides in general and of Agent Orange in particular was fairly limited. Exhaustive testing of the spray mechanisms by the USAF produced equipment that was designed to minimize drift and spray the targeted vegetation to perform its intended task of defoliation. The stated assumption of the USAF in Vietnam was that the jungle canopy was going to absorb the vast majority of the defoliant. A number of studies in the 1960s and 1970s confirmed that the half-lives of 2,4-D and 2,4,5-T were limited, measured in weeks rather than months or years.[14] Knowledge of the fate of dioxin was more limited, both by a lack of reliable case studies with known dioxin contamination and by the state of technology. To measure accurately the residue of dioxin in soil over a long period of time, it was necessary to measure samples at levels of one ppt rather than ppm or ppb, a procedure that was not even widely possible until the mid-1970s.[15] Some early studies, including those conducted during the NAS trips to Vietnam in the early seventies, found elevated dioxin levels ranging from 18 to 810 ppt in fish from areas that had been heavily sprayed. Although the samples were limited, the NAS scientists thought it possible, even likely, that dioxin was accumulating in the food chain.[16]

Given the little long-term analysis that had been done, the evaluations of the environmental fate of dioxin in soil done by the air force, based on herbicide test sites such as Eglin Air Force Base, were valuable. Those studies, conducted largely by Young, found very limited long-term residue and upheld a body of evidence that showed dioxin was subject to photodegradation, that is, sunlight rapidly accelerated its dissipation.[17] Into the 1970s the USAF continued to emphasize the limited long-term effects of dioxin in affected plants, soil, and animals. In its report on dioxin in 1978, prepared in conjunction with Operation Pacer HO, the USAF team led by Young summarized the matter as follows:

> Most TCDD sprayed into the environment during defoliation operations would probably photodegrade within 24 hours of application. Moreover, recent studies suggest that even within the shaded forest canopy, volatilization and subsequent photodecomposition of TCDD would occur. Since translocation into vegetation would be minimal, most TCDD that escaped photodegradation would enter the soil-organic complex on the forest floor following leaf fall. Soil chemical and microbial processes would further reduce

TCDD residues. Bioconcentration of the remaining minute levels of TCDD may occur in liver and fat of animals ingesting contaminated vegetation or soil. However, there are no field data available that indicate that the levels of TCDD likely to accumulate in these animals would have a biological effect.[18]

Longer-term evaluations in certain areas of Vietnam did not support these assumptions. Neither, however, did they completely discredit them. Heavily contaminated areas of central and southern Vietnam would indeed show significant bioaccumulation of dioxin into the food chain, but the sources of the contamination seemed to be not in areas where the herbicides were sprayed but in those where they were stored. Spread out over a large area, the herbicides did indeed break down largely as expected, if a bit more slowly. Concentrated amounts in smaller areas like the storage sites proved much more persistent.

The most notable work on dioxin hot spots was done by Hatfield Consultants, a Canadian environmental services firm that contracted with the Vietnamese government in the early 1990s to study the human and environmental impact of Agent Orange on the country. The hundreds of soil samples Hatfield collected from 1996 to 2004 as well as those from other teams showed elevated levels of TCDD in the soil, blood, breast milk, and wildlife in areas of Vietnam near former storage sites, particularly Bien Hoa, Da Nang, Phu Cat, and A Luoi (fig. 13). Schecter's team, for instance, found elevated TCDD levels in the blood of a limited sample of residents of Bien Hoa, many of whom were born long after Ranch Hand had ended. Some of the samples showed levels as high as 271 ppt, far above normal or "safe" levels. A control group from Hanoi showed far lower levels of dioxin in their blood. Schecter's team included the Vietnamese doctor and Agent Orange research pioneer Le Cao Dai and John Constable, who had been on the NAS team in the 1970s. The findings, which contradicted the presumed dissipation of dioxin levels over time, suggested that the TCDD from Agent Orange was, in fact, remaining in the environment much longer than expected and had likely entered the food chain.[19]

A few years later Schecter's team proposed seemingly conclusive evidence to back up their hypothesis, finding elevated dioxin levels in fish and poultry from the area surrounding Bien Hoa. On the basis of these findings the authors contended that the source of the ongoing contamination was a major storage site for Agent Orange, the Bien Hoa air base.[20] Not everyone was convinced. A later edition of the journal in which Schecter published his findings featured a lengthy letter to the editor from Truyet Mai, a member

Figure 13. Discarded barrels of White and Orange, Bien Hoa Air Base, 1969. (Courtesy National Archives, College Park, Maryland)

Decades after the Vietnam War, studies showed the most significant residual dioxin contamination from military herbicides to be not in locations that were heavily sprayed during the war, but at the sites of former air bases that served as storage sites for Agent Orange and other chemicals, including Bien Hoa, shown here littered with herbicide barrels in 1969.

of the Vietnamese–American Science and Technology Society of Orange County, California. Orange County is home to a large Vietnamese-American community known for its virulent anti-communism and its assertion that concerns about Agent Orange are nothing more than "communist propaganda."[21] Mai observed that Schecter's team had not shown conclusively that food was the "route of current intake" of dioxin and that sources of industrial contamination around Bien Hoa were the more likely suspects.[22]

Yet any doubt about the hot spot hypothesis had already been cast aside by the time of Mai's letter, thanks to the work of Wayne Dwernychuk and his team at Hatfield Consultants, first published in 2002. Like Schecter's team, Dwernychuk worked with Vietnamese scientists and never presumed that Agent Orange was the only source of dioxin contamination in central and southern Vietnam. In fact, as one of the Hatfield reports put it, "After over a decade of research, the 10-80 Division [of the Vietnamese Ministry of Health] and Hatfield Consultants have concluded that residual levels of wartime Agent Orange dioxin (TCDD) in soils of southern Vietnam are generally *at or below background levels* found in industrialized nations of North America. However, 10-80 Hatfield research has shown that significant hot spots of TCDD remain in select areas of South Vietnam."[23]

The key development in the hot spot hypothesis came in the isolated A Luoi valley in the central highlands above Da Nang. At this site, home to several special forces bases occupied by U.S. troops between 1963 and 1966, Dwernychuk's team found noticeably higher levels of residual dioxin than at other areas nearby that had been heavily and repeatedly sprayed but had not served as storage sites. By taking extensive samples of soil, sediment, fish and duck tissues, human blood, and breast milk, Hatfield and the 10-80 committee showed conclusively that dioxin was moving from heavily contaminated soil into sediment, where it entered the food chain through dioxin's location of choice: the fatty tissues of animals, especially fish and poultry, two staples of the Vietnamese diet. Once consumed by humans, the dioxin present in the animals could be passed on from parents, particularly through breastfeeding mothers, to children born decades after the war.[24]

The Hatfield research also showed that while the residual dioxin contamination was a threat to human and environmental health, it was a "manageable problem": "The principal concern today, regarding dioxin in the environment of Vietnam, is that people living near some of the former military installations continue to be exposed to dioxin. People born after the war are also at risk of contamination. Through the use of dioxin-laden herbicides, the Vietnam War has left a legacy of environmental contamination that continues to this day; however, with simple mitigation measures this problem can be addressed and the probability of exposure significantly reduced."[25] Unlike the ongoing uncertainty about particular effects on human health related to Agent Orange exposure, the scenario for at least seven major hot spots (and potentially several other, smaller ones) in Vietnam was fairly clear: dangerous levels of dioxin are present in the environment, humans are thereby at risk, but the knowledge about how to fix the situation exists. All that remained was the will and the financing to clean up the hot spots. These crucial ingredients, however, would prove just as elusive as scientific certainty.

Given the common refrain heard in Vietnam that testing for dioxin contamination in humans was prohibitively expensive (about one thousand dollars per person), the prospect of a major environmental remediation project, which would involve the excavation and incineration of contaminated soil and would cost, at the very least, tens of millions of dollars, would not be possible without, as the Hatfield report noted, "international cooperation and international financial assistance."[26] Such cooperation and assistance would have been most effective had they come from the United States, in light of its financial and technological resources, not to mention its responsibility for the presence of the dioxin reservoirs in the first place. But such assistance was not readily forthcoming.

The Agent Orange issue had been a sticking point throughout the long, painful process of postwar normalization between the United States and Vietnam. Ever since the war ended, the United States has maintained a consistent official line that there is no scientific evidence linking contemporary health concerns in Vietnam to the historic use of Agent Orange. U.S. Ambassador to Vietnam Michael Marine availed himself of the discourse of uncertainty in 2007 when he stated, "[I] cannot say whether or not I have myself seen a victim of Agent Orange. The reason for that is that we still lack good scientific definitions of the causes of disabilities that have occurred in Vietnam. . . . We just don't have the scientific evidence to make that statement with certainty."[27] The Vietnamese point to the gradual increases in benefits paid to American veterans suffering from conditions presumed to be caused by Agent Orange as well as to the extensive cleanup in places like Times Beach.[28] For more than twenty years the United States refused to accede to any linking of Agent Orange remediation and normalization, trade, or easing of the draconian regime of economic sanctions it enacted against Vietnam during and after the war; in fact, the issue is conspicuous by its near total absence in talks between the countries until the early twenty-first century, once Vietnam enjoyed normal trade relations with the United States and membership in the World Trade Organization.[29]

When President George H. W. Bush laid out his road map to normalization in 1991, for instance, he posited a mechanism for U.S. humanitarian aid to Vietnam but made no mention of Agent Orange or dioxin. When President Bill Clinton made his historic visit to Vietnam in 2000, leaders there pressed him for increased assistance for Agent Orange victims as part of a larger aid package, but the U.S. delegation agreed only to a joint scientific meeting.[30] The joint meeting took place in 2002, but talks over future research projects quickly fizzled out, largely because of the rather crass and now infamous "embassy memo" from the U.S. delegation in Hanoi, which became public in 2003. This sensitive but unclassified memo revealed that the U.S. embassy believed the lack of progress on Agent Orange issues reflected "the unwillingness of the GVN to allow its scientists to engage in genuinely transparent, open rigorous scientific investigation to determine the true extent of the impact of AO/dioxin on health in Vietnam. We believe that the GVN will attempt to control, disrupt, or block any research project that could potentially produce scientific evidence that refutes the GVN's allegations of broad, catastrophic damage to the health of Vietnamese citizens, especially birth defects."[31] Only in the second term of President George W. Bush did the two nations make some headway on remediation efforts. Bush and President Nguyen Minh Triet signed a carefully worded joint statement in 2006 that read in part, "Further joint efforts to address the environmental contamination near former dioxin storage sites would make a valuable contribution to the continued development of bilateral relations."[32] Building on that pledge, Congress appropriated a total of $3 million in both 2007 and 2009 for assistance with dioxin removal and "related health activities" in Da Nang. Most of the initial grant was directed to various nongovernmental organizations (NGOs) operating in Da Nang. Three years later, however, only a fraction of the original money had been released by the DOS, leaving the remediation efforts almost completely reliant on other sources of funding and international aid, including the United Nations, the Red Cross, and the Ford Foundation, which has maintained a multiyear, multimillion-dollar project related to Agent Orange.

Two sites, the Da Nang air base and the A Luoi valley, illustrate the dilemmas of environmental remediation in contemporary Vietnam, even when the money required is in hand. In late 2009 Hatfield completed its assessment and initial remediation plan for the Da Nang airport. As one of the main herbicide storage sites and one of the key locations of Operation Pacer IVY, the airport and the adjacent Lake Sen are among the most heavily contaminated hot spots in the country. Hatfield and the 10-80 committee deduced that the most common pathway for exposure was for the dioxin to move from the soil located under the former loading, mixing, and storage areas at the airport into the lake, from the lake into the fish (particularly the popular tilapia fish), and from the fish into humans. In 2007 the GVN erected a barrier around the lake, prohibiting fishing and other forms of public use. A sizable population still lives within a stone's throw of the lake (fig. 14). Although other sources contribute to the contamination and other potential exposure pathways exist, above all for those working at or living near the site, Hatfield's research has shown a reduction in human dioxin exposure since the lake closure.[33]

To make the site safe for the surrounding community and ecosystem, Hatfield recommended the construction of a secure, on-site landfill for the contaminated soil at the airport, consideration of incineration for the

Figure 14. Neighborhood adjacent to Lake Sen hot spot, 2008. (Photo Courtesy of William Richard Zeller Jr.)

Many of the dioxin hot spots at former air bases are in heavily populated areas. Residents of the neighborhood shown here, adjacent to the Da Nang airport hot spot, have shown significantly elevated dioxin levels, which scientists believe to be evidence that dioxin from Agent Orange has entered the food supply.

contaminated soil, and increasing the awareness of local residents to exposure pathways.[34] The on-site landfill, while a less costly option, has not yet proven to be a long-term solution. The operations during the Pacer HO and Times Beach episodes have proven that incineration is an effective way to eliminate the most dioxin in the safest manner, but it is also costly. The estimated cost of the cleanup at Da Nang has risen steadily over the past several years from $10 million to $17 million to more than $20 million. The United Nations Development Program estimated in 2010 that the cost of cleaning up three major hot spots—Da Nang, Bien Hoa, and Phu Cat—would be over $50 million.[35] In late 2010 the GVN and the USAID signed a memorandum of intent and began the long-overdue program of dioxin remediation at Da Nang. In June 2011, after years of wrangling, the United States and Vietnam launched the first phase of a joint cleanup effort at Da Nang. Having an estimated cost of $31 million to remove dioxin from up to seventy-one acres of land, the project is a large step forward, but the process had only just begun there in 2011 and has not yet started at other hot spots.[36]

The situation in the A Luoi valley is no less serious. Removed from the population and economic center of Da Nang, which is home to nearly one million people, the remote valley is populated largely by poor, rural residents, many of whom are members of various minority tribal groups that have strained relationships with the GVN. These people rely on farming as a means of survival, and there are no resources to support alternatives to breastfeeding among new mothers. Some human blood samples from the Hatfield research in the valley revealed dioxin levels as high as 15 ppt, nearly as high as those found in Da Nang (17 ppt). Soil samples taken by Hatfield found levels as high as 879 ppt.[37] At Da Nang, Bien Hoa, and other hot spots at former bases concrete casings and incinerators could easily be built on-site, but the excavation of soil in A Luoi is a far greater logistical and financial challenge. The remote location of the valley and the narrow, winding roads leading to it do not favor on-site incineration or excavation.

A Luoi is not alone among poor, rural hot spots. Although most of the attention with regard to dioxin remediation has focused on major sites near urban areas like Da Nang and Bien Hoa, Hatfield has identified more than two dozen other potential hot spots throughout central and southern Vietnam. These sites, nearly all of which are former air bases or storage sites, have not all been tested for elevated dioxin levels in the soil and the local population, but given the historical patterns of use and the findings from places like A Luoi, Hatfield considers them potential hot spots in need of further study.

Korea

While the situation in Vietnam remains, rightly so, at the center of attention for dioxin remediation, more recent revelations about the global scale of herbicide testing, production, and storage have expanded the list of sites that have become part of the global landscape of Agent Orange. Among the many sites that have shaped and been shaped by the contested legacies of Agent Orange is the demilitarized zone (DMZ) on the Korean Peninsula.

In 1999 a South Korean government investigation revealed that U.S. forces stationed in the area had sprayed more than twenty thousand gallons of Agent Orange along the DMZ. The spraying took place between 1968 and 1969, when the forces were on heightened alert following an incident in which North Korean forces attacked the *USS Pueblo* and proceeded to hold some eighty crew members hostage for nearly a year.[38] As tensions between the United States and North Korea rose, U.S. and South Korean forces were ordered to defoliate areas around the DMZ to protect against possible infiltration. Tens of thousands of U.S. troops were on active duty in Korea at the time, and records released by the DOD during the investigation show that as many as fifty thousand South Korean troops did most of the spraying by hand. The VA initially ruled that U.S. troops serving near the DMZ between April 1968 and July 1969 would be eligible for service-related disability claims presumed to be caused by exposure to Agent Orange. Under growing pressure from others who served in the affected area, in 2011 the VA expanded coverage to "any veteran who served between April 1, 1968, and Aug. 31, 1971, in a unit determined by VA and the Department of Defense to have operated in an area in or near the Korean DMZ in which herbicides were applied."[39]

Several aspects of the revelation of Agent Orange use in Korea and of the ensuing decade-long struggle for service-related benefits by veterans stationed in Korea at the time are striking. First, several early reports posed the issue in conspiratorial language, describing a coverup by the Pentagon, which, according to one report, "had previously claimed that Agent Orange was used only in Vietnam."[40] As the same article goes on to claim, however, there was never anything secretive about the program. Although many of the documents released by the Pentagon had been classified, classification is the rule rather than the exception when it comes to most military documents. The use of herbicides itself was not classified, as a Pentagon spokesperson noted at the time.[41] The default belief that the government is hiding information about Agent Orange, especially among many in the veterans' community, not only speaks to the lack of trust and sense of betrayal Vietnam veterans feel toward the government, but also relates to the shifting frameworks of knowledge and environmental awareness about herbicides and chemicals. Even during 1968–69, before the cancellation of Ranch Hand and the domestic ban on 2,4,5-T in the United States, most Americans, including those serving in the military, believed that herbicides were safe; they had little reason at the time to think otherwise. Given what was widely known about chemical herbicides in general and Agent Orange in particular by 1999 and 2000, however, the idea that the government would openly use such chemicals, exposing its own soldiers and allied troops, no longer fit the traditional narrative confines of the Vietnam-era military experience. Government conspiracies, however, did. A final major source of tension in the story about Agent Orange use in the DMZ has to do with who was ultimately responsible for the spraying. When the story broke, the United States and South Korea traded swipes in the media over who had ordered or requested the spraying operations. The South Korean government initially admitted that its troops did the majority, if not the entirety, of the spraying but claimed that the operation was undertaken at the request of the U.S. government. The DOD and DOS countered the claims, pointing to records in the Korean report that show the Korean government's active involvement in the decision. The Pentagon did, however, acknowledge the "spotty" nature of the records in question.[42]

Relations between the United States and South Korea were already strained at the time because of the recent revelations about the massacre of civilians at No Gun Ri by soldiers of the United States Army just after the start of the Korean War. The disclosure that dozens, perhaps hundreds, of unarmed Korean civilians had been killed by U.S. troops, first revealed in a series of Associated Press articles, was also published in 1999, not long before the release of the DMZ Agent Orange report.[43] Bracketed by the memories of two divisive and bloody wars, then, the incidents posed a number of dilemmas for the veterans involved as well as for their respective governments, which had to grapple with both the historical records and the contemporary diplomatic, political, and economic realities of U.S.–South Korea relations.

Many of these same tensions would be on display a few years later, when South Korean veterans of the Vietnam War filed two lawsuits: the first demanded compensation from the U.S. government and was eventually thrown out of federal court; the second targeted Dow and Monsanto, seeking $500,000 per claimant from them. The suits were the culmination of a slow but long-simmering accretion in activism among Korean veterans of the war. Ironically, although Korean troops were by far the largest group of allied soldiers outside of U.S. and ARVN forces, totaling over three hundred thousand, they did not participate in the original veterans' class action suit, as Australian and New Zealand troops had. Still, the Korean Vietnam Veterans' Association noted the similarities between the experiences of Korean soldiers and those of other nations: "We were no different," Hwang Myung Chul, vice-president of the association told the *New York Times* in 1992. "We wanted the media to relate our pain and difficulties but they were told not to. The Government virtually ignored our problems. That is what we face now."[44] High on the list of veterans' complaints was the lack of support for Agent Orange claims compared to other allies. The South Korean defense ministry declared that no illnesses had been detected among its Vietnam veterans, an assertion the veterans sharply disputed. Speaking in Hanoi in

2006, Kim Jung Wook, secretary general of the Korean Disabled Veterans Association, accused the Korean government of suppressing information about Agent Orange "in order to maintain its political power and not to stimulate the U.S." Frustrated by a lack of progress on the political front, they eventually sought legal remedies instead.[45]

Some 10,000 people joined the initial class of the suit against the chemical companies, and the initial claim totaled more than $5 billion. In each of the cases filed by the Korean veterans, both the South Korean and U.S. governments attempted, unsuccessfully, to define the issue as a political and diplomatic matter rather than a legal one. Although the eventual size of the class shrank, the plaintiffs won their case, and in January 2006 the Appeals Court of the Republic of Korea ordered the two companies to pay more than $60 million to 6,795 people, the awards to range from $6,000 to $46,000.[46] The case was especially meaningful because it is the only ruling since the original veterans' class action suit that has found in favor of Vietnam veterans, although the court did not support every allegation of the plaintiffs, notably rejecting that of second-generation birth defects. The companies appealed to the Supreme Court of South Korea. Although it is not clear how the companies would prevail under international law even if the decision were upheld at the Supreme Court, they both have broad business interests in the country that have already been negatively affected by the case.[47]

While the human cost of the Korean spraying remains to be seen, particularly to the South Korean veterans who used hand applicators, there is not much cause for concern about the environmental health of the DMZ itself. The relatively low volume of spray, the short duration of the operations, and the absence of long-term storage of the herbicides leave little chance of the presence of grave dioxin contamination of the landscape. In fact, the DMZ has become something of an unusual, Cold War–era nature preserve. The environmental historian Lisa Brady explored the issue in an essay entitled "Life in the DMZ: Turning a Diplomatic Failure into an Environmental Success" (2008). Since it has been kept essentially free of human action and development, the DMZ is, according to Brady, "one of the Cold War's greatest successes." Although it remains riddled with land mines, the area has become a haven for native plants and animals as well as migratory birds. Brady goes so far as to suggest that the shared environmental concern about the DMZ might lead to improved relations between North and South Korea and could serve as "an internationally recognized and supported symbol of peace and preservation."[48] Sadly, not all of the unintended consequences of war turn out as well as the DMZ, particularly when Agent Orange is involved.

Even more recently, new revelations have surfaced about the use of Agent Orange on the Korean Peninsula. In the spring of 2011 three U.S. soldiers who had been stationed at Camp Carroll in southeastern South Korea alleged that they helped bury perhaps as many as 250 drums containing Agent Orange at the base in the late 1970s. All three of the veterans were now ill with conditions they believed to be linked to Agent Orange. The United States Army launched an inquiry into the matter immediately after Korean news outlets picked up the story.[49] Investigations done in the early 1990s had revealed that chemical drums had indeed been buried on the base at some point and that several dozen tons of contaminated topsoil had been excavated from the base in the early 1980s. Those inquiries did not conclude that the chemicals included Agent Orange; nor did they document what became of the drums. A study in 2004 revealed trace amounts of dioxin in soil, but the levels were deemed not to be a threat to the surrounding area.[50]

While the army sprang into action to complete its investigation, the South Korean government worked equally quickly to assuage the fears of local citizens and to forestall anti-American protests.[51] The allegations about Agent Orange at Camp Carroll led to investigations at other U.S. bases in the country, which were also found to be polluted, although not by dioxin from Agent Orange. In June 2011 the South Korean government pumped two thousand gallons of groundwater believed to be polluted with benzene, toluene, and xylene out of the water supply near Yongsan Garrison, just outside of Seoul. The initial investigations at Camp Carroll also found pollution in the groundwater supply, likely caused by runoff from the base, but no elevated levels of dioxin and no trace of Agent Orange.[52] The investigations in Korea will continue and likely be repeated at U.S. bases around the world. Such inquiries will probably not find appreciable residual levels of dioxin, but these cases are reminders that Agent Orange is only one among many pollutants potentially connected to the hundreds of U.S. overseas military bases.[53]

New Zealand

Perhaps more than any other country associated with the prosecution of the war in Vietnam, New Zealand has had the most difficult time coming to terms with the legacies of herbicidal warfare at home and abroad. New Zealand continued to produce 2,4,5-T for domestic use long after it was banned in the United States, England, Australia, and Canada. Activists and veterans continued to oppose its use and to press for increased compensation to New Zealand veterans and Vietnamese victims of Agent Orange. Some of these efforts were the result of the perseverance of committed activists working against entrenched financial interests. Others reveal a sense of national guilt about the nation's treatment of its veterans and its responsibility for the war in Vietnam as well as ambivalence about its relationship with the United States.

The new century in New Zealand brought with it renewed concerns about dioxin. New studies of dioxin levels in New Plymouth residents continued to surface, as did renewed allegations that IWD supplied Agent Orange for use in Vietnam, while veterans' groups continued to agitate for greater compensation from the government. Studies in New Plymouth proved inconclusive for elevated risks of cancer or birth defects, but a potential bombshell related to IWD dropped in 2005, when Harry Duynhoven, who represented New Plymouth in Parliament, revealed that he had new information that "the ingredients of Agent Orange" were shipped from New Plymouth's State of Taranaki to Subic Bay U.S. forces base in the Philippines. According to the Agence France Presse headline, the statement by Duynhoven represented the New Zealand government's view, confirming that it supplied Agent Orange, which was far from true. In fact, Duynhoven's supposed evidence consisted of little more than the anecdotal claims of retired veterans, similar to the unsubstantiated allegations made in the late 1980s that led to the parliamentary inquiry in 1990 that found no evidence that IWD supplied Agent Orange, 2,4,5-T, or 2,4-D to U.S. forces.[54]

During the initial investigation into whether or not IWD had produced Agent Orange for use in Vietnam, the divisiveness of the war returned to the forefront of New Zealand politics. When the issue of Agent Orange was reopened by the parliamentary inquiry, many of the wounds of those who had protested New Zealand's involvement in the war were reopened also. Former prime minister Robert Muldoon, of the conservative National Party, which had strongly supported the war, labeled trade minister and Labour Party member Michael Moore, who supported the production inquiry, a hypocrite: "You were marching against [veterans] when they were [in Vietnam]," Muldoon charged, "and now you're supporting them." In response Moore noted that Muldoon must have been "the only person in the Western world who still believes that adventure was correct."[55] Years later, after Duynhoven's allegations, the government released to the public additional documents from the inquiry, none of which called into question the results. Critics were not mollified; some remained "absolutely convinced" that the government was hiding something.[56]

Many New Zealand veterans shared those sentiments. Following the trajectory of their American and Australian counterparts, whose developments they followed closely, veterans in New Zealand continued to press the government for more accountability and greater benefits as knowledge about Agent Orange and dioxin grew. Veterans of many countries feared that even minimal exposure during their service increased the likelihood of birth defects in their children. In 2001 *Investigate* magazine ran a story detailing allegations of improper toxic waste disposal at the IWD plant in New Plymouth. The piece also touched on veterans' concerns, however, linking government denials about safety and exports from IWD to government's refusal to acknowledge the wider range of health claims expressed by veterans. The piece quoted John Moller, the president of VVANZ, who said veterans' frustration level was so high they might "do a Timothy McVeigh," referring to the American who destroyed the Oklahoma City Federal Building in 1995, killing 168 people: "Passions are running so high that [Moller] and his colleagues have had to work 'damned hard' recently to persuade dying veterans whose children have also been affected by dioxin-related deformities, 'not to take the law into their own hands. These guys have had enough. They're being cheated and lied to by the politicians and the bureaucrats.'"57 The Ministry of Defence reacted by commissioning two new inquiries: the Reeves report, which focused on the overall health of veterans and their children and on the question of whether Agent Orange might be to blame; and the McLeod report, chaired by Deborah McLeod, dean of the Otago University School of Medicine in Wellington, which reviewed all available international research related to the health effects of veterans and dioxin exposure and attempted to interpret the literature in "a New Zealand context," given, as the final report described, "the very limited potential New Zealand troops had for exposure to Agent Orange." Both reports essentially determined that Agent Orange was not the cause of any adverse health effects in veterans or their families. "In this context, and given the small increased risks found in studies of very exposed populations," McLeod wrote, "the conclusion reached by this appraisal of the literature is that there is no evidence that exposure to chemicals in Vietnam has affected the health of the children of New Zealand Vietnam Veterans."58 While both reports explored the health effects on children of New Zealand veterans involved in Operation Grapple, a series of twenty-one atmospheric nuclear tests conducted by the United Kingdom between 1952 and 1958, the majority of the report and nearly all of the controversy it generated focused on Vietnam veterans.

The uproar over the McLeod report centered on its use of the phrase "New Zealand context." The introduction notes that "the information available to the authors was that ANZAC Forces generally served in Phuoc Tuy Province, where there was no aerial spraying."[59] Although the McLeod committee apparently was given that information, it was wrong. Most ANZAC forces did serve in Phuoc Tuy province, but, as earlier government investigations in New Zealand and Australia showed, aerial as well as perimeter base spraying had occurred there. Not long after the report, former army officers produced a wealth of data to reinforce the point, including maps of spray missions and the extensive anecdotal evidence of veterans who had been sprayed directly. The data suggested that as much as 1.8 million liters of herbicide—Orange, White, and Blue—had been sprayed over Phuoc Tuy between 1965 and 1968.[60] In the view of many, this gross oversight tainted the report and inflamed those who saw a concerted governmental coverup.

Regardless of the increased likelihood of exposure, however, the McLeod report was essentially correct that the voluminous body of literature examined reveals "no consistent positive association between exposure to Agent Orange or a range of chemicals or pesticides and any specific birth defect," except for spina bifida.[61] The American Ranch Hand Study and Australian studies confirmed this, but skeptics were not convinced. McLeod defended the report before a parliamentary committee, focusing on the scientific findings rather than on the false claims regarding exposure. She also drew attention to the negative effects that the heated debate over this and other reports related to Agent Orange and Vietnam veterans would have, making other scholars and scientists less likely to undertake such studies. "I certainly would think very seriously about being involved in any controversial topic," McLeod told the committee, "because of the amount of time—unfunded time—that we have had to spend on this project and, in particular, the abusive and offensive manner in which debate has been carried out on this topic."[62]

By seeming to bolster charges that the government was not accurately accounting for the history and danger of Agent Orange, the McLeod report paved the way for a final round of reports, negotiations, and a potential resolution of the long-standing tensions between New Zealand veterans and their government. In October 2004, Parliament launched yet another inquiry into the issue, this one conducted by the Health Committee and chaired by Steve Chadwick. Designed explicitly to reevaluate the Reeves and McLeod reports in light of the evidence of troop exposure, the Chadwick report included extensive testimony from veterans that was integrated into the report narrative and included verbatim in the published report in hundreds of pages of appendices.⁶³ The committee concluded that New Zealand troops were exposed to Agent Orange and to "a toxic environment." On these grounds the report then made a major departure from previous government positions: "In line with the Australian and the United States governments, we accept that service in Vietnam is evidence of likely exposure to defoliants and other, possibly toxic, chemicals. We recognize the length of time it has taken to publicly acknowledge Vietnam veterans' exposure to a toxic environment, and the frustration this has caused for veterans and their families."⁶⁴ "To satisfy these concerns, and in an attempt to finally resolve this issue," the report recommended increased health monitoring and benefits for veterans and their families (including the addressing of "unanticipated health outcomes" related to service in Vietnam) and "recognition of their achievements and acknowledgement of past wrongs."⁶⁵

Over the next two years the Department of Veterans' Affairs developed a compensation package for veterans, one that was strengthened by a report in 2006 purporting to show "significant genetic damage" among Vietnam veterans in New Zealand.⁶⁶ The final compensation package, totaling more than NZ$30 million, provided free health examinations for registered Vietnam veterans and one-time payments between NZ$20,000 and NZ$40,000 to veterans suffering from a variety of illnesses. Some veterans voiced displeasure at the terms, saying the package did not go far enough in covering other conditions, but by late 2007 more than forty-six hundred had registered with the program.⁶⁷

In May 2008 the government, through an official statement made by the Crown—the British monarch who technically still directs the country's defense forces—issued an official apology, which was read by Prime Minister Helen Clark and broadcast throughout the country: "The crown extends to New Zealand Vietnam veterans and their families an apology for the manner in which their loyal service in the name of New Zealand was not recognized as it should have been, when it should have been, and for inadequate support extended to them and their families after their return home from the conflict." Acknowledging the crucial role of Agent Orange in bringing veterans' issues to the forefront, the Crown's statement cited the "toxic environment" described in the Chadwick report, noting that "the failure of successive governments and their agencies to acknowledge the exposure of veterans to dioxin contaminated herbicides . . . exacerbated the suffering of veterans and families."⁶⁸ The remarks were especially poignant in that they were made by Prime Minister Clark, who not only led the Labour government, which had opposed New Zealand's involvement in the war and removed its troops in 1972, but also had been active in the antiwar movement in the late sixties and early seventies. According to news accounts of the address, veterans wept as the two major parties of New Zealand came together to recognize their shortcomings in dealing with veterans.[69] And yet, as in the case of Australian and American veterans and of every issue related to the legacies of the Vietnam War, closure refused to come easily. "The apology covered everything veterans expected," one veteran told the Wellington *Dominion Post*. The article added, "But unresolved issues relating to Agent Orange remained."[70]

Canada

While the New Zealand case refused to go quietly despite multiple high-level public investigations, the Canadian encounter with Agent Orange seemed to come out of nowhere. In 2005 Louise Elliot, a reporter for the Canadian Broadcasting Company (CBC), broke the story that more than 1.3 million liters (about 343,000 gallons) of Orange, Purple, and White and other combinations of herbicides were sprayed at Canadian Forces Base Gagetown in New Brunswick from 1956 through 1984, when the Canadian government banned 2,4,5-T. In the ensuing investigations, documents from the Department of National Defence revealed that more than three hundred thousand Canadian troops served at Gagetown during that twenty-eight-year period.[71] As part of the global testing apparatus of the U.S. military, Gagetown served as a test case for a variety of herbicides; as part of the global network of local and scientific knowledge about dioxin, it linked Canadian veterans and civilians with other alleged victims of Agent Orange around the world.

The Liberal government set up a fact-finding review, which quickly became politicized during the run-up to the federal election in January 2006 when conservatives accused the Liberal party of attempting a coverup. Campaigning in New Brunswick just before the election, the Conservative leader, Stephen Harper, promised locals that his government would provide free medical testing along with compensation to any victims. Bundled with promises of improved highway infrastructure, Harper's words were dismissed by the CBC as mere wooing of voters in the maritime province.[72] Harper's party won the election, but the new prime minister would quickly discover the challenges of navigating the choppy political waters of Agent Orange. The uproar surrounding the investigation was fueled by incorrect reports in the media, some of which misstated the relationship between Agents Orange and Purple and primed the public to expect significant levels of dioxin to still be in the soil twenty to thirty years after the spraying took place.[73] Many people were shocked when the initial probe reported in August 2006 that the levels of dioxin around Gagetown were low and of no threat to the community.[74] The government declared the base to be safe and suggested that the risks to personnel in the 1960s had been overstated. A series of additional reports released in 2007 found that the area around Gagetown did not have an elevated risk for a variety of cancers associated with exposure to dioxin. In fact, the study showed, Gagetown had lower cancer rates on average than the rest of New Brunswick.[75] Still, faced with new charges of a whitewash, the Harper government pressed ahead with a compensation package for those who believed their illnesses to be caused by dioxin exposure.

Announced in September 2007, the CA$96-million package offered a one-time payment of CA$20,000 to "veterans and civilians who worked on or lived within five kilometres of the base between 1966 and 1967, and only those who have illnesses associated with Agent Orange exposure, including Hodgkin's disease, lymphoma, respiratory cancers, prostate cancer, and type 2 diabetes, as determined by the U.S. Institute of Medicine."[76] The Agent Orange Association of Canada, which had formed in 2005, shortly after the Gagetown story broke, called the package "ridiculously inadequate" and criticized the government for focusing only on personnel present in 1966 and 1967. The government estimated that more than four thousand people would be eligible for claims under the program, but seventeen hundred were already involved in a class action suit filed against the government and the chemical manufacturers of the herbicides. The lawsuit, which covered a larger expanse of time at the base, swelled by several hundred claimants after the announcement of the compensation package. Tony Merchant, the lawyer for the plaintiffs, repeatedly claimed that the total number affected by the Gagetown testing might well exceed four hundred thousand.[77]

The case gathered additional class members but it failed to gain momentum, running into stiff, well-funded opposition from the government and the chemical companies. Legal teams for the companies used the growing diversity of the class against it, arguing that, unlike the U.S. veterans' class action suit in the 1980s, the very different circumstances of potential exposure for the various groups included in the Canadian class made it nearly impossible to try as a single case. In April 2010 an appeals court sided with the companies, concluding that "in light of the timeframe involved, the large number of people, the size of the base, and the different chemicals used, the proposed common issues would be insignificant when compared to the large number of individual inquiries necessary to resolve the claim. Accordingly, there would be no judicial economy in certifying a class action."[78] A year later the Supreme Court of Canada refused to hear the case, effectively ending the class action phase of the lawsuit. While individual cases would still be allowed to come forward, they would effectively have to start from scratch.

As was true of so many aspects of the transnational saga of Agent Orange, the issues in the Gagetown case, historical, scientific, and legal, all revolved around uncertainty. Neither the claimants nor their lawyers could, except in rare circumstances, show clear exposure to Agent Orange. All the soil samples and spectrometers in the world could not provide the type of evidence and causality that Western legal systems rely on, and the studies appeared to work against, rather than in support of, the claimants. The question of uncertainty was brought home in no uncertain terms by the final environmental report on Gagetown, issued in April 2007. It found that personnel who had worked directly with herbicides might have experienced elevated risks, "and, as such, potentially unacceptable health risks," during their time at Gagetown. Most others, however, including hunters, farmers, bystanders, and people from the larger community surrounding the base, were not believed to have been exposed to dangerous doses. Working within a traditional model of exposure, whereby "risk equals exposure times toxicity," the report cited the difficulty, if not impossibility, of determining any exact findings for claimants: "The level of uncertainty resulting from the recreation of activities, some of which occurred more than 50 years ago, coupled with the uncertainties inherent in standard forward-looking risk assessment, is very large. As a result, the expectations regarding the level of precision that this risk assessment exercise can produce should be limited."[79] Limited results, while probably the most accurate in terms of historical and scientific methodologies, were the least likely to satisfy those who believed Agent Orange was to blame for the conditions they and their loved ones experienced.

One chapter in the Canadian story was not even fully closed before another one opened. In early 2011 a new story broke in the media, claiming Agent Orange was used to clear brush alongside railroad tracks in Ontario in the 1970s. Unsurprisingly, news stories confused 2,4-D and 2,4,5-T—which were legal and registered and had been sprayed alongside railroad tracks and roads in Ontario and elsewhere—with Agent Orange, the herbicide built and delivered according to military specifications in the 1960s.[80] The Ontario government responded quickly, admitting that a mixture of 2,4-D and 2,4,5-T had indeed been used throughout the province in the seventies and eighties until the federal government banned 2,4,5-T in 1985. The issue quickly became a political football amid calls for a government probe. Readers of the CBC website flooded the page with comments alternately sharing their stories of Agent Orange exposure, blaming the current government for the situation, and demanding compensation. Hundreds of calls lit up the hotline set up by the province, although the government was forced to admit it had far more questions than answers.[81] The same could be said of nearly every other community affected by Agent Orange and its associated dioxin, even forty years removed from Operation Ranch Hand. Nowhere was this truer than in Vietnam.

Vietnamese Victims' Lawsuit

All of the additional research and attention drawn to hot spots in Vietnam bolstered the latest round of legal activity surrounding Agent Orange: the Vietnam victims' lawsuit. In December 2003 a group of Vietnamese scientists, lawyers, and activists launched the Vietnam Association for Victims of Agent Orange/Dioxin (VAVA), an umbrella group whose aim was to provide resources for Vietnamese believed to be suffering from the consequences of dioxin exposure caused by Agent Orange and to seek justice from the United States and the chemical manufacturers of the herbicide. Although the U.S. government would not agree to be sued, VAVA filed its lawsuit against the chemical companies in January 2004. Using the Alien Tort Claims Act to advance its case, the initial claim held that the companies had been in "violation of international law and war crimes, and under the common law for products liability negligent and intentional torts, civil conspiracy, public nuisance and unjust enrichment, seeking many damages for personal injuries, wrongful death and birth defects and seeking injunctive relief for environmental contamination and disgorgement of profits."[82] In an open letter to the American people in August 2004, VAVA argued that the lawsuit was a last resort, the United States having provided "no positive response" to the "goodwill" of the Vietnamese people. "We affirm with you that the Vietnamese people have never had any sense of hatred for the American people, who themselves have also written important pages of the history of hard struggle for independence and freedom," the letter read. "It must be understand [sic] that the use of toxic chemicals is a brazen violation of international laws and constitutes an act of war crime and betrayal of the ideals of the 1776 Declaration of Independence."[83] In his Declaration of Independence for the Democratic Republic of Vietnam on September 2,

1945, Ho Chi Minh had also alluded directly to the American declaration. Sixty years later VAVA positioned itself as the latest in a long line of Vietnamese efforts to appeal to the better angels of the American people, their government, and their legal system, only to become the latest to be disappointed with the government and the legal system.

The case was heard in the Eastern District Court of New York, presided over by Jack Weinstein, who had overseen the veterans' case in the seventies and eighties. The preliminary phase included extensive oral arguments and a plethora of amicus briefs, including one from the Department of Justice of the Bush administration, which argued that a ruling in favor of VAVA would infringe upon the president's powers as commander-in-chief. The brief also took strong exception to the implication by VAVA's legal team that the use of herbicides in Vietnam constituted a war crime under international law, putting the United States in the company of, among others, the Nazis:

> Finally, the Nuremberg and Eichmann cases are readily distinguishable in a more fundamental way. Those cases all involved conduct so abhorrent as to be readily identifiable as violative of international legal norms and all standards of civilized human conduct. By contrast, as discussed above, there was much debate regarding the wisdom and legality of the United States' use of chemical herbicides in Vietnam. Whatever one may conclude about the 1925 Geneva Protocol's intended applicability to chemical herbicides, it is beyond peradventure that the use of such herbicides constitutes conduct of a qualitatively different nature from the mass murder and slavery engaged in by the Nazis.[84]

While Judge Weinstein rejected most of this defense by the government, he nevertheless found little legal basis for VAVA's case and dismissed the suit in March 2005.[85]

In a typically voluminous decision, Weinstein denied that the herbicides were poisons and that, as such, they should not be considered chemical weapons: "Agent Orange and the other agents used for the purposes [of the motion under consideration] should be characterized as herbicides and not poisons. While their undesired effects may have caused some results analogous to those of poisons in their impact on people and land, such collateral consequences do not change the character of the substance for present purposes."[86] Weinstein also upheld the "contractor defense" offered by the chemical companies, arguing that since the companies produced the materials in question for the U.S. government according to contract, they too were immune from legal action. The companies' immunity had been a major issue in the veterans' class action suit, allowing Weinstein to leverage a settlement. In fact, according to the legal scholar Peter Schuck, the author of *Agent Orange on Trial*, the contractor defense actually strengthened during the appeals process of the veterans' suit. As the case moved through the appeal process, the defense morphed into a "military contractor defense" that strengthened the chemical companies' position in future litigation, making it, in Shuck's words, less vulnerable to "new scientific findings on the etiology of the veterans' devastating illnesses."[87] In yet another example of how the American legal system is designed to reward the type of evidentiary uncertainty on which the chemical companies thrive, despite the fact that knowledge about dioxin and its effects on human health grew exponentially in the twenty years between the veterans' settlement and the dismissal of the Vietnamese victims' lawsuit, the manufacturers of Agent Orange found themselves far more insulated from the threat of a successful lawsuit.

As VAVA's case made its way through the appellate system, Weinstein's dismissal was repeatedly upheld. Press releases from VAVA described Weinstein's decision and subsequent appellate rulings as "irrational, unusual, and unfair," pointing to the list of conditions compiled by the U.S. Department of Veterans' Affairs believed to be linked to Agent Orange, the compensation packages offered to U.S. veterans for such conditions, and, most notably, Weinstein's own role in fashioning the settlement in the veterans' class action suit.[88] These issues, however, were not at stake in the VAVA suit, which dealt primarily with issues of international law. In early 2008, VAVA asked the U.S. Supreme Court to hear the case. The VAVA legal team thought the court might elect to hear the case since a number of conservative justices had previously expressed concerns about the Alien Claims Tort Act and might be looking for an opportunity to strike it down, but in February 2009 the court officially declined to hear the case, effectively ending the Vietnamese victims' lawsuit.

Recounting their personal views of the case at VAVA headquarters in Hanoi in 2008, several leaders of the group expressed their profound disappointment in the outcome of the case. Nguyen Trong Nhan, the president of VAVA and a former minister of health, reiterated that they began the lawsuit to bring attention to the issue in the face of ongoing intransigence by the U.S. government and the chemical companies. They knew it would be a difficult challenge, but in light of what veterans in the United States, Australia, and New Zealand had won in their case, along with more recent developments in Korea and New Zealand, they believed they had public opinion and the law on their side. While public opinion was mixed, the law was definitely not on VAVA's side.[89] The law professor Luu Van Dat recalled that he considered the case a "test of the independent judiciary" enshrined in the American Constitution, a test he believes the court failed. At both the district level in New York and the appellate level, Dat believed the court "ignored the truth" of the case, which, in the view of VAVA, was that the United States and the chemical companies were in clear violation of international law. Easily slipping into the voice of a prosecutor, Dat sharply contested the idea that the herbicides in question were not poisons and that the damage done by Agent Orange and the other herbicides was tantamount to "collateral damage."[90] To Dat, this argument ran counter to logic, to the facts, to a larger sense of morality, and to the point that the United States should, at the very least, be responsible for what it did to the land and people of Vietnam during the war. Yet morality had little, if any, place in the legal case against Dow, Monsanto, and the other thirty plus entities that now represented the original manufacturers of Agent Orange.

The response of the U.S. courts notwithstanding, the question of responsibility looms large among not only VAVA's leadership, but also the entire VAVA apparatus, which in 2009 had offices in fifty of Vietnam's fifty-eight provinces, many of which had a number of district-level offices as well. Visiting several of these offices in 2008, I was struck by the degree to which even district-level VAVA staff were informed about the intricacies of the legal case and the technical information about Agent Orange and dioxin. In Vinh Long province, southwest of Ho Chi Minh City, the province chief and VAVA representatives told me that the area was home to more than six thousand victims of Agent Orange, ranging from the mildly to the severely disabled, many of whom were what the Vietnamese refer to as third-generation victims, meaning they are the grandchildren of those exposed during the war. VAVA focuses on the children believed to be suffering from conditions related to dioxin exposure, attempting to support families caring for high-needs children. The group prioritizes basic needs of victims, starting with housing, food, and medicine, but whenever possible provides support through physical therapy and rehabilitation and, eventually, education and job training.[91]

Strict standards for determining exposure and medical or scientific tests for elevated dioxin levels do not exist in the countryside. "Vietnam is a poor country," the VAVA staff in Vinh Long told me. "We have suffered long from war and have many needs. The government has a small budget, and we rely on outside aid for support." Even in a remote area of the province, the talk of responsibility was never far below the surface. "The U.S. government has a responsibility," the province chief said in an interview. "It ordered the spraying of Agent Orange, and it should negotiate with the chemical companies to help provide assistance." He added, in a refrain heard often in Vietnam, that it was strange that the government would pay American victims of Agent Orange but ignore what he called "the Vietnamese laboratory," which remains full of victims.[92]

To learn how people get counted as Agent Orange victims, I visited Quang Tri province, just south of the former demilitarized zone and the site of very heavy spraying, in 2008. Le Kim Tho, the chairman of the Quang Tri VAVA association, told me the province was home to more than 15,000 victims, one of the highest rates in the country. Of these, 7,021 were second-generation victims and another 256 were third-generation.[93] Tho showed me the handwritten ledgers in which he kept the listings of those designated as victims, all of whom were receiving some form of assistance from the government through VAVA. Broken down by area, the ledger showed 8,202 families receiving assistance in Quang Tri's nine districts, several of whom had offered testimony in support of the legal case, which the provincial office had followed very closely. The staff there knew not only the exact number of corporate defendants in the case, but also each of the central tenets in Judge Weinstein's decision, which was on appeal at the time of our meeting. "They say dioxin is not a poison under international law," Tho noted, "but the companies knew of the dangers of dioxin." He described how the companies neglected the processes they knew could limit the dioxin content of herbicides, choosing instead to make more in less time by producing it at higher temperatures, which resulted in higher dioxin levels. "We don't think of the U.S. as an enemy," Tho said, but it has a responsibility "to help resolve this humanitarian issue."[94] As I explored the provincial maps indicating the locations of families receiving assistance, the VAVA staff kept turning the pages of the booklets in front of me back to pictures of children in the area born with horrific birth defects. "How do you determine which children are Agent Orange victims?" I asked. Tho explained that they receive requests from families who believe their children have an Agent Orange–related condition. Upon visiting the family, the VAVA staff makes the determination, adding the name of the child in the ledgers. Lack of resources and other priorities rule out medical testing, and no determination of dioxin exposure is made. When asked about this lack of evidence, Tho again shifted the conversation back to responsibility, noting that as recently as the previous year the United States had made available $3 million. "Why so much money if you do not accept responsibility?" he wondered out loud. "Three million dollars is a very small number compared to [the compensation provided to] American soldiers. And the money is for soil cleanup at Da Nang and Bien Hoa, not for families." VAVA consistently noted that helping families and children was its top priority; cleaning up the environment was secondary.[95]

VAVA has long maintained that there are more than 3 million Agent Orange victims in Vietnam, but it has never revealed how it arrived at that estimate. The Stellmans' article in *Nature* estimated that between 2.1 and 4.8 million people in Vietnamese were sprayed by Ranch Hand missions, a number that is generally accepted by VAVA and by various Vietnamese government agencies.[96] VAVA informed the Congressional Research Service (CRS) in late 2008 that "2.1 to 4.8 million Vietnamese were exposed to Agent Orange during the war and at least three million suffer serious health problems due to that exposure." The Ministry of Labor, War Invalids, and Social Affairs estimates that around 365,000 veterans, their children, and grandchildren suffer from dioxin-related conditions.[97] The state-run Vietnam news agency is prone to hyperbole and misstatement, which further calls into question the accuracy of the numbers. In early 2011 a story about renewed engagement between the U.S.–Vietnam Dialogue Group on Agent Orange dioxin noted that "between 1962 and 1971, U.S. troops sprayed more than 80 million litres of AO/Dioxin on one-fourth of southern Viet Nam's land area. *At least 4.8 million* Vietnamese people were exposed to the toxic chemical." The Stellmans' numbers suggest that around 73 million liters of herbicides were sprayed over central and southern Vietnam, 46 million of which were Agent Orange/dioxin. Furthermore, 4.8 million was the upper limit of the range given in the study, but that number is now regularly used by the Vietnamese government as the *minimum* number of victims.[98] Even sympathetic Western scientists like Schecter think such overstatements do more harm than good in drawing attention to the issue and in searching for other possible causes and solutions to conditions in Vietnam.[99]

On the basis of interviews with VAVA representatives at the national, provincial, and district levels and of the documentation, the number 3 million seems to be a rough estimate reflecting the range in the Stellman study. If Quang Tri is, in fact, one of the provinces most heavily populated with Agent Orange victims, and its official VAVA estimate in 2008 was 15,485, that number does not extrapolate to 3 million nationwide. Even assuming 15,000 as the average number of victims per province, multiplying by fifty-eight provinces amounts to 870,000 victims. Assuming that the 3 million includes those who have already died as a result of Agent Orange–related conditions, it is still nearly impossible to arrive at the number of 3 million.

The accounting is also likely attributable to the fact that the list of diseases considered by the Vietnamese government to be caused by dioxin from Agent Orange is more expansive and inclusive than the American list. In the late 1990s and early 2000s the Vietnamese Red Cross also developed such a list, and the most notable difference is the inclusion in Vietnam of

"Reproductive Abnormalities," a phrase that is far more vague than the specific list of abnormalities used by the VA.[100] The issue of birth defects and their relationship to dioxin exposure has been the most controversial aspect of the scientific, legal, and historical debates over Agent Orange.

Reports of birth defects and miscarriages by women in southern Vietnam helped draw attention to the potential dangers of Operation Ranch Hand in the 1960s.[101] While the Bionetics studies of 1968 helped show the dangers of dioxin as a teratogen in mice, the link remained unproven for humans. Throughout the lawsuits of the seventies and eighties, stories of birth defects among Vietnam veterans commonly appeared in the American press, but the VA and other agencies continued to resist allowing claims in these areas even as they expanded in others. Well into the twenty-first century birth defects remain among the most common images associated with Agent Orange but among the least proven potential consequences of dioxin exposure in humans. VAVA claims that the "rate of severe congenital malformations in herbicide-exposed populations is 2.95%, compared to .74% in non-exposed populations," but in view of the scarce documentation of exposure or even of exposure likelihood scenarios as well as of the few resources for testing, it is unclear how VAVA determines "exposed" versus "non-exposed" populations.[102]

As of 2011 the IOM included only spina bifida and anencephaly as birth defects with proven links to Agent Orange and dioxin exposure. Vietnamese scientists, whose studies have often been dismissed by their Western counterparts as lacking scientific rigor, have consistently pushed this issue only to find their claims refuted. For example, in 2006 a team led by Anh Ngo of the University of Texas Health Science Center published a meta-analysis of published findings concerning the association between Agent Orange and birth defects. The authors reviewed twenty-two studies published between 1966 and 2002, thirteen of which were by Vietnamese scientists working in Vietnam. The analysis was the first published article to include "unpublished data from Vietnamese civilians and sprayed civilians," which were given to the team by the 10-80 Committee in Vietnam. Ngo's review concluded that exposure to Agent Orange "is associated with a statistically significant increase in the risk of birth defects." The study did acknowledge that most of the Vietnamese studies were "methodologically weaker" than their counterparts and suggested that additional large-scale studies were needed "to further elucidate the aetiology of the Agent Orange and birth defects relationship."[103] The conclusion, however, was clear: Agent Orange exposure resulted in an increased likelihood of birth defects.

When the study was published in the *International Journal of Epidemiology* it was accompanied by commentary from Schecter and Constable, both of whom have done extensive research in Vietnam and both of whom are sympathetic to the plight of the Vietnamese people. "There is no doubt about the toxicity of the dioxin contaminant of Agent Orange," they write. "This dioxin has resulted in serious health effects in humans," they continued, citing their own considerable research in the field. Calling Ngo et al.'s article admirable, Schecter and Constable nevertheless faulted its "inclusive approach." Most of the Vietnamese studies included in the meta-analysis had not undergone peer review, and several were taken from the First International Conference on Agent Orange and Dioxin, held in Ho Chi Minh City in 1983, during which, Schecter and Constable point out, "Western scientists in attendance were joined unanimously by their Vietnamese colleagues at the conference in concluding that, although several of these papers were suggestive, or even very suggestive, none of them proved, to international standards, a connection between herbicide exposure and unfortunate outcomes of pregnancy, including congenital abnormalities." While supporting the authors' call for additional research that might one day construct a firmer foundation for the links, Schecter and Constable concluded that "to date, the answer is, at best, scientifically equivocal and, at worst, without valid positive scientific evidence."[104]

Scientific uncertainty has not kept the issue of birth defects from headlining media stories about Agent Orange. Photos of horribly deformed children accompany nearly every major newspaper feature about Agent Orange.[105] The preserved fetuses of deformed babies allegedly caused by exposure are on display throughout Vietnam, sometimes just down the hall from young children currently believed to be third- or fourth-generation victims, often despite any records of where the children have come from. During congressional hearings in 2008 entitled "Our Forgotten Responsibility: What Can We do to Help Victims of Agent Orange?" committee chairman Eni Faleomavaega of American Samoa and countless witnesses stated for the record the link between Agent Orange exposure and a variety of birth defects for which there exists no clear scientific link. When a DOS representative testified that "there is not comprehensive scientific information about how many of those birth defects were caused by X, Y, or Z, but what we have been doing is saying, what we can do is help, and what we ought to do is help and provide assistance to people, even without scientific evidence of what exactly caused it," Faleomavaega started to interrupt and eventually replied, "I did not mean to interrupt you, but when you say that not enough scientific evidence has proven that there was presence of dioxin in this Agent Orange, the substance that we used to defoliate the forests, the jungle there in

Vietnam, I thought we had moved beyond that already."[106] The parade of witnesses that followed testified to the persistence of third-generation birth defects among the Vietnamese population. Buried deep in the appendices of the published transcript of the hearing was a prepared statement by Schecter qualifying the statements linking Agent Orange and unspecified birth defects while calling for additional studies and preventative care.

The Double-Standard of Proof

Caregivers, advocates, and policymakers in Vietnam have repeatedly stressed the seeming contradiction of denying aid and support to Vietnamese victims of Agent Orange while awarding what seemed to most Vietnamese an embarrassment of riches to Americans. Although the actual per family payout from the class action suit in the United States was seen as meager by most in the class, Vietnamese regularly pointed to the $180 million award. Adjusting for the per family rate or the actual cost of care, let alone attending to the subtleties of the chemical companies not admitting guilt as part of the settlement, often failed to translate as effectively to the Vietnamese as the sheer size of the package. Similarly, the Vietnamese contrast the increasingly expanded package of benefits for veterans with their own limited resources.

As the VAVA lawsuit made its way through the American legal system, many U.S. veterans continued to press their case as well. In 2003 the Supreme Court agreed to hear the case of Daniel Stephenson, who, like many others, believed he had been either unaware or inadequately represented during the settlement in 1984. The high court divided evenly after the recusal of Justice John Paul Stevens, who lost his son, a Vietnam veteran, to cancer in 1996. The tie vote left in place an appellate ruling that allowed veterans to continue to pursue damages outside the boundaries of the class action.[107] That case was sent back to Judge Weinstein, who had rejected the veterans' claim to begin with. When Weinstein rejected the case on largely the same grounds as the Vietnamese victims' suit, the Supreme Court eventually declined to hear that case as well, making the original class action suit of 1984 the only successful legal case in American courts against the manufacturers of Agent Orange.

Similarly, despite Vietnamese views about the extent and scope of the VA benefits package, many veterans continued to feel that the U.S. government was ignoring or dismissing their claims and their calls for a more inclusive benefits regime. The Agent Orange Act of 1991 granted the secretary of veterans' affairs leeway in determining service-related disabilities for Vietnam veterans, but profound tensions endured, not only between veterans and the government but also between the decision to allow service-related disability claims for a particular condition and the scientific case for linkage between that condition and Agent Orange. When the IOM published its first report, *Veterans and Agent Orange,* in 1994, as mandated by the Agent Orange Act of 1991, the IOM review committee devoted several chapters to the Agent Orange controversy over the past two decades and dozens of pages summarizing the methodologies employed in determining its recommendations to the VA on the associations between various conditions and Agent Orange exposure. The result, while thorough in its analytic rigor, produced sentences like the following: "Hence, in a strict technical sense, the committee could not prove the absence of any possibility of a health outcome associated with herbicide exposure. Nevertheless, for some outcomes examined for which there was no evidence consistent with an association, there was limited or suggestive evidence consistent with no association, and the committee was able to conclude *within the limits of the current resolving power of the existing studies* that there is no association with herbicide exposure."[108] On the issue of birth defects and other reproductive health issues, the committee wrote, "Given the large uncertainties that remain about the magnitude of potential risk from exposure to herbicides in the occupational, environmental, and veterans studies that have been reviewed, effects of information bias in these studies, and the lack of information needed to extrapolate from the level of exposure in the studies reviewed to that of individual Vietnam veterans, it is not possible for the committee to quantify the degree of risk likely to have been experienced by Vietnam veterans because of their exposure to herbicides in Vietnam."[109] While such statements reflected the state of medical and scientific research on these issues, they offered little comfort to veterans and family members who believed their miscarriages or the conditions afflicting their ill children had been caused by Agent Orange. Some of the same information was published by the VA in the more accessible, easily distilled form of *Agent Orange Review,* but the overall message to attentive readers was the same throughout most of the 1990s and the early twenty-first century: except for the small and slowly expanding list of conditions identified by the VA as qualifying for service-related disabilities, there was no scientific basis for associating Agent Orange and a variety of conditions, including most birth defects.[110]

Some in the veterans' community continued to hold out hope for the Ranch Hand study, which, despite the cloud of controversy that seemed to hang over it, was the only long-term epidemiological study of a reasonably sized sample of people known to have been exposed to dioxin through Agent Orange. Beginning around 2000, evidence from the Ranch Hand study began to suggest a possible link between herbicide exposure and diabetes. In *Agent Orange Update* for December 2000 the VA announced it had approved a presumptive service connection for Type 2, or adult-onset, diabetes. This was a major development for Vietnam veterans, as nearly one-fifth of all veterans receiving treatment in VA facilities at the turn of the century had been diagnosed with diabetes.[111] The new regulations did not fully take effect until the summer of 2001, but nevertheless this was one of the major changes made possible by the Agent Orange Act of 1991, and it demonstrates the degree to which the politics of uncertainty had shifted in favor of veterans after the 1980s.

Prior to 1999 the VA officially held there was "inadequate/insufficient evidence to determine whether an association existed" between Agent Orange and diabetes. After a review of the updated scientific literature, however, the IOM committee in 2000 switched its recommendation, claiming there was now "limited/suggestive evidence of an association between exposure to the herbicides used in Vietnam or the contaminant dioxin and Type 2 diabetes."[112] This understated shift in language was all that was required, under the Agent Orange Act, to allow the secretary of veterans' affairs to qualify Type 2 diabetes as a service-related disability for Vietnam veterans. Far removed from the burden-of-proof days of the late 1970s, "limited" evidence was now enough to earn even the navy electrician who spent only hours in Vietnam a service-related disability benefit for his diabetes. Perhaps most remarkable is that, buried in the summary of recent epidemiology, mortality, and morbidity studies, there was even less evidence of a link between exposure and diabetes than the "limited/suggestive" description might have suggested: "Presently available data allow for the possibility of an increased risk of Type 2 diabetes in Vietnam veterans. It must be noted, however, that these studies indicate that the increased risk, if any, from herbicide or dioxin exposure appears to be small. The known predictors of diabetes risk—family history, physical inactivity, and obesity—continue to greatly outweigh any suggested increased risk from wartime exposure to herbicides."[113] In other words, by 1999 veterans were beginning to show a greater incidence of diabetes, but little, if any, evidence linked those with diabetes and those claiming exposure to Agent Orange. In the next several years, particularly in the first years of the administration of Barack Obama, the list of presumptive benefits grew rapidly to include such conditions as prostate cancer and Parkinson's disease.[114] Few veterans in 1979 would have believed such policies would be in place thirty years later, but they represented a realization of what they, their fellow veterans around the world, and countless civilians and activists had long called for: placing the burden of proof on those denying links between Agent Orange, dioxin, and their possible effects on the health of humans and the environment rather than on those who believed they were suffering from those effects. The shift demonstrates the power of the veterans' lobby in Washington. Through the sustained efforts of veterans' groups, it has become exponentially easier for Vietnam veterans to receive service-related benefits even though with the passage of time it has gotten exponentially more difficult to equate exposure to Agent Orange with specific medical conditions.

And yet the ongoing story of Agent Orange, Vietnam veterans, and their governments well into the twenty-first century remains one based on distrust. When the Ranch Hand study was allowed to expire as scheduled in 2006, it represented the termination of probably the most expansive, detailed study of its kind. The study resulted in limited expansion of benefits for veterans, helping to determine the association between herbicide exposure and spina bifida, drawing attention to the possible link between Agent Orange and elevated risk for Type 2 diabetes. But because of its limitations—the Ranch Handers by definition were not typical Vietnam veterans—the study was plagued by questions about whether or not the results could be extrapolated to combat veterans who served in the field, let alone to those with a lower likelihood of exposure. Additionally, a GAO report to Congress in 2000 found the USAF ineffectual in communicating (or failing to communicate) its findings, limitations, and data to the public.[115]

Equally contested was what constituted Vietnam-era service. While veterans who served in the RVN in the sixties and seventies were increasingly able to obtain service-related benefits, others who served in the military during that time were not. This discrepancy became a highly contentious issue as the global scope of Agent Orange became known. When the revelations about the use of Agent Orange in the Korean DMZ became common knowledge, a major push for expanded benefits began. The DOD's report in 2006 on the testing, evaluation, and storage of herbicides led veterans who had served at the more than thirty sites included on the list, places like Johnston Island, Puerto Rico, and Thailand, to make similar demands.[116] But VA regulations were not as flexible for veterans who did not serve on the ground in Vietnam. Unlike those who served in-country, Vietnam-era veterans who believed they were exposed to Agent Orange even though they did not set foot in Vietnam had to prove to the VA that they were exposed, a task certainly no easier for veterans in 2007 than for combat veterans only years removed from their service.[117]

One group that embodied the gray area of Vietnam-era service is the Blue Water navy men who served off the coasts of Vietnam. Unlike the Brown Water navy veterans, who had been in-country, few Blue Water veterans ever set foot in South Vietnam and were thus outside the boundaries of the VA's presumptive benefits regime. A Blue Water veteran, Jonathan Hass, sued the VA, claiming he had been unfairly denied benefits for his diabetes because he had not served in-country. Haas told the Associated Press in 2010 that he remembered seeing clouds of chemicals drift out over the harbor and "engulf" his ship.[118] In 2006 a U.S. Court of Appeals for Veterans' Claims ruled in favor of Haas, holding that Blue Water veterans should be entitled to presumptive benefits; the VA overturned the decision on appeal. In 2009 the Supreme Court refused to hear the case, leaving the Blue Water veterans dependent on proposed legislation, such as the Agent Orange Equity Act of 2009, which would have included in presumptive benefits any veteran who served in Vietnam's "inland waterways, ports, and harbors, waters offshore, and airspace above," who served on Johnston Island during the period of Operations Pacer IVY and Pacer HO, or who received the Vietnam Service Medal or the Vietnam Campaign Medal, as Haas had. The bill never got out of committee.[119]

The Blue Water veterans have kept their fight alive. In 2010 the Blue Water Navy Vietnam Veterans Association self-published its own study attempting to make the case for its members. The Da Nang Harbor report argued that

> ports, bays and harbors became collection points for the residuals of herbicide that washed from the inland country via rivers and streams to the sea. Da Nang Harbor had additional herbicide input to its water, given that a major air base for the herbicide spraying project was less than a mile from the harbor's edge. Vast amounts of diluted and undiluted herbicides, along with their contaminants, entered the harbor waters at Da Nang. This provided a continuous replenishment of vaporizing herbicide, including Agent Orange and its byproduct dioxin, one of the most toxic substances known. When ships from offshore entered Da Nang Harbor, they were surrounded by a floating mass of dioxin-laden herbicide clinging to oil, particulates, and debris on the water's surface. The crews of these ships were exposed to the toxic atmosphere that was more likely than not contaminated with dioxin molecules arising from surface waters around their ships.[120]

Leaving aside the gaps in its scientific analysis, the Da Nang report makes a strong case for Blue Water veterans' inclusion in the benefits regime; the arguments reveal more about the nature of that regime than about the legitimacy of the group's contentions. Even if the soundness of the scientific analysis of exposure through vaporization and downstream particle accumulation is debatable, the Blue Water veterans presented a damning indictment of the "boots on the ground" version of what constituted a Vietnam veteran. Under the VA's guidelines, they noted, several navy and air force pilots who were shot down over North Vietnam never "had the opportunity to walk on the landmass of South Vietnam" and would thereby not be Vietnam veterans.[121]

Yet the inventive rhetorical argument in no way evidences, as the report states, that Blue Water veterans had "no escape from direct exposure" because of the presence of dioxin in Da Nang harbor. The logical extension of the Blue Water position, according to John Wells, a spokesman for the group, is that "if you were on board a ship in Territorial Seas of RVN, you were exposed."[122] This argument reveals less about the strength of the Blue Water veterans' evidence than about the fundamental weaknesses of most veterans' claims about Agent Orange. Although not a convincing explanation of how Blue Water vets were inevitably exposed to dioxin, the study does imply that they were, in many ways, no less likely to have been exposed than the hundreds of thousands of veterans who served "in the rear with the gear," as the saying goes, working desk jobs in Saigon and elsewhere. The results of the IOM inquiry in 2011 into Blue Water veterans' claims confirmed their deficiencies, finding insufficient evidence to determine exposure. "Given the lack of measurements taken during the war and the almost 40 years since the war," the IOM report noted, in language strikingly reminiscent of the exposure debates of the 1980s, "this will never be a matter of science but instead a matter of policy." Blue Water veterans and the American Legion were predictably outraged at the decision, but the headline at the *New York Times* summed up the situation perfectly: "Inconclusive Agent Orange Study Is Conclusive Enough for Vet Groups."[123]

Conclusion

In the early twenty-first century Vietnam veterans from all branches faced a new hurdle in their quest to maintain and expand service-related benefits from the VA. Despite having a strong ally at the VA in Gen. Eric Shinseki, a deficit-conscious Congress scrutinized the secretary during the Obama administration, focusing on the skyrocketing cost of veterans' benefits during the economic crisis that began in 2008. Ironically, it was a Vietnam veteran, Jim Webb, the Democratic senator from Virginia and former secretary of the navy, who led the charge against the expansion of Agent Orange benefits under President Obama. After placing a hold on funding for the VA, Webb, at a hearing of the Senate Veterans Affairs Committee in September 2010, grilled Shinseki on the cost of the programs and the scientific evidence supporting the expanded list of conditions covered. Although most of the committee members, Republicans and Democrats alike, supported the expanded coverage and couched their backing of it in terms of honoring the service of American veterans, Webb pointed out that under the new program the VA would be inundated with hundreds of thousands of veterans' claims, at a cost in the billions of dollars over the next decade alone. Shinseki maintained that his decisions were based "upon careful consideration of the scientific and medical evidence" available to him, but Webb pressed both Shinseki and the scientists present on how they determined that TCDD was a causal agent in these conditions. Neither Shinseki nor the scientists, any more than their predecessors of ten or twenty years earlier, could satisfactorily answer that question. Speaking instead of "associations," as dictated by the Agent Orange Act of 1991, the panelists made the case as simply as they could, citing "the credible evidence for the association" between service in Vietnam and increased risk for diabetes, hypertension, and Parkinson's disease.[124] The Agent Orange Act of 1991 makes no mention of the cost of benefits. By giving the benefit of the doubt to veterans and to the VA, the act limits Congress's ability to intervene in the determination of benefits.

At the time of the hearing featuring Webb and Shinseki, diabetes was by far the most common condition of veterans on VA disability, accounting for more than 270,000 of the 1 million plus veterans receiving service-related payments. The payments for diabetes alone were expected to cost more than $850 million in 2011. The additional claims for heart disease, Parkinson's, and certain types of leukemia alone will likely total more than $42 billion by 2021, according to the VA, which is currently spending $34 billion annually for disability benefits for *all* veterans.[125] Haas, who was eventually granted a full disability claim for his diabetes, thought that other vets, especially Blue Water veterans, were still "getting screwed," while the Ranch Hand veteran Jack Spey sided with policymakers like Webb, arguing that Ranch Handers, like all Vietnam veterans, were simply getting older. "It's a bunch of

B.S," Spey said in 2010. "We're all going to die someday." That year, erectile dysfunction, often linked to diabetes, was one of the ten most common disability claims of Vietnam veterans.[126]

Viewed in the light of the expanding benefits being awarded by the VA, the frustration of the Vietnamese with the obstinacy of the U.S. government in acknowledging the link between Agent Orange and health concerns in Vietnam is understandable. The science is largely the same: there is certainly "limited" and "suggestive" evidence of a link between exposure to Agent Orange and a variety of health problems in those living near hot spots like Da Nang, Bien Hoa, and A Luoi. The standard, however, is patently different. American diplomats tell the Vietnamese people there "is no established scientific link between" Agent Orange and a particular condition, yet VA publications describe the same links as being "suggestive," all the while acknowledging the lack of good evidence linking the condition to herbicide exposure at all.

It is no surprise that the United States assists its veterans who prosecuted the war in Vietnam far more generously than it does the citizens of Vietnam, many of whom were not alive during the war. It is no surprise either that Vietnamese soldiers, civilians, and government representatives point out the double standard employed to avoid dealing with the question of U.S. responsibility for ongoing dioxin-related problems in Southeast Asia. The politics of uncertainty have always been fluid and subject to mobilization by various constituencies in the service of various projects and agendas. While it is certainly true that dioxin still presents all of its victims, real or imagined, with more questions than answers, it is just as true that the caregivers of the children in the orphanages in Da Nang could work miracles with a tiny fraction of the money the VA pays out in claims each year.

The initial complaint in the class action lawsuit filed on behalf of Paul Reutershan and other veterans read that the chemical companies who manufactured Agent Orange acted in "wanton and reckless disregard of the public health, safety and welfare and the health, safety and welfare of the deceased plaintiff herein and *all those others so unfortunate as to have been and now to be similarly situated at risk,* not only during this generation but during those generations yet to come."[127] While such arguments clearly do not hold much weight in American courtrooms, given the flexible standard of proof for exposure and causation now in place for Vietnam veterans who claim disability from Agent Orange, even the most cynical and hardened observer would be hard-pressed to see the children of Vietnam and judge that they should not be included among "all those others so unfortunate."

CONCLUSION: AGENT ORANGE AND THE LIMITS OF SCIENCE AND HISTORY

In the central highlands of Vietnam, near some of the areas most heavily defoliated during the war, a thick, invasive grass thrives along open hills where the trees refuse to return even now, forty years after they were destroyed. They call it American grass. Standing in a gnarled patch in the A Luoi valley is Phung Tuu Boi, who, at barely five feet tall, seems as though he will be overtaken by the renegade growth. Since the war ended in 1975 Boi has waged a battle to regrow Vietnam's lush forests. He has worked with Australian colleagues to plant acres of acacia trees, which can grow under harsh conditions and are a profitable source of lumber, in heavy demand from companies like Ikea. Boi has also developed one of the most innovative responses to dioxin contamination in Vietnam: he calls it a green fence. This fence, made of trees covered with sharp, cactus-like needles, is meant to keep both humans and animals out of hot spots like the one here in the A Luoi valley, which sits nestled in the highlands above Da Nang. The fence is made primarily of acacia and gleditschia, a type of locust tree imported from Australia that produces fruit that can be harvested, sold, and processed by locals. The long-term success of the green fence won't be known for some time, but Boi is the first to admit that it cannot help the many villagers still living in the A Luoi valley, who have been farming the land for years, potentially disturbing dioxin that has been in the soil since the early 1960s.[1]

Boi and thousands of others like him throughout Vietnam and throughout the world are not necessarily looking for answers; they are looking for solutions. Despite the absence of clear, scientific evidence linking Agent Orange to conditions like birth defects, they press on with the goals of mitigating the levels of dioxin around hot spots, helping those less fortunate than themselves who cannot otherwise care for themselves or their families, and drawing increased attention to the issues related to Agent Orange.

Most people around the world who have encountered Agent Orange are looking for answers. They want to know if a specific condition experienced in a particular body is caused by exposure to a specific agent. It is still, and will for some time be, beyond the scope of both scientific and historical analysis to answer such questions. There are simply too many variables, too many unknowns. As I said in the introduction, it is beyond the scope of this book and beyond my abilities as a scholar to provide such answers. What I have attempted instead is to offer a broader sense of understanding of issues surrounding Agent Orange as it was developed, used, encountered, experienced, imagined, and negotiated at different places and in different times. In doing so, however, I believe I have provided sufficient basis to posit preliminary answers to the most pressing questions about the history and legacies of Agent Orange that I posed in the introduction.

1. How could the United States and its allies do such a thing?

How could the United States and its allies do such a thing? is the question about Agent Orange that is the simplest to answer historically and the one least often addressed in historical terms. Many critics of Operation Ranch Hand and contemporary Agent Orange activists attribute the program of herbicidal warfare to the evils of American imperialism. For those who became disillusioned about their government during the Vietnam War, Agent Orange has served as a screen onto which they can project those feelings. Given the level of destruction wrought on the Vietnamese people and landscape during the period of direct U.S. military involvement, it was easy for many to believe the worst about the herbicide program. Similarly, given the challenges faced by Vietnam veterans in dealing with the government, particularly the VA, in the postwar period, it became a given for many that the government was lying to them and withholding crucial information. And although there was no coverup about the effects of Agent Orange, the U.S. government could have responded earlier and with greater urgency to the concerns raised by veterans and others. American veterans were right to view themselves as victims of the Vietnam War; they were given an impossible task in a brutal war that lacked popular support. Upon returning home, most still did not enjoy the support of their government. The VA became a symbol of this distrust. So, too, did Agent Orange.

But intentions matter to historians, perhaps particularly so in the case of historians of foreign relations; and there is simply no evidence that the policymakers in the White House, the DOS, and the military believed the herbicides, including Agent Orange, were dangerous to human health and therefore deliberately used them as weapons to inflict direct harm on people. In chapters 2 and 3 I argued that policymakers were guilty of, among other things, arrogance and shortsightedness. Again and again they sought to impose the will of the United States through superior resources and technology, seeking military solutions to a political problem and treating Southeast Asia as a laboratory for various counterinsurgency strategies, including the weaponization of herbicides. When these proposed solutions continually failed, military and political leaders repeatedly ignored evidence about those failures in favor of less convincing evidence that the programs were working. There is no excuse for what the United States did to the Vietnamese people during the war, including experimenting with herbicidal warfare. But it does not follow that every decision made about the prosecution of that war was necessarily evil by design. Rather than immoral, the decision to weaponize herbicides was amoral. Acknowledging this does not simply whitewash history or turn the war itself into the supposed noble cause championed in retrospect by many politicians and revisionist historians. It does require one to attempt to see the decisions from the standpoint of those who made them fifty years ago and more.

Standing figuratively in the shoes of those who advocated and implemented Operation Ranch Hand requires one to imagine a world in which the bipolar logic of the Cold War reigned supreme, the threat of Communism was deemed very real, and technology and chemicals were believed to be potential solutions to an array of problems ranging from garden weeds to guerilla warfare. As the historical record shows, policymakers in Washington debated the politics of the chemical war at length, but their initial doubts about the program were not about the effects of the chemicals on human health or the environment; they were about appearances and public opinion. The Kennedy administration did not consider defoliation and crop destruction tantamount to biological warfare, but it actively worried about the specter of germ warfare publicized by communist propaganda. Even in the DOS, where opposition to the program was the strongest, critics framed their objections in political, rather than moral, ethical, or humanitarian terms. Those who worried about the effects of crop destruction on Vietnamese civilians did so because of the hunger that would ensue, not because of the contamination of food supplies. Yet even so, the misgivings of such figures as Dean Rusk and Roger Hilsman were owing to the politics of crop destruction, not to the act itself. They thought longer and harder about crop destruction than about defoliation, but the fact remains that even then they did not ask questions about the effects on human health because few Americans were asking such questions at the time. Like nearly every American citizen in the early 1960s, Rusk, Hilsman, McNamara, and Kennedy believed herbicides to be safe for human contact. There is simply no evidence to the contrary.

To this day the only evidence offered in support of the view that the U.S. government knew about the dangers of Agent Orange is the letter written in 1988 by James Clary to Admiral Zumwalt and included in the Zumwalt report of 1990. As I discussed in chapter 4, however, this letter, which claimed that "because the material was to be used on the 'enemy,' none of us were overly concerned," was written decades after the fact and has never been substantiated by contemporary documentation. Should not the burden of historical proof be on those who advocate conspiracies and coverups and continue to claim that the military knew about the dangers of herbicides early on? What is more likely, that the dangers were well known at the time and yet no trace of such knowledge can be found in the historical record? or that Clary's letter, like other reflections on herbicidal warfare that came years, sometimes decades, after the war is guided by a contemporary rather than a historical understanding of Agent Orange?

Indeed, this is the key insight to be gained by addressing these questions historically. As I have made clear, the rise of environmentalist thinking about chemicals, the body, nature, and the state fundamentally changed the ways in which people thought about the dangers of herbicides in general and Agent Orange in particular. By the time Ranch Hand ended, the dangers of herbicides and dioxin were much more widely known, but that knowledge was not available to policymakers in the early 1960s. Arguably, in light of the Bionetics study and the work of the AAAS, Ranch Hand should have been ended earlier than it was, but, considering the discussion in chapters 4 and 5, those who make that argument must account for how unstable knowledge about dioxin was well into the 1980s and 1990s. While it is regrettable that the warnings were not heeded sooner, it is also, in the context of the time, understandable. It took decades, for example, to change the public's perceptions about the dangers of cigarette smoking, and even then, people everywhere, knowing full well of the risks, continue to smoke. Using this historical approach, one might speculate about what future generations will say about the current moment. Might they not wonder why the United States used depleted uranium in the production of armaments, or why people drank water and fed children from plastic bottles?

My final point here is that the question of knowledge, agency, and decision making should separate government officials and the chemical companies. As the legal battles of the seventies and eighties divulged, the major manufacturers of Agent Orange knew about the presence and potential danger of dioxin in 2,4,5-T prior to the introduction of that toxin and also of U.S. troops into Southeast Asia. Just as no evidence supports claims that American policymakers knew about the dangers of herbicides to human health, there is, relatedly, no evidence that the chemical companies explicitly warned anyone in the government, including the military, about those dangers. While no one, historians included, knows for certain what lurks deep in the archives, it is difficult to imagine such documents not coming to light over the past several decades. The evidence at this point supports the view that American war planners had no knowledge of the potential effects of Agent Orange on the health of its many victims, even if the chemical companies that produced it did.

2. Should the use of Agent Orange be considered chemical warfare?

Was the use of Agent Orange chemical warfare? This is likely the most difficult question to address historically. In many ways it is related to the previous question of the intentions of policymakers: if those who designed the war were not aware of the dangers of the chemicals, how could their actions be considered chemical warfare? Many in the Kennedy administration discussed this question openly in the early 1960s, exploring both the possibility that defoliation and crop destruction could be considered chemical warfare and, more often, that they would likely be described as such by critics and communists around the world regardless of the law. While even critics in the DOS acknowledged that the military use of herbicides was not prohibited by international law, however, the actual use of herbicides, as opposed to the intended use, is also at issue. As is the case in the debates over knowledge and intention, the chemical warfare issue is not simply an academic debate. Just as the question of what the government knew and when it knew it was a key component in determining the course of the veterans' class action suit in the 1980s, the question of violating international law was one of several important factors in the dismissal of the VAVA case more recently. Here again, the historical record offers some useful context and guidance.

There was no clear prohibition against the military use of herbicides at the time, but, in addition, the Kennedy and Johnson administrations believed they were operating within international law because the chemicals were not known to be a threat to human health and, moreover, because U.S. and ARVN forces were using the chemicals against natural targets, such as forests, trees, and crops, not against "the physical person of the enemy," as prohibited by chemical warfare conventions. The last point, however, unlike the first two, is not nearly resolved.

As I argued in chapter 3, it would not have been uncommon for policymakers to distinguish the physical person of the enemy from the environment in which that person was living, working, eating, and fighting. Thinking about a sovereign human body as moving through a distinct natural environment was the norm in the United States and other Western nations in the early 1960s. As much as one might now, through the lens of some four or five decades of environmentalist thinking, view the relationship between human bodies and their environments as more permeable and integrated, that worldview was not available to the policymakers who designed and implemented Operation Ranch Hand.

Yet it became obvious early in that operation that the physical person of the enemy was likely being sprayed directly and repeatedly, along with large numbers of Vietnamese civilians. The psyops efforts of U.S. and ARVN forces made clear that "friendly forces" regularly "inhaled" the chemicals and discussed the safety of the herbicides. I described in chapter 2 that Ranch Hand missions regularly destroyed civilian crops near populated areas, dramatically increasing the chances of direct spraying of persons, whether civilian or combatant. A further complication is that while Agent Blue became the weapon of choice for crop destruction, other agents, including Orange and Purple, both of which contained 2,4,5-T and its dioxin contaminant, were also used on occasion. And yet even in the case of crop destruction and even when the military used Agents Purple or Orange, rather than Blue, to destroy crops, it would be easy to argue that the physical person of the enemy, or of civilians for that matter, was not the intended target of the attack. But the use of chemicals on the physical person of the enemy patently occurred regardless of the intent of policymakers, military commanders, and individual pilots.

[244]

The question of chemical warfare thereby leaves one squarely in the gray area that occupies so much space in the long history of Agent Orange. A good case can be made either way, and although the legal argument that the weaponization of herbicides should be considered chemical warfare has not held up well in court, it was far from the only factor in the dismissal of the VAVA case. History can add context to the issue, and it necessarily complicates the views of anyone who believes that the case can be made on either side. If historical analysis can identify policymaker's intentions, however, it cannot hand down a watertight verdict on this matter. Calling the spraying of herbicides chemical warfare is not a helpful strategy for those seeking compensation and obscures, rather than illuminates, its historical realities. I remain convinced that it is a less than useful way to think about the chemical war and its legacies.

3. What can and should be done for U.S. veterans, Vietnamese victims, and others around the world who believe they are suffering as a result of Agent Orange?

If historical analysis has helped illuminate and complicate understanding of the history and legacies of Agent Orange, it has been overshadowed by the many contributions of science. Whereas until recently historians largely ignored Agent Orange and dioxin as subjects of study, scores of scientists and medical practitioners have been working on them for decades. The dramatic shift in attitudes, policy, and knowledge about chemicals, bodies, and the environment over the past half century is owing not only to shifts in social and cultural attitudes but also to remarkable advancements in scientific understandings of the interaction of those different actors and agents. And yet, as I show in chapters 3, 4, and 5, the production of these bodies of knowledge and the shifting frameworks they helped shape have not always translated into answers, let alone comfort, for the victims and alleged victims of Agent Orange. What, if anything, do history and science suggest about what can and should be done for them?

I have argued throughout this book that there are few, if any, concrete answers about Agent Orange. I have also held that when history and science have failed to yield answers, decisions about exposure, resources, and benefits have been overtly political. When confronted with a reluctant and at times obstinate bureaucracy, U.S. veterans forced a change in the VA benefits regime through political action guided by their powerful congressional allies. In the face of government agencies that unfailingly used the discourse of uncertainty to claim the safety of 2,4,5-T, activists in New Zealand continued to organize politically to shift the burden of proof for safety claims away from everyday citizens and onto the states and chemical manufacturers.

In Vietnam the situation is more complicated, given the local, national, and transnational relationships at work. Vietnamese victims for years struggled to obtain scarce resources within their nation, as their government refused to make the issue of Agent Orange so visible that it would jeopardize the chances of normalization with the United States. The brutal economic sanctions regime led by the United States ruled out most recourses Vietnamese victims had through binational and international frameworks. Even since the end of the embargo and the normalization of relations change has been slow to come. As the United States continues to reject the claims of Vietnamese veterans and civilians, and as American courts dismiss the cases brought by VAVA, justice seemingly continues to elude Vietnamese victims of Agent Orange. The attention brought to the issue by the Vietnamese lawsuit, however, and by the blatant, transparent double standards of the U.S. government on the issues of exposure and causality has made possible transnational political action and fundraising that are beginning to bring much-needed resources to the people of Vietnam. Despite the paltry sums committed by the U.S. government, increasing amounts of private American aid and NGO money are going to families and individuals who need help and to the glaring problem of cleaning up known dioxin hot spots like Da Nang and Bien Hoa.

Yet VAVA and its allies around the world risk alienating potential sources of support when they offer dubious, unreliable, unsubstantiated claims regarding the number of Vietnamese victims. Surely reasonable people can agree that despite the fact that any U.S. veteran who served in Vietnam and now has diabetes is eligible for service-related benefits, not every Vietnam veteran with diabetes was therefore exposed to Agent Orange and its associated dioxin. Similarly, millions of children in Vietnam suffer from a variety of birth defects. The absence of scientific consensus notwithstanding, likely some of them, not least those from known hot spots who have conditions widely recognized as related to dioxin exposure, can be described as victims of Agent Orange. But in orphanages throughout the country, from the Red River to the Mekong Delta, the assumption is that most, if not all, of these children are so characterized; victimhood is the rule rather than the exception. Drawing attention to the needs of children, especially those with disabilities whose families are unable to care for them, is laudable, but overstating the case does not help the cause. By its practice of continually increasing the number of victims without explaining the source of its data, VAVA hinders those for whom it advocates.

By calling into question the nature and reliability of the information, such claims not only distract from the very real needs of genuine victims and others with serious disabilities, but also potentially divert additional resources from the search for other potential causes and solutions. If nearly every child in Vietnam born with a birth defect is presumed to be an Agent Orange victim, those with the knowledge and the resources to address such issues are increasingly likely to stop looking for alternative explanations. What if other factors are contributing to these conditions? What if there are dietary, medical, or environmental causes unrelated to Agent Orange at work? What if there are other solutions to the problem? Acknowledging such possibilities does not require a return to the punitive politics of uncertainty, in which an absence of proof constitutes an outright denial of connections and causality, accompanied by a denial of claims. Neither history nor science can supply the type of certainty to support or reject these claims, but without some demonstrable base of evidence grounded in historical and scientific documentation, the fates of victims of Agent Orange, in Vietnam and around the world, are even more likely to be driven by the ebbs and flows of politics and the resource flows that accompany them.

Raising the question of accuracy and evidence does not absolve the U.S. government and the chemical manufacturers of their moral, ethical, and historical responsibilities. The continuation of a legal structure that rewards states and corporations and punishes citizens by citing scientific uncertainty raises the likelihood that chemical companies will never be held accountable for their role in ongoing dioxin-related concerns in Vietnam. Despite the most recent contributions by the United States to dioxin remediation in Da Nang, the scenario is eerily reminiscent of the period immediately after the embargo was lifted in 1994. American firms flooded Vietnam with advertising and consumer goods while trolling for cheap sources of labor; meanwhile Vietnamese exports to the United States were greeted with questionable tariffs, charges of environmental contamination (such as catfish contaminated with dioxin), and demands that the Vietnamese government improve its human rights record. The result was, as I have argued elsewhere, a continuation of the same asymmetries of power that have long defined Vietnamese–American relations.[2] For years the United States participated in the diplomatic process, committed a shamefully small amount of resources, and continued to dictate to the Vietnamese the conditions under which progress will continue. Even if the cost of the final remediation plan for all known hot spots runs to $50 million, that is the equivalent of about $20 million less that what the United States was spending *per day* in 2011, on the war in Afghanistan, or about 0.04 percent of funds appropriated for the wars in Iraq and Afghanistan between 2001 and 2010.[3] The $32 million committed to Da Nang is a sign of progress, but much more aid is needed, particularly for rural areas like A Luoi, where the cleanup will be more difficult and more costly.

If the history of Agent Orange teaches one anything, it is that there are distinct limits to the types of answers that scientific or historical analysis can provide. I know this to be true in part because I have spent the better part of the past decade immersed in that history, and at the end of this part of the journey I feel as though I have barely scratched the surface. I have learned so much and yet feel as though I know so little. I have had my original assumptions about Agent Orange turned upside down so often I can barely remember what they were. Throughout it all I have met many people—veterans, civilians, scientists, writers, and artists among them—whose *lives* have been turned upside down by Agent Orange. Whether I believe their stories or not and whether or not those stories are supported by historical and scientific evidence, they are nevertheless part of the global history of Agent Orange. As I said in the introduction, my goal is to provide context, not closure. I have attempted to bring some balance to the history and legacies of the chemical war, to seek out some likely explanations in the middle ground long abandoned in the polarized politics of Agent Orange. No amount of historical context will provide comfort to the children of Da Nang and A Luoi, to aging American veterans, and to the millions around the world who feel, rightly or wrongly, they and their loved ones are victims of the chemical war. Their search for answers will continue.

So, too, will mine.

Notes

The following abbreviations are used in the notes:

ALY	Alvin L. Young Collection on Agent Orange, National Agricultural Library, Beltsville, Maryland
ANZ	Archives New Zealand, Wellington, New Zealand
AOCRF	Agent Orange Claims Resolution Files, NARA Record Group 341/190
AOLF	Agent Orange Litigation Files, NARA Record Group 341/190
FRUS	*Foreign Relations of the United States*
MACJ3-06	Military Assistance Command, Vietnam, Surface Operations Division: Chemical Warfare Operations, NARA Record Group 472/270
MACJ3-09	Military Assistance Command, Vietnam, Chemical Operations Division: Herbicide Operations NARA Record Group 472/270
MACV–CORDS	Military Assistance Command, Vietnam—Civil Operations and Revolutionary Development Support, NARA Record Group 472/270
NARA	National Archives II, College Park, Maryland
NYT	*New York Times*
RG	Record Group
TTU	Vietnam Archive, Texas Tech University
WP	*Washington Post*

INTRODUCTION

1. Mike Baker, "Diabetes Now Tops Vietnam Vets' Claims," Associated Press Report, August 30, 2010. Baker, who reviewed several veterans' files under a Freedom of Information Act Request, notes in the article that there "is no record" of the veteran's journey but that he was nevertheless able to make a successful claim based on it. Although the Veterans Administration became the Department of Veterans Affairs in 1989, I will use the abbreviation VA throughout to avoid confusion and to reflect the term used by most veterans and policymakers.
2. Interview with author, April 20, 2008.
3. The full names for 2,4-D and 2,4,5-T are 2,4-Dichlorophenoxyacetic acid and 2,4,5-Trichlorophenoxyacetic acid, respectively. 2,3,7,8-tetrachlorodibenzo-para-dioxin, more commonly known as 2,3,7,8-TCDD, or simply TCDD, is one of dozens of toxins known collectively as dioxins. Its name is derived from the location

of the chlorine atoms on the molecule (positions 2, 3, 7, and 8), and their position relative to the benzene and oxygen atoms. The configuration of these components on the TCDD molecule makes 2,3,7,8-TCDD by far the most toxic form of dioxin, thousands of times more toxic than other polychlorinated dioxins. TCDD can be produced by a number of processes, including the manufacture of herbicides such as 2,4,5-T. Throughout this book, unless otherwise noted, when I refer to dioxin or to Agent Orange and its associated dioxin, I am referring to 2,3,7,8-TCDD. Similarly, while it is critical to understand that Agent Orange and TCDD are distinct and noninterchangeable things, for reasons of brevity I will refer to *Agent Orange* in most sections of the book without always qualifying it as Agent Orange and its associated dioxin, a more accurate but also a more cumbersome phrase. For more on the makeup, history, and characteristics of dioxins and of TCDD, see Alastair Hay, *The Chemical Scythe: Lessons of 2,4,5-T and Dioxin* (New York: Plenum Press, 1982).

4. For instance, George Herring's *America's Longest War: The United States and Vietnam 1950–1975*, one of the most widely used textbook accounts of the war, has only two *sentences* about the use of herbicides in Vietnam. Herring, *America's Longest War* (1979; repr. New York: McGraw-Hill, 1996), 168. Marilyn Young's *The Vietnam Wars* (New York: Harper-Perennial, 1991) devotes only about three paragraphs to the issue, two of which appear in the epilogue (see 82, 190–91, 325–26). Even Ronald Frankum's *Like Rolling Thunder: The Air War in Vietnam 1964–1975* has only eight pages on defoliation (Lanham, MD: Rowan and Littlefield, 2005).

5. For the traditional military approach, see William Buckingham, *Operation Ranch Hand: The Air Force and Herbicides in Southeast Asia, 1961–1971* (Washington: Office of Air Force History, 1982); and Paul Cecil, *Herbicidal Warfare: The Ranch Hand Project in Vietnam* (New York: Praeger, 1986). Cecil's is perhaps the best-known work on the subject. Cecil is a veteran of Operation Ranch Hand, however, and, like Buckingham, is overly sympathetic to the U.S. military and now somewhat outdated in his discussions of science and medical developments in the field. For the advocacy approach, see Fred A. Wilcox, *Waiting for an Army to Die* (1989; repr. Cabin John, MD: Seven Locks Press, 2011); and Jock McCulloch, *The Politics of Agent Orange* (Richmond, Australia: Heinemann Books, 1984). More recently Wilbur Scott's *Vietnam Veterans since the War: The Politics of PTSD, Agent Orange, and the National Memorial* (Norman: University of Oklahoma Press, 2004) has taken this school in a much more constructive direction, illuminating the issues of veterans through historical sociology.

6. The first of the significant new works, a dissertation by the anthropologist Diane Fox, who must be recognized as a major pioneer in the field of Agent Orange studies, is in many ways a work of advocacy as much as of scholarship, reconstructing powerful narratives about exposure to Agent Orange thirty years or more after the war. Second, the historian Thi Phuong-Lan Bui's dissertation combines environmental history with ethnography. Both Fox's and Bui's reconstructions of Agent Orange narratives, particularly those of Vietnamese women, are incredibly important contributions to the field. Both works, however, fail at times to account for change over time, and the different connotations Agent Orange has in the twenty-first century than it did in the 1960s and 1970s. Diane Niblack Fox, "'One Significant Ghost': Agent Orange, Narratives of Trauma, Survival, and Responsibility," (Ph.D. diss., University of Washington, 2007); and Lan Thi Phuong Bui, "When the Forest Became the Enemy" (Ph.D. diss., Harvard University, 2003).

7. David Zierler, *The Invention of Ecocide* (Athens: University of Georgia Press, 2011). With this major contribution, Zierler joins a group of historians whose work crosses disciplinary boundaries, combining diplomatic and environmental history with the history of science and technology. Several works in this area have helped illuminate the politics of chemicals, both foreign and domestic, including Thomas Dunlap's *DDT: Scientists, Citizens, and Public Policy* (Princeton: Princeton University Press, 1981) and, perhaps most famously, Edmund Russell's *War and Nature: Fighting Humans and Insects with Chemicals from World War I to Silent Spring* (New York: Cambridge University Press, 2001). Also see Richard Tucker and Edmund Russell, eds., *Natural Enemy, Natural Ally: Toward an Environmental History of War* (Corvallis: Oregon State University Press, 2004). As will be clear throughout much of the first three chapters, I am indebted to Zierler's work. His chapters on the development of herbicides, the initial militarization of them, and particularly the public debates over them in the late 1960s and early 1970s are quite simply the best work yet available on these subjects.
8. I have made use of a variety of collections from these archives, but one bears special mention: the Alvin L. Young Papers on Agent Orange at the National Agricultural Library (NAL). Young, who is an important part of the history of Agent Orange, is a retired air force colonel as well as a Ph.D. in environmental toxicology. I interviewed Young for this book and made heavy use of his archival materials, many of which have been digitized by the NAL staff. Young is a controversial figure, in part because of his connections both to the military and to Dow Chemical, for which he consulted on occasion, but also because he has long been seen by critics as an apologist for the use Agent Orange. As I attempt to make clear in the chapters that follow, Young is hardly the monster that many of his critics make him out to be. Regardless of how one feels about him, however, his papers are an invaluable source for historians, who have all but ignored them. Aside from being personally connected to a number of important developments in the history of Agent Orange, he has meticulously documented seemingly every study ever done related to herbicides, dioxin, and the effects of exposure on human health.
9. In addition to the works cited above, such as Scott's excellent *Vietnam Veterans Since the War*, see Christian Appy's *Working Class War* (Chapel Hill: University of North Carolina Press, 1993) and *Patriots: The Vietnam War Remembered by All Sides* (New York: Viking, 2003); Myra MacPherson's *Long Time Passing: Vietnam and the Haunted Generation* (New York: Anchor Books, 1984); Jerry Lembcke's *The Spitting Image: Myth, Memory, and the Legacy of Vietnam* (New York: New York University Press, 2000); and Patrick Hagopian's *The Vietnam War in American Memory: Veterans, Memorials, and the Politics of Healing* (Amherst: University of Massachusetts Press, 2009).
10. I am indebted to Clark Dougan for helping me to clarify the ideas and language included in this passage.
11. For a fascinating take on this, see Naomi Oreskes and Erik Conway, *Merchants of Doubt: How a Handful of Scientists Obscured the Truth on Issues from Tobacco Smoke to Global Warming* (New York: Bloomsbury Books, 2010).
12. For more on the "science wars" and the Sokal affair, see the special issue of *Lingua Franca* (July 1996) available at http://linguafranca.mirror.theinfo.org/9607/mst.html; Andrew Ross, ed., *Science Wars* (Durham: Duke University Press, 1996); and

Keith Parsons, ed., *The Science Wars: Debating Scientific Knowledge and Technology* (New York: Prometheus Books, 2003).

13. For more on the relationship between Agent Orange, herbicides, and chemical warfare, see Zierler, *The Invention of Ecocide*, chap. 8.
14. Despite this view, I have chosen not to place the word *victims* in quotation marks throughout the book. In part, this is the result of an attempt to keep the narrative as uncluttered as possible. It also reflects the uncertainty surrounding the ability to determine with any real accuracy who is and is not a victim of Agent Orange. Since the history presented here is in part the story of how individuals and groups who consider themselves victims of the chemical war made sense of their experience, it seemed unnecessary to continue to impose my verdict in specific cases.

CHAPTER ONE. ONLY YOU CAN PREVENT FORESTS

1. RAND interview twenty, ALY, series II, box 14, folder 128, p. 9. For background on the RAND interviews, see chapter 2, notes 32, 34. Because of the low altitude and relatively low speeds at which the C-123 aircraft were forced to fly to achieve effective spray techniques, strafing the area with ground fire became more common as the U.S./RVN forces retaliated against ground fire from the Front.
2. RAND interview twenty, 10.
3. Odd Arne Westad has been eloquent in writing about the Cold War as, in part, a "contest of narratives," in which the U.S. and Soviet camps attempted to win over the hearts and minds of other nations, above all, the recently decolonized nations of Asia and Africa. Westad, *The Global Cold War: Third World Interventions and the Making of Our Times* (Cambridge: Cambridge University Press, 2006).
4. The classic text on this is David Halberstam's *The Best and the Brightest* (New York: Random House, 1972). On the administration's technocratic mindset, also see James Gibson, *The Perfect War: The War We Couldn't Lose and How We Did* (New York: Random House, 1986), and the recent literature on modernization theory and its chief architect and proponent in the administration, Walt W. Rostow. Highly useful among the latter are Michael Latham, *Modernization as Ideology: American Social Science and "Nation Building" in the Kennedy Era* (Chapel Hill: University of North Carolina Press, 2000); David Engerman et al., eds., *Staging Growth: Modernization, Development and the Global Cold War* (Amherst: University of Massachusetts Press, 2003); Nils Gilman, *Mandarins of the Future: Modernization Theory in Cold War America* (Baltimore: Johns Hopkins University Press, 2004); and David Milne, *America's Rasputin: Walt Rostow and the Vietnam War* (New York: Hill and Wang, 2009).
5. The past decade has witnessed a veritable onslaught of literature reassessing Diem and the Republic of Vietnam from a variety of perspectives. Among the many recent works in this area are Phillip Catton, *Diem's Final Failure: Prelude to America's War in Vietnam* (Lawrence: University of Kansas Press, 2002); Seth Jacobs, *America's Miracle Man in Vietnam: Ngo Dinh Diem, Religion, Race and U.S. Intervention in Southeast Asia, 1950–1957* (Durham: Duke University Press, 2004); Edward Miller, "Vision, Power, and Agency: The Ascent of Ngo Dinh Diem, 1945–54," *Journal of Southeast Asian Studies* 35, no. 3 (October, 2004): 433–58; Jessica Chapman, "Staging Democracy: South Vietnam's 1955 Referendum to Depose Bao Dai," *Diplomatic History* 30, no. 4 (September, 2006): 671–703; James Carter, *Inventing Vietnam: The*

United States and State Building, 1954–1968 (New York: Cambridge University Press, 2008); and Matt Masur, "Exhibiting Signs of Resistance: South Vietnam's Struggle for Legitimacy, 1954–1960," *Diplomatic History* 33, no. 2 (April 2009): 293–313.

6. William Turley, *The Second Indochina War: A Concise Political and Military History*, 2nd ed. (New York: Rowan and Littlefield, 2009), 63. Turley's estimates are based on examinations of both American and Vietnamese sources. He notes that in 1964 U.S. intelligence put "the total number of revolutionary armed forces" in South Vietnam at 106,000, while the revolutionary forces themselves put the tally at 140,000. These numbers reflect main forces, regional forces, and guerilla militias. Throughout this and the next chapter I use the term *Front* interchangeably with NLF, to describe the National Liberation Front. Other recent works, including David Hunt's *Vietnam's Southern Revolution* (Amherst: University of Massachusetts Press, 2009), have argued that these terms should not be seen as interchangeable in that various decentralized liberation front groups appeared in the south prior to the formation of the NLF itself (see Hunt, 29–43, for more on this). Although this is an important distinction, it does not bear on the discussions offered here, so I continue to use the terms interchangeably for stylistic reasons.

7. For the "Plan of Action," see George Herring, ed., *The Pentagon Papers*, abridged ed. (New York: McGraw-Hill, 1993), 44–49.

8. Cecil, *Herbicidal Warfare*, 25.

9. William Buckingham, *Operation Ranch Hand: The Air Force and Herbicides in Southeast Asia, 1961–1971* (Washington: Office of Air Force History, 1982), 10. Also quoted in David Zierler, *The Invention of Ecocide* (Athens: University of Georgia Press, 2011), 59.

10. The history and scientific development of Agent Orange and other commercial and tactical herbicides have been well documented and need not be reconstructed fully here. For more on this background, see Buckingham, *Operation Ranch Hand*, 1–8; Cecil, *Herbicidal Warfare*, 1–20; and Zierler, *The Invention of Ecocide*, chap. 3. Zierler's chapter, "Agent Orange before Vietnam," is well researched and provides valuable historical and scientific context for understanding the development of these herbicides.

11. Zierler, *The Invention of Ecocide*, 38.

12. "Chemical Control of Vegetation," U.S. Army Biological Laboratories Report, December 11, 1962, AOLF, box 183, folder 1953, no. 1953-6, pp. 3–4.

13. For a partial list, see "Chemicals Used in Military Operations during the Vietnam War," Michigan Agent Orange Commission, available at www.gmasw.com/chemlist.htm.

14. Gale Peterson, "The Discovery and Development of 2,4-D," *Agricultural History* 41, no. 3 (July 1967), 243–53. Also quoted in Cecil, *Herbicidal Warfare*, 19n22.

15. *Vegetation Control Testing, Vietnam*, internal film produced by the United States Army, available in NARA RG 340, nos. 1184–88.

16. "Information from Department of Defense (DoD) on Herbicide Tests and Storage Outside of Vietnam, available at www.publichealth.va.gov/docs/agentorange/dod_herbicides_outside_vietnam.pdf.

17. *Vegetation Control Tests, Vietnam*.

18. Buckingham, *Operation Ranch Hand*, chap. 4, "Early Evaluations and Expanded Operations," esp. 46–55.

19. Zierler, *The Invention of Ecocide*, 52.
20. National Security Action Memorandum No. 115: "Defoliant Operations in Viet Nam," AOLF, box 194, no. 1945-248.
21. Buckingham, *Operation Ranch Hand*, 38.
22. Ibid., 49.
23. "Review and Evaluation of ARPA/OSD 'Defoliation' Program, in South Vietnam," April 1962, AOLF, box 187, no. 1945-670, p. 49.
24. Buckingham, *Operation Ranch Hand*, 50.
25. Ibid., 67.
26. Zierler, *The Invention of Ecocide*, 81–82. This debate connects directly to the larger debate over whether or not President Kennedy planned to withdraw U.S. forces from Vietnam had he not been assassinated in November 1963. Popularized perhaps most famously by films like Oliver Stone's *JFK* (1991) and Koji Masutani's *Virtual JFK: Vietnam If Kennedy Had Lived* (2008), the argument has received the most scholarly attention from the historian Fred Logevall in his *Choosing War: The Lost Chance for Peace and the Escalation of War in Vietnam* (Berkeley: University of California Press, 1999). James Blight, Janet Lang, and David Welch, eds., *Virtual JFK: Vietnam If Kennedy Had Lived*, (Lanham, MD: Rowman and Littlefield, 2010), serves as a companion piece to Masutani's film and bases a good deal of its evidence on Logevall's work.
27. Alvin Young, "The Military Use of Herbicides in Vietnam," in *Agent Orange and Its Associated Dioxin: Assessment of a Controversy*, ed. A. Young and G. M. Reggiani (Amsterdam: Elsevier Science Publishers, 1988), 20. As specific notes below will indicate, I am indebted in this section to the work of Young, whose exhaustive re-creation of the procurement, shipment, and distribution of tactical herbicides, first published in the chapter cited above, is based on a combination of military records and interviews with Ranch Hand personnel.
28. Young, *The History, Use, Disposition, and Environmental Fate of Agent Orange*, 42–44.
29. D. A. Craig, "Use of Herbicides in Southeast Asia," Historical Report, (San Antonio: Air Logistics Center, Kelly AFB, 1975), cited in Young, *The History, Use, Disposition, and Environmental Fate of Agent Orange*, 46.
30. Young, "The Military Use of Herbicides in Vietnam," 20.
31. Young, *The History, Use, Disposition, and Environmental Fate of Agent Orange*, 78.
32. Stellman et al., "Extent and Patterns of Usage of Agent Orange and Other Herbicides in Vietnam," *Nature* 422 (April 17, 2003), 685.
33. Young, "Military Use of Herbicides in Vietnam," 22.
34. MACV, "Lessons Learned: Accidental Herbicide Damage," AOLF, box 183, no. 1945-13.
35. "Defoliant Damage in Da Nang City," March 25, 1969, MACJ3-09, box 18.
36. Young, "Military Use of Herbicides in Vietnam," 20.
37. Young, T*he History, Use, Disposition, and Environmental Fate of Agent Orange*, 82.
38. In *History, Use, Disposition, and Environmental Fate of Agent Orange*, for instance, Young notes on pages 40–41 the safety procedures developed by air force staff; on page 80 he writes that "most of the personnel involved in the initial handling of the herbicides were Vietnamese military. However, a USAF flight mechanic was responsible for ensuring that the aircraft was properly loaded and the spray system functional."

39. The HERBS tapes are a database of records for Operation Ranch Hand spray missions, created in 1980 as a tool to help determine the likelihood of exposure of U.S. personnel on the ground in Vietnam. The strengths and limitations of these records are discussed in chapters 4 and 5. The original HERBS files can be located in the Alvin Young Collection at the National Agricultural Library in Beltsville, MD, series II, nos. 00108 and 00109. These files are also accessible digitally through the ALY collection website.
40. "Herbicide Damage to Vegetable Plots Vicinity Da Nang Air Base," October 31, 1968. MACVJ3-09, box 18.
41. Ibid., 3; Robert Darrow, "Report of Trip to Republic of Vietnam, 15 August–2 September, 1969," September 23, 1969, ALY, series II, no. 00271, pp. 17–18. For more on herbicide dumps and the difficulty in determining the history and risks associated with them, see Stellman et al., "Extent and Patterns of Usage of Agent Orange and Other Herbicides in Vietnam," 685.
42. "Memorandum: Defoliant Damage in Da Nang," Lt. Col. Harold Kinne to General Wheelock, March 29, 1969, MACVJ3-09, box 18, 2; ALY, series II, no. 00211.
43. "Herbicide Damage to Vegetable Plots Vicinity Da Nang Air Base," 3.
44. "Memorandum: Defoliant Damage in Da Nang," 8.
45. Darrow, "Report of Trip to Republic of Vietnam," 19.
46. "Memorandum: Reported Herbicide Damage, from Commanding General to First Coastal Zone Adviser, Naval Advisory Group," October 3, 1968, ALY, series II, no. 00209.
47. For more on long-term consequences, see chapter 5, and "Comprehensive Assessment of Dioxin Contamination in Da Nang Airport, Viet Nam: Environmental Levels, Human Exposure and Options for Mitigating Impacts," report by Hatfield Consultants, Vancouver, CA, November 2009. Available at www.hatfieldconsultants.com.
48. "Chemical Control of Vegetation," Fort Detrick Report, December 11, 1962, AOLF, box 191, no. 1953-6, p. 4
49. "Proceedings of the First Defoliation Conference," Fort Detrick, MD, June 1963, AOLF, box 184, no. 1945-256, p. 3.
50. "Proceedings of Second Defoliation Conference," Fort Detrick, MD, August 1964, AOLF, box 184, no. 1945-257; Young, *The History, Use, Disposition, and Environmental Fate of Agent Orange*, 32–33.
51. Young, *The History, Use, Disposition, and Environmental Fate of Agent Orange*, 34–35.
52. "Proceedings of Second Defoliation Conference," Fort Detrick, MD, August 1964, AOLF, box 184, no. 1945-258.
53. "Defoliation Operations in Laos," MACJ3-06, box 1, folder 201-1.
54. Stellman et al., "The Extent and Patterns of Usage of Agent Orange and Other Herbicides in Vietnam," 685.
55. "Memo: COMUSMACV to SECDEF, SECSTATE (December 1969)," MACJ3-06, box 1, folder 201-1. Andrew Wells-Dang has also written about herbicide use in Laos and Cambodia. His generally well-documented essays are located at www.ffrd.org/indochina/agentorange.html#Cambodia and Laos. Wells-Dang's work is an especially good starting point for examining the little that is known about the Cambodian missions. Some of the best evidence of herbicide use in Cambodia comes from accidental sprayings, particularly the diplomatic row over the spraying

of Kampong Cham in 1969. For more on that incident, in addition to Wells-Dang, see "Report of Cambodian Rubber Damage," December 11, 1969, ALY, series VI, subseries 1, no. 03124.
56. "The History of the US Department of Defense Programs for the Testing, Evaluation, and Storage of Tactical Herbicides," available at www.dod.mil/pubs/foi/reading_room/TacticalHerbicides.pdf.
57. "The History of the U.S. Department of Defense Programs for the Testing, Evaluation, and Storage of Tactical Herbicides," 26.
58. Alvin Young and Michael Newton, "Long Overlooked Historical Information on Agent Orange and TCDD following Massive Applications of 2,4,5-T-Containing Herbicides, Eglin Air Force Base, Florida," *Environmental Science and Pollution Resolution* 11, no. 4 (2004): 214–21.
59. A summary of these studies can be found in "Long Overlooked Historical Information on Agent Orange and TCDD."
60. Cited in Young, *The History, Use, Disposition, and Environmental Fate of Agent Orange*, 239, 258. The ALY collection—particularly series II and series VI, subseries 2 and 3—has extensive documentation on both the history of Eglin and the long-term environmental monitoring there.
61. "Chemical Defoliation of Northern Tree Species," Technical Memorandum 145, Fort Detrick, Maryland (October 1968), ALY, series I, no. 00030, pp. 2–3.
62. "The History of the U.S. Department of Defense Programs for the Testing, Evaluation, and Storage of Tactical Herbicides," 37–40.
63. Chris Arsenault, *Blowback: A Canadian History of Agent Orange* (Blackwood, Nova Scotia: Fernwood Publishing, 2009), 3, 10, 13. Louise Elliot, "In Depth: Agent Orange and Agent Purple," Canadian Broadcasting Corporation, available at www.cbc.ca/news/background/agentorange.
64. Arsenault claims there is "ample evidence" that U.S. scientists, for example, knew that the herbicides in questions were dangerous to humans and animals. Yet in support of this he offers only the "secret 1990 memo" from Adm. Elmo Zumwalt to the VA, summarizing the statement, long after the war was over, of one scientist, James Clary, who had worked with the herbicides. To this day, this is the only piece of evidence that any American scientific, military, or civilian leaders had any knowledge of the toxicity of Agent Orange prior to the late 1960s. I discuss the memo and the issue of prior knowledge by U.S. officials and the chemical manufacturers of herbicides at greater length in chapters 4 and 5.
65. "Memorandum: Substitution of Orange for Purple," November 7, 1963, AOLF, box 195, no. 1977-12.
66. Ibid.
67. Marilyn Young, *The Vietnam Wars, 1945–1975* (New York: Harper Collins, 1991), 191.
68. Buckingham, *Operation Ranch Hand*, 133.
69. Ibid.
70. "Letter, Dow to USAF RE: Herbicide Shortages," AOLF, box 183, folder 1922, nos. 1922-15, 1922-16.
71. Buckingham, *Operation Ranch Hand*, 133.
72. "Pesticides Used in Vietnam Hostilities and Their Use in Australian Agriculture: A Comparative Study," (August 1981), ANZ, ABQU632/W4552, file 2, "1981–82."
73. The report is based on the Australian War Memorial Series Records RG 181

(Canberra), "Herbicide Files." ANZ Foreign Affairs and Defence Committee Inquiries – Manufacturing of Agent Orange in New Zealand, ANZ, ABGX 4731, file B: Submissions to Committee.

74. Letter, John Moller, VVANZ, to MoH [Minister of Health], March 14, 1989, ANZ, Department of Health Records, Poisons-Substances-Agent Orange, ABQU632, W4552, file 5, "1983–1992." "Inquiry into Agent Orange," *The Dominion* (NZ), April 6, 1989, Foreign Affairs and Defence Committee Inquiries—Manufacturing of Agent Orange in New Zealand, ANZ, ABGX 4731, file A: Correspondence; John Dux and P. J. Young, *Agent Orange: The Bitter Harvest* (Sydney: Hodder and Stoughton-Australia, 1980).

75. IWD, Submission to Parliamentary Inquiry, ANZ ABGX 4731, file B: Submissions to Inquiry.

76. Ibid.

77. "Exports of New Zealand Products: Chemical Materials and Products—Weedkillers," Customs Department Records, ANZ, ABGX 4731, file B: Submissions to Inquiry.

78. Memo, Embassy to Defence Department, July 20, 1967, "Defence Department Memos," NZ Parliamentary Inquiry, ANZ, ABGX 4731, file A: Correspondence.

79. Memo, Turner to Minister of Defence, July 14, 1967, "Defence Department Memos," NZ Parliamentary Inquiry, ANZ, ABGX 4731, file A: Correspondence.

80. V. R. Johnson, Submission to Parliamentary Inquiry, ANZ, ABGX 4731, file B: Submissions to Inquiry.

81. B. R. Thomas, "Submission to Inquiry," ANZ, ABGX 4731, file B: Submissions to Inquiry.

82. "Correspondence, Rowe to 2,4,5-T Manufacturers," March 19, 1965, TTU, Burch Collection, box 1, folder 1, p. 1. This issue is discussed at length at the beginning of chapter 4.

83. Gibson, *The Perfect War*, 77–80.

84. For recent work on the strategic hamlet program and nation building in Vietnam, see Phillip Catton, "Counter-Insurgency and Nation Building: The Strategic Hamlet Programme in South Vietnam, 1961–1963," *International History Review* 21 (1999): 918–40; Carter, *Inventing Vietnam*, 117–28; David Elliot, *The Vietnamese War: Revolution and Social Change in the Mekong Delta, 1930–1975*, concise ed., (New York: M. E. Sharpe, 2007), 164–65, 175–76; and Gibson, *The Perfect War*, 82–86. For more recent work on the McNamara line and its antecedents, see Seymour J. Deitchman, "The 'Electronic Battlefield' in the Vietnam War," *Journal of Military History* 72, no. 3 (July 2008): 869–87; and Antoine Bousquet, *The Scientific Way of Warfare: Order and Chaos on the Battlefields of Modernity* (New York: Columbia University Press, 2009).

85. Quoted in Zierler, *The Invention of Ecocide*, 45.

86. Paul Cecil and Alvin Young, "Operation Flyswatter: A War within a War," *Environmental Science and Pollution Resolution* 15 (2008): 3.

87. Cecil, *Herbicidal Warfare*, 92–93.

88. "Memo: Transfer of C123 Aircraft," May 2, 1962, AOLF, box 184, no. 1945-293.

89. Cecil and Young, "Operation Flyswatter," 5–6.

90. Ibid.

91. Ibid., 6–7; Young, *The History, Use, Disposition, and Environmental Fate of Agent Orange*, 114; Arthur Westing, *Ecological Consequences of the Second Indochina War* (Stockholm: SIPRI/Stockholm International Peace Research Institute, 1976).

92. Darrow, "Report of Trip to Republic of Vietnam," 10. Also cited in Young, *The History, Use, Disposition, and Environmental Fate of Agent Orange*, 114.
93. Paul Cecil and Alvin Young have made this claim in several publications, often referencing their own work as evidence in support of this. More discussion of the issue of exposure to Agent Orange is offered below in chapter 5. For the Cecil and Young references, see "Operation Flyswatter," 6.
94. "Vietcong Fleeing Huge Forest Fire," *NYT*, April 14, 1968, A1.
95. Buckingham, *Operation Ranch Hand*, 142; "Vietcong Fleeing Huge Forest Fire."
96. "Forest Fire as a Military Weapon," Final Report of the U.S. Forest Service to the Advance Research Project Agency, June 1970, MACV-CORDS, box 102, p. A-6.
97. James Brown report, cited in Buckingham, *Operation Ranch Hand*, 49n11.
98. See, for instance, Lan Thi Phuong Bui, "When the Forest Became the Enemy: The Legacy of American Herbicidal Warfare in Vietnam." Lan describes the American pursuit of "hypervisibility" by the military in Vietnam, which led the Kennedy administration to approve herbicidal warfare, waging war against the jungle itself.
99. Buckingham, *Operation Ranch Hand*, 111.
100. "Operation Sherwood Forest: MACV Command History, April 6, 1964," available at TTU; (www.vietnam.ttu.edu/virtualarchive/), item no. 2130506001, pp. 16–17.
101. Buckingham, Operation *Ranch Hand*, 112.
102. "Operation Sherwood Forest," 17.
103. "Forest Fire as a Military Weapon," iii.
104. "Operation Pink Rose: Final Report," U.S. Department of Agriculture-Forest Service, May 1967, MACV-CORDS, box 102, p. 3. For more on Project EMOTE and weather modification, see Peter Caplan, "Weather Modification and War," *Bulletin of Concerned Asian Scholars* 6, no. 1 (January–March 1974); and "Weather Modification," Hearings before the Senate Subcommittee on Oceans and International Climate, Committee on Foreign Relations, March 20, 1974 (Washington: Government Printing Office, 1975).
105. "Forest Industries of the Republic of Vietnam" USFS Report, April 1967; Preliminary Report on the Forest Industry in Vietnam," USFS Report, January 1968, AOLF, box 205, no. 1993.
106. Howard Benton, "The Forest Service and Herbicides," U.S. Department of Agriculture Report (October 1970), 2; ALY, series VI, no. 03710; James Lewis, "Gallery: Smokey the Bear in Vietnam," *Environmental History* 11, no. 3 (July 2006), 599.
107. "Defoliation Project Request 20-69(U)," January 17, 1966, AOLF, box 187, no. 1945-701; "Memorandum: Burning of Chu Pong Mountain," March 15, 1966, AOLF, box 187, no. 1945-699.
108. "Summary Report: Operational Test Burning of Chu Pong Mountain," March 25, 1966, AOLF, Box 187, no. 1945-582.
109. "Summary Report: Operation Test Burning of Chu Pong Mountain [Attachments]," 4.
110. "Operation Pink Rose," 6, 9.
111. Ibid., 10–12, 86.
112. "Operation Pink Rose," 58.
113. Ibid., 74.
114. Ibid.
115. "Memorandum: Dr. W. G. McMillan to General Westmoreland," February 12, 1967, reprinted as appendix III in "Operation Pink Rose," 117.

116. "Weather Modification," Hearings before the Senate Subcommittee on Oceans and International Environment, March 20, 1974 (Washington, DC: GPO, 1975), 92–93.
117. Ibid., 88, 123.
118. Stellman et al., "The Extent and Patterns of Usage of Agent Orange," 685.
119. Lewis, "Smokey the Bear in Vietnam," 601.
120. Ibid., 603. For more on the role of USFS at home and abroad, see Lewis, *The Forest Service and the Greatest Good* (Durham, N.C.: Forest History Society, 2005); and Terry West, "USDA Involvement in Post World War II International Forestry," in *Changing Tropical Forests: Historical Perspectives on Today's Challenges in Central and South America,* ed. Harold Steven and Richard Tucker (Durham, N.C.: Forest History Society, 1992), 277–99.
121. "Prohibiting Military Weather Modification," Hearings before the Subcommittee on Oceans and International Environment, Senate Foreign Relations Committee, July 26–27, 1972 (Washington: Government Printing Office, 1974), 22.
122. Buckingham, *Operation Ranch Hand,* 109.

CHAPTER TWO. HEARTS, MINDS, AND HERBICIDES

1. "Memorandum, Director, USIA, to the President's Assistant for National Security Affairs," August 16, 1962, *FRUS,* 1961–63, vol. 2: Vietnam, 1962, no. 266.
2. Russell Betts and Frank Denton, "An Evaluation of Chemical Crop Destruction in Vietnam," report prepared for ARPA (Santa Monica, CA: RAND Corporation, October 1967); and Anthony Russo, "A Statistical Analysis of the U.S. Crop Spraying Program in South Vietnam," report prepared for ARPA (Santa Monica, CA: RAND Corporation, October 1967).
3. William Buckingham, *Operation Ranch Hand: The Air Force and Herbicides in Southeast Asia, 1961–1971* (Washington: Office of Air Force History, 1982); and Paul Cecil, *Herbicidal Warfare: The Ranch Hand Project in Vietnam* (New York: Praeger, 1986). For a sampling of advocacy works, see Fred Wilcox, *Waiting for an Army to Die* (Potomac, MD: Seven Locks Press, 1989), and Jock McCulloch, *The Politics of Agent Orange* (Richmond, Australia: Heinemann, 1984).
4. L. Goure, A. Russo, and D. Scott, "Some Findings of the Viet Cong Motivation and Morale Study," Memorandum RM-4911-2-ISA/ARPA, (Santa Monica: RAND Corporation, February 1966), 9.
5. Among the many works which have pointed out the problems with this distinction during the Vietnam War, some particularly effective examples are Chris Appy, *Working Class War: American Combat Soldiers in Vietnam* (Chapel Hill: University of North Carolina, 1993), 153–59; Sam Adams, *War of Numbers: An Intelligence Memoir* (South Royalton, VT: Steerforth Press, 1994); and David Elliot, *The Vietnamese War: Revolution and Social Change in the Mekong Delta, 1930–1975* (New York: M. E. Sharpe, 2003).
6. Cecil, *Herbicidal Warfare,* 26.
7. "Use of Defoliants in Vietnam," Memo, Assistant Director Far East to the Director USIA, November 17, 1961, *FRUS,* 1961–63, vol. 1: Vietnam, 1961, no. 265.
8. "Defoliant Operations in Vietnam," Memo, SECSTATE to JFK, November 24, 1961, AOLF, box 184, no. 1945-53, p. 1.
9. "Chemical Defoliant, Spraying Equipment and Aircraft, Vietnam," Memo, McNamara to JCS, November 7, 1961, cited in Buckingham, *Operation Ranch Hand,* 16n23.

10. For another example of DOS opposition to crop destruction, see the memo from Roger Hillsman to Averill Harriman, "Re: Crop Destruction in Vietnam," July 28, 1962, quoted in Buckingham, *Operation Ranch Hand*, 72n13.
11. "Use of Defoliants in Vietnam."
12. Buckingham, *Operation Ranch Hand*, 21.
13. "Telegram from the Department of State to the Embassy in Vietnam," December 14, 1961, *FRUS*, 1961–63, vol. 1: Vietnam, 1961, no. 361.
14. "Memorandum, Johnson to Rostow," November 17, 1961, *FRUS*, 1961–63, vol. 1: Vietnam, 1961, no. 264.
15. "Review and Evaluation of ARPA/OSD 'Defoliation' Program," AOLF, box 186, no. 1945-670, pp. 46–52.
16. Buckingham, *Operation Ranch Hand*, 69.
17. Memorandum: Hilsman to Harriman, July 28, 1962, *FRUS* 1961–63, vol. 2, no. 250. For more on Hilsman's views on herbicidal warfare and the early years of the Vietnam War in general, see his *To Move a Nation: The Politics of Foreign Policy in the Administration of John F. Kennedy* (Garden City, NY: Doubleday Books, 1967).
18. "Chemical Crop Destruction: Vietnam," JCS to McNamara, July 28, 1962, *FRUS* 1961–63, vol. 2, no. 251. Also cited in Buckingham, *Operation Ranch Hand*, 72.
19. "Chemical Crop Destruction: Vietnam."
20. "Chemical Crop Destruction, South Vietnam," McNamara to Kennedy, August 8, 1962, *FRUS* 1961–63, vol. 2, no. 262.
21. "Vietnam: Project for Crop Destruction," Rusk to JFK, August 23, 1962, *FRUS* 1961–63, vol. 2, no. 270. Also see Buckingham, *Operation Ranch Hand*, 74.
22. "Viet-Nam: Project for Crop Destruction." For the earlier drafts of language that made its way into the Rusk memo, see, in particular, "Crop Destruction," Deputy Assistant Secretary of State for Far Eastern Affairs Rice to Assistant Secretary of State Harriman," August 2, 1962, *FRUS* 1961–63, vol. 2, no. 256.
23. "Memorandum of the Substance of Discussion at a Department of State–Joint Chiefs of Staff Meeting," August 24, 1962, *FRUS* 1961–63, vol. 2, no. 271.
24. "Memorandum of Conversation: Situation in Viet Nam," September 19, 1962, *FRUS* 1961–63, vol. 2, no. 285.
25. "Memorandum of Conversation between President Kennedy and the Vietnamese Secretary of State at the Presidency (Thuan)," September 25, 1962, *FRUS* 1961–63, vol. 2, no. 292.
26. Buckingham, *Operation Ranch Hand*, 102.
27. Kastenmeier to Kennedy, March 7, 1963, AOLF, box 187, no. 1945-626.
28. Bundy to Kastenmeier, March 1963, AOLF, box 187, no. 1945-627. For more on the exchange, see Buckingham, *Operation Ranch Hand*, 82–83.
29. Christopher Sellers, "Body, Place and the State: The Makings of an 'Environmentalist' Imaginary in the Post–World War II U.S.," *Radical History Review* 1999, no. 74 (Spring 1999): 31–64.
30. For two recent examples of this ethnographic approach, see Diane Niblack Fox, "One Significant Ghost": Agent Orange, Narratives of Trauma, Survival, and Responsibility" (Ph.D. diss., University of Washington, 2007); and Lan Thi Phuong Bui, "When the Forest Became the Enemy" (Ph.D. diss., Harvard University, 2003).
31. The Chieu Hoi program began in 1963 as a tool to encourage defections from the NLF to the RVN. For more on the Chieu Hoi program, see J. A. Koch, "The Chieu

Hoi Program in South Vietnam, 1963–1971," report prepared for ARPA (Santa Monica, CA: RAND Corporation, January 1973).

32. The interviews were carried out by RAND employees, the vast majority of whom were South Vietnamese academics, and many of whom were also born in the North. Despite the limitations and problems associated with these sources, they shed light on the reactions among the NLF, the party leadership, and local villagers to the use of herbicides. The interviews in Series H (for Herbicides), mostly conducted in 1966, contain reactions from both captured soldiers and defectors under the Chieu Hoi program. For more on the interview program and related reports, see W. Phillips Davison, "User's Guide to the RAND Interviews," report prepared for ARPA (Santa Monica, CA: RAND Corporation, 1972), available on the RAND website, www.RAND.org/pubs/reports/R1024/ (last accessed February 10, 2007); and David Hunt, *Vietnam's Southern Revolution: From Peasant Insurrection to Total War* (Amherst: University of Massachusetts Press, 2009), especially his extended discussion of the RAND interviews, "Appendix A: The Uses of a Source," 225–34.

33. The interviews include several subjects from, among other places, My Tho province in the Mekong Delta, Binh Dinh province in the Central Coastal region, and Tay Ninh province, in the Parrot's Beak region near the Cambodian border.

34. RAND interview twenty, ALY, series II, box 14, folder 128, p. 9. Note: All RAND interviews here refer to Series H of the RAND collection. The full series is located in series II, boxes 13 and 14 of the ALY collection, in sequential folders 108–47. I will hereafter refer to these as "RAND interview X," followed by the page number; RAND interview five, 9; RAND interview eleven, 6; RAND interview twelve, 8. Several research libraries also have the full RAND series available on microfiche.

35. "VC Preventive Measures Against Chemical Warfare," MACJ3-09, box 1, folder 1, pp. 28–29.

36. NLF Central Committee A.302, "Directive to Executive Agents Committee Provincial Level Cadres on Poisonous Substances," August 1962, HVWM (Captured Documents Collection, Joiner Center, University of Massachusetts-Boston), File W, cited in Bui, "When the Forest Became the Enemy," 45n82. The recipe for purging suggested in this directive, according to Bui, was to eat "corn, pumpkin, green tea, manioc, two spoons charcoal powder with one liter of water." Another suggested two spoonfuls of water mixed with ash, followed by one spoonful of iron sulfate and water every five to fifteen minutes until vomiting commenced.

37. Bui, "When the Forest Became the Enemy," 49.

38. Military Section of Chau Thanh District, "Directive to Village Level of NLF on Actions to be taken to Counter Enemy Defoliation Campaign," May 1962, HVWM, file W, cited in Bui, "When the Forest Became the Enemy," 44n81.

39. RAND interview twenty-one, 6.

40. RAND interview eight, 6, 8; RAND interview twenty-three, 10.

41. RAND interview five, 9.

42. RAND interview five, 9; RAND interview eleven, 6; RAND interview seven, 3.

43. RAND interview two, 2.

44. RAND interview thirty-nine, 4; RAND interview thirteen, 5–6. This finding was confirmed in the RAND morale study and is discussed in greater detail below.

45. RAND interview seven, 6.

46. RAND interview eighteen, 9.

47. "Effectiveness of Defoliants and Chemicals Dropped by Allied Aircraft Over VC Controlled Areas in Quang Ngai and Binh Dinh Provinces," NIC Interrogation Report, June 1967, MACJ3-06 box 18, folder 501; "Intelligence Reports," MACJ3-06 box 2, folder 201, p. 1-F, 1-G.
48. Hunt, *Vietnam's Southern Revolution*, 117–19.
49. Elliot, *The Vietnamese War*, 1174, 1275.
50. RAND interview seven, 2; RAND interview thirty-nine, 3.
51. RAND interview ten, 10; RAND interview thirty-nine, 6; RAND interview thirty-four, 3.
52. RAND interview sixteen, 3.
53. Cited in "Crop Destruction Operations in RVN During CY 1967," Scientific Advisory Group Working Paper 20-67, AOLF, box 189, folder 1945-685, p. D2.
54. This is a pattern reinforced in other studies of the village war, such as Eric Bergerud's *Dynamics of Defeat: The Vietnam War in Hau Nghia Province* (Boulder: Westview Press, 1991), 70–71; Jeffrey Race, *War Comes to Long An: Revolutionary Conflict in a Vietnamese Province* (Berkeley: University of California Press, 1972); and Hunt, *Vietnam's Southern Revolution*, esp. chaps. 10–12.
55. RAND interview thirty-four, 3.
56. RAND interview eighteen, 3.
57. RAND interview eighteen, 2.
58. "VC Preventive Measures against Chemical Warfare," 28–31.
59. "MACV Bulletin no. 17391: Enemy Documents," October 20, 1968, MACJ3-06, box 18, folder 501.
60. RAND interview thirty-four, 2; RAND interview eighteen, 4.
61. RAND interview, eighteen, 4.
62. For more on this, see Robert Frankum, *Like Rolling Thunder* (Boston: Rowman and Littlefield, 2005), 54, 58.
63. RAND interview five, 9.
64. RAND interview thirteen, 5, 8.
65. RAND interview twenty, 15.
66. RAND interview twelve, 10.
67. RAND interview thirty-nine, 6.
68. RAND interview twelve, 9.
69. RAND interview twelve, 11.
70. Bui, "When the Forest Became the Enemy," 44.
71. For more on the strategic hamlet program, see chapter 1, note 85.
72. This is also part of the area explored in painstaking detail by Elliot in *The Vietnamese War* and by Hunt in *Vietnam's Southern Revolution*.
73. "Visit to Dinh Tuong Province for Defoliation Target Reconnaissance," April 7, 1964, MACJ3-09, box 4, pp. 1–2.
74. "Memorandum: Proposed Defoliation in Dinh Tuong," May 6, 1964, MACJ3-09, box 4.
75. "Notice: Announcement of the Dinh Tuong Province," April 9, 1964, MACJ3-09, box 4.
76. "Letter from Province Chief Le Qui Ky, April 9, 1964," MACJ3-09, box 4.
77. "USOM Weekly Report, Dinh Tuong Province, May 15, 1964," MACJ3-09, box 4.
78. Ibid.

79. "Memorandum: Proposed Defoliation in Dinh Tuong"; letter from Province Chief Le Qui Ky, April 9, 1964, MACJ3-09, box 4.
80. "Civilian Affairs and Psyops," MACJ3-09, box 2, folder 2-25. Also see Cecil, *Herbicidal Warfare*, 39.
81. "Brother Nam Has Questions about Defoliant Chemicals," MACJ3-09, box 6, folder 20-59. Other cartoons in this series vary the story about Nam's family and ambushes by the NLF but conclude with nearly identical versions of the last three frames about compensation and Nam's newfound understanding of the safety of defoliants. I am indebted to Molly O'Connell and Diane Fox for their help with the translation.
82. "Province Chief's Leaflet, No. 2," MACJ3-09, box 4, folder 20-18.
83. For more on the Cambodia incident, see Typescript: "Report of Cambodian Rubber Damage" (December 11, 1969), ALY, series II, no. 03124; box 10, "Cambodian Herbicide Spraying and Crop Destruction" in MACJ3-06, box 2, folder 2-25; and Andrew Wells-Dang, "Agent Orange in Cambodia: The 1969 Defoliation in Kampong Cham," available at www.ffrd.org/Agent_Orange/cam1969.htm.
84. "COMUSMACV to JCS, RE: Claims for Crop and Tree Damage," January 2, 1970, MACJ3-06, box 6, folder 2. For the full text of the Foreign Claims Act, see 10 USC §2734.
85. "Claims against the United States," November 30, 1968, MACJ3-06, box 18.
86. "Herbicide Policy Review," August 20, 1968, ALY, series VI, no. 3120, pp. 32–33. The value of the RVN piaster fluctuated wildly during the brief existence of the republic and served as an ongoing source of tension between U.S. and RVN leaders. In 1970 the Nixon administration finally convinced the RVN to revalue the currency relative to the U.S. dollar to combat rampant inflation, from 118 piasters to 1 U.S. dollar to 275 piasters. It is thus hard to compare monetary values across this period. For present purposes I estimate that during the period 1967–70, 1 U.S. dollar was equivalent to about 200 piasters, roughly the average of the two formal points of 118 and 275.
87. "GVN Herbicide Claims Program," "Payment of Herbicide Claims," and "Chart for Indemnification Payments," MACJ3-06, box 6, folder 3; "COMUSMACV to JCS, RE: Claims for Crop and Tree Damage," 2.
88. "An Evaluation of Chemical Crop Destruction in Vietnam," 17–18.
89. "Popular Opinions about Defoliation in Long Khanh Province," CORDS Report, December 1967," MACJ3-06, box 18, p. 3.
90. "Policy Review of the US/GVN Herbicide Program," December 28, 1967, MACJ3-06, box 10.
91. "Herbicide Policy Review," 31–32.
92. The article, written by Jim Lucas, appeared in the *WP* on May 27, 1964. The citation comes from Buckingham, *Operation Ranch Hand*, and 94n11.
93. Buckingham, *Operation Ranch Hand*, 96 (emphasis added).
94. "Request for Defoliation," March 11, 1964, MACJ3-06, box 4, folder 2-20.
95. "Evaluation of Herbicide Operations in the Republic of Vietnam as of 30 April 1966," Combined Intelligence Center, Vietnam (July 1966), ALY, series VII, box 113, no. 3115, p. 3 (hereafter cited as CICV Report [1966]).
96. Cecil, *Herbicidal Warfare*, 39.
97. CICV Report (1966), 4–5.
98. Ibid., 6. Also cited in Buckingham, *Operation Ranch Hand*, 121.

99. CICV Report (1966), 7.
100. Ibid.
101. Ibid., 9.
102. Ibid.
103. Ibid., 13.
104. "Viet Cong Motivation and Morale Study," ix.
105. Ibid., 7.
106. Ibid., 9–10.
107. Mai Elliot, *RAND in Southeast Asia: A History of the Vietnam War Era* (Santa Monica, CA: RAND Corporation, 2010), viii, x. While Elliot's study was commissioned and published by RAND, it is nevertheless a critical and probing history, the best study to date of RAND's involvement in the Vietnam War.
108. "An Evaluation of Chemical Crop Destruction in Vietnam," ix.
109. Ibid., xi.
110. Ibid., xii.
111. Ibid., 14.
112. Ibid., 17.
113. Ibid., 18.
114. "A Statistical Analysis of the U.S. Crop Spraying Program in Vietnam," ix, x; 20, 23, 32.
115. Buckingham, *Operation Ranch Hand*, 133–34.
116. "Herbicide Crop Destruction Operations in South Vietnam," COMUSMACV to CINCPAC, December 1967, MACJ3-06, box 10.
117. MACSA/Griggs to COMUSMACV, December 14, 1967, AOLF, box 189, folder 1945-694, p. 1 (emphasis added to both).
118. Elliot, *RAND in Southeast Asia*, 227–28. The Russo quote comes from an interview conducted by Elliot in 2004.
119. "Crop Destruction Operations in RVN during CY 1967," 19 (emphasis added).
120. "Effectiveness of Defoliants and Other Chemicals Dropped by Allied Aircraft over VC Controlled Areas in Quang Ngai and Binh Dinh Provinces," NIC Report, April 12, 1967, MACJ3-06, box 18, folder 501.
121. "Attitudes Toward Defoliation: Anti-Americanism," Survey Team Report from Cu Chi District, Hau Nghia Province, May 19, 1967, MACJ3-06, box 18, folder 501.
122. "MACV Bulletin No. 8796: Enemy Documents," January 5, 1968, MACJ3-06, box 18, folder 501.
123. "DIA Interrogation Report: Defoliation Mission in Binh Duong Province," April 14, 1969, AOLF, box 192, no. 1962, p. 4. For other examples supporting this finding, see, in the same record group, "DIA Interrogation Report: Effects of Defoliation in Hau Nghia Province," July 11, 1969, p. 3; and "Crop Destruction in Khanh Hoa Province, June 14, 1969," p. 3.
124. "CG, III Corps to COMUSMACV," November 1967, MACJ3-06, box 10.
125. Ibid.
126. "CG, IFFORCEV to COMUSMACV," November 1967, MACJ3-06, box 10.
127. R. N. Pesut and W. P. Virgin, "Defoliation-Incidents Correlation Study," Report to ARPA/Project Agile, April 1, 1967, p. iv. As of the summer of 2009, this document had been withdrawn from the recently declassified CORDS-Analysis Division files at NARA II (RG 472/270/78/33-35, box 81) for reasons that the archivists at College Park were unable to explain to me. However, the same document was, at the time, available

for purchase as microfilm from the Defense Technical Information Center (DTIC. mil, Accession Number AD0383296), and listed as "approved for public release" as late as April 2010. The citations here refer to the page numbers in the microfilm.

128. "Defoliation-Incidents Correlation Study," 54, 57.
129. JSCM-719-67, JCS to SECDEF, "Review of Crop Destruction Operations in Vietnam," December 29, 1967, AOLF, box 189, no. 1945-689, pp. 1–2.
130. Denton to Tom Thayer, Director of SEA Programs, Office of Assistant SECDEF for Systems Analysis, February 14, 1968, AOLF, box 189, no. 1945-686.
131. Elliot, *RAND in Southeast Asia*, 236.
132. Ibid., 436–41. For more on the relationship between Russo, Ellsberg, and the Pentagon Papers, see Daniel Ellsberg, *Secrets* (New York: Viking Books, 2002), 290–301.
133. For a sampling of these memos, see George Herring, ed., *The Pentagon Papers*, abridged ed. (New York: McGraw Hill, 1993), 194–207. For McNamara's own take on his departure, see his *In Retrospect: The Tragedy and Lessons of Vietnam* (New York: Random House, 1995), 311–17. McNamara's account begins tellingly, "I do not know to this day whether I quit or I was fired." The book makes no mention of Agent Orange, Operation Ranch Hand, or crop destruction.
134. "Herbicide Policy Review," August 20, 1968, ALY, series VI, no. 03120, p. 17.
135. "Herbicide Policy Review," 1.
136. For more on the decision and military opposition, see Buckingham, *Operation Ranch Hand*, chap. 9, "Ranch Hand Ends Its Work."
137. "Corona Harvest: Defoliation Operations in Southeast Asia, A Special Report," (March 1970) AOLF, box 188, no. 1945-180, p. 26.

CHAPTER THREE. INCINERATING AGENT ORANGE

1. "Press Release: Home Use of 2,4,5-T Suspended," ALY, series VIII, subseries 1, no. 05170; "Final Environmental Statement: Disposition of Orange Herbicide by Incineration," Department of Air Force (November 1974), ALY, series II, box 3, no. 0094, p. 1 (hereafter "Final EIS"); "U.S. Curbs Sales of a Weed Killer," *NYT*, April 16, 1970.
2. "Table: Agent Orange in Vietnam," ALY, series II, box 197, no. 05548. One of the major problems in assessing the levels of dioxin exposure as a result of contact with Agent Orange is that the dioxin levels varied so widely from producer to producer. Dow Chemical, which held the second largest market share for the product behind Monsanto, produced Agent Orange that contained comparatively low levels of dioxin. Dow supplied over four million gallons for military use between 1965 and 1970, the average dioxin content of which was estimated at between .05 and .24 ppm, according to military records. Conversely, Diamond Alkali produced fewer than seven hundred thousand gallons, but its average dioxin content was between 8 and 14 ppm, well beyond acceptable risk levels. Peter Schuck, *Agent Orange on Trial* (Cambridge: Harvard University Press, 1986), 86–87.
3. These estimates are also taken from Stellman et al., "The Extent and Patterns of Usage of Agent Orange and Other Herbicides in Vietnam," *Nature* 422 (April 17, 2003), 681–87. For more on Russell Bliss, see Alistair Hay, *The Chemical Scythe: Lessons of 2,4,5-T and Dioxin* (New York: Plenum Press, 1982), 113–32; and Aaron Wildavsky and Brendan Swedlow, "Dioxin, Agent Orange, and Times Beach," in *But Is It True?: A Citizen's Guide to Health and Safety Issues,* ed. Aaron Wildavsky (Cambridge: Harvard University Press, 1995), 374–94.

4. "Dioxin: Quandary for the 80s," *St. Louis Post-Dispatch*, Special Section, November 14, 1983, 6. Barry Commoner, "The Political History of Dioxin," keynote address at the Second Citizens' Conference on Dioxin, St. Louis, Missouri, July 30, 1994, available at www.greens.org/s-r/078/07-03.html.
5. Christopher Sellers, "Body, Place and the State: The Makings of an 'Environmentalist' Imaginary in the Post–World War II U.S.," *Radical History Review* 74 (Spring 1999), 58.
6. Zierler, *The Invention of Ecocide* (Athens: University of Georgia Press, 2011), esp. chaps. 6–8.
7. Zierler, *The Invention of Ecocide*, chap. 8.
8. "Final EIS," 1.
9. "Disposition of Herbicide Orange," March 12, 1971, AOLF, box 183, no. 1945-237.
10. "The Geneva Protocol of 1925," hearings before the Committee on Foreign Relations, United States Senate, 92nd Congress, First Session, March 1971 (Washington: Government Printing Office, 1972). For the definitive account of this debate, see Zierler, *The Invention of Ecocide*, chap. 8.
11. "Notes on JCS Plan for Disposal of Herbicides," AOLF, box 195, no. 1971-36, p. 1.
12. Buckingham, *Operation Ranch Hand*, 184.
13. For more on the transformations in environmentalist thinking and environmental policy in the postwar world, see Craig Colten and Peter Skinner, *The Road to Love Canal: Managing Industrial Waste Before EPA* (Austin: University of Texas Press, 1996); and J. R. McNeil's *Something New Under the Sun: An Environmental History of Twentieth Century World* (New York: Norton, 2000).
14. *U.S. Environmentalism since 1945: A Brief History with Documents*, Steven Stoll, ed. (Boston: Bedford St. Martin's, 2007), 19, 23. Joan Hoff's *Nixon Reconsidered* (New York: Basic Books, 1994) is a well-researched attempt to recover this and other often-overlooked legacies of the Nixon administration. For Hoff's take on Nixon's environmental record, see pages 22–27. I am indebted to David Kieran and Jeremi Suri for helping me to clarify the significance of Nixon's actions and motivations.
15. Rick Perlstein, *Nixonland: The Rise of a President and the Fracturing of America* (New York: Scribner, 2008), 460–61, 517.
16. "Disposition of Herbicide Orange: Memorandum, SECDEF to Chair, JCS," September 13, 1971, AOLF, box 183, no. 1945-242.
17. "Fact Sheet: Status of Herbicide Orange," March 7, 1972, AOLF, box 195, no. 1971-35.
18. According to Col. Alvin Young, *Pacer* is a "USAF term for logistical movements"; Pacer IVY thus referred to the movement of the inventory (InVentorY) of Orange. Pacer HO referred to the final shipment and incineration of Herbicide Orange (HO). Alvin Young, *The History, Use, Disposition, and Environmental Fate of Agent Orange* (New York: Springer Books, 2009), 123.
19. Ibid., 128.
20. Quoted in ibid., 131–32.
21. Ibid., 135. For more on Hurricane Camille, see the U.S. Department of Commerce Preliminary Report on Hurricane Camille (November 1969), available at www.nhc.noaa.gov/archive/storm_wallets/atlantic/atl1969-prelim/camille/TCR-1969Camille.pdf; and Ernest Zebrowski and Judith Howard, *Category Five: The Story of Camille, Lessons Learned from America's Most Violent Hurricane* (Ann Arbor: University of Michigan Press, 2005).

22. Alvin Young, "Dilemma for Disposal of Agent Orange," ALY, series VI, subseries 3, no. 03899, pp. 3–4.
23. "Summary of Report on Disposal of Herbicide Orange in Utah," Lyman Olson, Director of Public Health, to Governor Rampton, April 30, 1973; AOCRF, box 1, folder 3; Young, "Dilemma for Disposal of Herbicide Orange"; "Dioxin in Utah and Vietnam," ALY, series VI, subseries 3, no. 03085; Rampton to Billy Welch, May 7, 1973, AOCRF, box 1, folder 3.
24. A concise sample of the responses to the proposed action by the State of Texas are included in the Final EIS, appendix L, esp. pp. L9–L20. Additional memos from a variety of state agencies can be found in "Technical Report: Incineration of Herbicide Orange by Incineration," appendix D5, pp. 65–70. In an interesting parallel to this story, the Deer Park facility did become the final point of reprocessing and destruction of the Americans' supplies of napalm in 2001, although again not without controversy, as local residents and others in the path of the napalm inventory's transit from San Diego raised safety concerns. For more on this, see "Relegating Napalm to Its Place in History," *Los Angeles Times*, April 1, 2001.
25. William Spain, "Yes, In My Backyard: Tiny Sauget, Illinois, Likes Business Misfits," *Wall Street Journal*, October 3, 2006. According to its official, published corporate history, the founder John Queeny named both the town and the company Monsanto after his wife, Olga Mendez Monsanto. Dan J. Forrestal, *The Story of Monsanto: Faith, Hope, and $5000: The Trials and Triumphs of the First 75 Years* (New York: Simon and Schuster, 1977), 13.
26. "Subject: Disposition of Orange by Incineration," Douglass Thornberry to Robert Seamans (March 1, 1972), Final EIS, appendix L, p. 16.
27. "Draft Environmental Impact Statement Reply," EPA to USAF, March 8, 1972, included in appendix C, "Technical Report: Incineration of Agent Orange," USAF Environmental Health Laboratory, March July 1972, ALY, series II, no. 00179, pp. 56–60. The stench in the air over the Sauget area was well known to locals and would later be immortalized in the song "Sauget Wind," by the southern Illinois-based band Uncle Tupelo ("They're poisoning the air for personal wealth") on their release *Still Feel Gone* (Rockville Records, 1991).
28. Young, "Dilemma for Disposal of Herbicide Orange," 5.
29. "Ranchers Would Use Banned Army Defoliant," *San Antonio Light*, April 17, 1972; located in AOCRF, box 1, folder 6; Armed Services Procurement Regulation 24-101.2, p. 2407; located in AOCRF, box 1, folder 6.
30. Correspondence: Newton to Young, September 22, 1972, AOLF, box 183. For more on Newton's work, see "Oregon Reforestation Trials with Herbicide Orange, Reports, Correspondence and Pictures, 1971–1980," ALY, series VI, subseries 3, no. 03759.
31. Col. C. G. MacDonald to Bob Sikes, AOCRF, box 15, folder 9.
32. Correspondence, AID to Blue Spruce Company; "Memorandum for the Record, USAF Supply and Maintenance, April 28, 1972," AOCRF, box 4, folder 2. The African weed control proposal is discussed in "Domestic Use of Herbicides," Charles Minarik to Commander, SAAMA, December 6, 1968, AOCRF, box 1 folder 6.
33. "Domestic Use of Herbicides."
34. Richard Patterson, "Modification and/or Destruction of Defoliants," presentation to Armed Forces Pest Control Board, March 1971, AOCRF, box 1, folder 1.

35. Ibid.
36. I am indebted to David Zierler for helping me to clarify several parts of this chapter, including this powerful, symbolic connection.
37. Final EIS, 2.
38. Ibid.
39. Ibid., 14.
40. This information is culled from the various drafts of the EIS issued by the air force prior to, during, and after operation Pacer HO. The most thorough and updated summary of Johnston Atoll can be found in the draft EIS of January 2004 produced for the termination of the air force mission on Johnston. United States Air Force, January 2004. The draft EIS had been available online through the EPA (epa.gov) through 2004 but appears to have been removed as of early 2011. A copy of the draft can be found in the Hawaii State Library, Honolulu (call number H 623.047 EA).
41. Draft EIS, Termination of the Air Force Mission, Johnston Atoll, 3–48. The DTRA was actually formally established within the DOD in 1998, consolidating a number of agencies that had long worked on chemical, nuclear, and biological weapons testing and monitoring. The agency tasked with the initial inspections on Johnston would most likely have been the Defense Atomic Support Agency, or DASA.
42. Draft EIS, Termination of the Air Force Mission, Johnston Atoll, 3–49.
43. For more on the ship's specifications, see "At Sea Incineration of Herbicide Orange Onboard the M/T Vulcanus," ALY, series 6, subseries 3, no. 03967.
44. "Marine Protection, Research and Sanctuaries Act of 1972," 33 U.S.C. §1401, text available at www.epa.gov/history/topics/mprsa/index.htm.
45. *Ocean Incineration: Its Role in Managing Hazardous Waste*, U.S. Office of Technology Assessment, OTA-O-313 (Washington: Government Printing Office, 1986), 181; Young, *The History, Use, Disposition, and Environmental Fate of Agent Orange*, 143.
46. 50 U.S.C. §1518, available at http://uscode.house.gov/download/pls/50C32.txt.
47. "The Applicability of 50 U.S.C. 1512–1518 to the Sale or Destruction of Agent Orange," Memorandum for the Assistant General Counsel, Installations, USAF Office of General Counsel, November 26, 1974, AOCRF, box 9, folder 6.
48. Final EIS, appendix L, p. L3.
49. "Harrison County Board of Supervisors Resolution Demands Removal of Herbicides from Naval Base (April 1974), AOCRF, box 2, folder 14; for a sample of similar letters and the mayor's letter, see AOCRF, box 2, folder 14.
50. Inouye to Seamans, September 6, 1973; AOCRF, box 8, folder 3; Col. Howes to Inouye, September 26, 1973; AOCRF, box 8, folder 2.
51. Young, *The History, Use, Disposition, and Environmental Fate of Agent Orange*, 144.
52. "Herbicide Orange Site Treatment and Environmental Monitoring: Summary Report and Recommendations for Naval Construction Battalion Center, Gulfport, MS," November 1979, ALY, series II, no. 0187, 11.
53. Young, *History, Use, Disposition, and Environmental Fate of Agent Orange*, 145–47. Young's source regarding the medical information is J. A. Calcagni, "Evaluation of Medical Examinations on Workers with Possible Exposure to Herbicide Orange during Project Pacer HO," USAF Occupational and Environmental Lab Report, November 1979, ALY, series III, subseries 3, no. 01854.
54. Instructions included in "Gulfport, MS Survey: Notes, 1976," ALY, series VI, subseries 3, no. 03932.

55. Final EIS; see also "Reports, Correspondence, Notes: Project Pacer HO, 1972–1982." ALY, series II, box 5, no. 03787.
56. "Herbicide Orange Motion Picture Records," NARA RG 341.5.2.
57. Firth to Proxmire, August 17, 1973, AOCRF, box 2, folder 13.
58. "A. Orange on Johnson [sic] Atoll," TTU, Bud Harton Collection, accessed electronically, no. 168300010714.
59. "Memorandum: Drum Disposal," May 1, 1973, ALY, series VI, subseries 3, no. 03804.
60. "Comments on Draft Environmental Statement—Disposition of Orange Herbicide by Incineration," Final EIS, appendix L, pp. 21–32.
61. Ibid., 25.
62. Thalken's notes are included in three images housed in the "Miscellaneous Records Related to the Herbicide Orange Project" collection, NARA RG 341/190, box 5.
63. Young, *History, Use, Disposition, and Environmental Fate of Agent Orange*, 150.
64. "Herbicide Orange Site Treatment and Environmental Monitoring: Summary Report and Recommendations," esp. pp. iii–iv, 27–30; Air Force Engineering and Services Laboratory, "Herbicide Orange Monitoring Program: Interim Report," December 1982, available at handle.dtic.mil/100.2/ADA143260.
65. Young, *History, Use, Disposition and Environmental Fate of Agent Orange*, 296.
66. "Land Based Environmental Monitoring at Johnston Island: Disposal of Herbicide Orange: Final Report for Period 1977–1978," USAF/OEHL, ALY, series VI, subseries 3, no. 03984; "Johnston Island Herbicide Orange Storage Site Monitoring Project," USFA/OEHL, ALY, series VI, subseries 3, no. 04010; "Progress Report to USAF on Degradation of Herbicide Orange and TCDD in Herbicide Sites on Johnston Island and Gulfport, Mississippi," Flammability Research Center, University of Utah (May 17, 1979), AOLF, box 19; Young, *History, Use, Disposition, and Environmental Fate of Agent Orange*, 272–76, 297–99; "Draft Environmental Impact Statement, Termination of the Air Force Mission Johnston Atoll Airfield."
67. Wildavsky and Swedlow, "Dioxin, Agent Orange, and Times Beach"; also see "Dioxin: Quandary for the '80s."
68. "Dioxin, Quandary for the '80s," 6.
69. Quoted in "Dioxin, Agent Orange and Times Beach," 115.
70. On the veterans' lawsuit and the legal fallout from the case, see Schuck, *Agent Orange on Trial*. I discuss the case in greater detail in chapter 4.
71. "Jury Awards $58 million to 47 Railroad Workers Exposed to Dioxin," *NYT*, August 27, 1982.
72. "Town Struggles with Toxic Legacy," *WP*, January 10, 1983. Also quoted in Wilbur Scott, *Vietnam Veterans since the War: The Politics of PTSD, Agent Orange, and the National Memorial* (Norman: University of Oklahoma Press, 2004), 174.
73. Quoted in "Poison or Not?" *WP*, January 14, 1983. Reprinted in Scott, *Vietnam Veterans since the War*, 174.
74. Bliss quoted in the documentary *Times Beach: Toxic Ghost Town* (Oakland: Video Project Productions, 1994).
75. "Dioxin Threat Puts U.S. Under Fire," *NYT*, December 26, 1982; "In Dioxin-Tainted Town, No 'Welcome' Signs," *NYT*, January 10, 1983; "Town Struggles with Toxic Legacy."
76. "Orange Lining in Dioxin Cloud," *St. Louis Post-Dispatch*, February 28, 1983. "Environmental News," December 8, 1982 (Washington: EPA, 1982), ALY, series IV, box

85, nos. 02113, 02114; "Town Struggles with Toxic Legacy," *St. Louis Globe*, December 9, 1982.

77. "Dioxin: Quandary for the '80s," 3. Examples of conflicting headlines in the *Post-Dispatch* in the introduction to this section include "FDA Reports Finding Dioxin in Cow Near Verona" (January 21, 1983), and "First Analysis Wrong: No Dioxin After All in Cow's Tissue" (February 1, 1983); "Scientist Calls Times Beach Safe, Says He'd Move There" (July 22, 1983), and "EPA Calls Dioxin Most Potent Material" (July 24, 1983).

78. "The Dioxin Legacy," *NYT* editorial, July 18, 1983. Also see, "No A.M.A. Vote on Dioxin 'Witch Hunt,'" letter to the editor from Raymond Scalettar, M.D., *NYT*, July 29, 1983; "Concern Growing over Unclear Threat of Dioxin," *NYT*, February 15, 1982; "EPA Allows Sale of Herbicides with Significant Dioxin Levels," *WP*, February 24, 1983; "Dioxin: Quandary for the '80s," 22–23.

79. "Concern Growing over Unclear Threat of Dioxin."

80. According to two accounts, Bliss's claim that he was unaware that the waste oil contained hazardous chemical waste was belied by the fact that he was paid by NEPACCO to remove the waste. Under Missouri law at the time, haulers paid manufacturers for nonhazardous oil, while manufacturers paid haulers to remove hazardous waste. "Ignorance Is Bliss," *Economist*, February 12, 1983, 42; "Dioxin: Quandary for the '80s," 8.

81. "Ignorance Is Bliss." For the bumper sticker, see "Dioxin, Quandary for the '80s," 8; "In Dioxin-Tainted Town, No 'Welcome' Signs," *NYT*, January 10, 1983; "Dioxin Experts Get Unfriendly Greeting at Missouri Meeting," *NYT* January 30, 1983.

82. Among the many works that touch on these topics from a variety of perspectives, see Michael Schudson, *Watergate in American Memory* (New York: Basic Books, 1993); Robert Buzzanco, *Vietnam and the Transformation of American Life* (Hoboken, NJ: Blackwell, 1999); Marc Hetherington, *Why Trust Matters: Declining Political Trust and the Demise of American Liberalism* (Princeton: Princeton University Press, 2006); Natasha Zaretsky, *No Direction Home: The American Family and the Fear of National Decline, 1968–1980* (Chapel Hill: University of North Carolina Press, 2007); and Michael Allen, *Until the Last Man Comes Home: POWs, MIAs, and the Unending Vietnam War* (Chapel Hill: University of North Carolina Press, 2009).

83. "Decade-Old Dioxin Has Turned Times Beach Scared and Angry," *WP*, January 10, 1983.

84. "Town Struggles with Toxic Legacy."

85. "Missourians Forced Out by Dioxin Are Impatient, Distrustful, and Worried," *NYT*, October 27, 1983.

86. "A Few Stay on in Lonely Times Beach, MO," *NYT*, June 12, 1983; "Missourians Forced Out by Dioxin Are Impatient, Distrustful, and Worried"; "Dioxin Level High at a Missouri Site," *NYT*, June 21, 1983.

87. "US Wins Suit on Cleanup of Dioxin," *NYT*, February 1, 1984. "Settlement Nearing in Dioxin Case," *St. Louis Post-Dispatch*, February 13, 1990.

88. "On Site Incineration at the Times Beach Superfund Site," final report of the EPA Office of Solid Waste and Emergency Response, Technology Innovation Office, available at www.clu-in.org/products/costperf/incinrtn/default.htm.

89. "Settlement Nearing in Dioxin Case." "It's War! Battle Lines Drawn for Dioxin Burner," *St. Louis Post-Dispatch*, May 27, 1990; "U.S. Officials Say Dangers of Dioxin

Were Exaggerated," *NYT* August 15, 1991; "Dioxin Scare Now Called Mistake: Citing Studies, Health Officials Conclude Times Beach Evacuation Wasn't Needed," *St. Louis Post-Dispatch,* May 23, 1991.
90. Robert Dewhirst, "Economic Development, Energy, and the Environment in Missouri," in *Missouri Government and Politics,* revised ed., ed. Richard Hardy, Richard Dohm, and David Leuthold (Columbia: University of Missouri Press, 1995), 293.
91. "On Site Incineration at the Times Beach Superfund Site."
92. Ibid.; "Dioxin Incinerator Emissions Exposure Study: Times Beach, Missouri," *Chemosphere* 40 (2000): 1063–74.
93. Bruce Wildblood-Crawford, "Environmental (In)Justice and 'Expert Knowledge': The Discursive Construction of Dioxins, 2,4,5-T, and Human Health in New Zealand, 1940–2007" (Ph.D. diss., University of Canterbury (Christchurch, NZ), 2008). For more on the grasslands revolution in New Zealand, the narrative scope of which is a source of much debate among environmental historians working in and on New Zealand, see Brooking et al., "The Grasslands Revolution Reconsidered," in *Environmental Histories of New Zealand,* ed. Pawson and Brooking (Melbourne: Oxford University Press, 2002), 169–82; and Gordon Winder, "Grassland Revolutions in New Zealand: Disaggregating a National Story," *New Zealand Geographer* 65, no. 3 (December 2009): 187–200.
94. Wildblood-Crawford, "Environmental (In)Justice and 'Expert Knowledge,'" esp. chap. 4, "The Progressive Discourses of Big Science and the Chemical Promise," 76–102.
95. For several excellent examples, see Zierler, *The Invention of Ecocide,* chap. 3.
96. "War on the Farm" (1960), New Zealand Film Unit, ANZ RV166.
97. For more on Seveso and the immediate fallout, see Arnold Schecter, ed., *Dioxins and Health* (New York: Plenum Press, 1994), esp. 538–39, 555–57, 587–626; Michael Gough, *Dioxin, Agent Orange: The Facts* (New York: Plenum Press, 1986); G. M. Reggiani, "The Seveso, Italy, Episode: July 10, 1976," in *Agent Orange and Its Associated Dioxin: Assessment of a Controversy,* ed. A. L. Young and G. M. Reggiani (New York: Elsevier Books, 1988), 227–69; A. L. Young, Typescript: "Seveso Scientific Review, June 9–16, 1983," ALY, series IV, subseries 4, no. 02341. This subseries of the ALY collection contains numerous studies from around the world on the Seveso incident.
98. For more on the Alsea studies, see Gough, *Dioxin, Agent Orange: The Facts,* 139–48; *Veterans and Agent Orange: Health Effects of Herbicides Used in Vietnam* (Washington: National Academies Press, 1994), 42–43; and ALY, series III, subseries 1 and 2, which contain the original Alsea studies and dozens of critiques and assessments of them.
99. "Correspondence: Leo and Suzanne Neal to MoH [Minister of Health]," March 21, 1979," Poisons: Substances—2,4,5-T, ANZ, ABQU 632, no. 49667: 1978–79.
100. "Uncertainty about Use of Herbicides," *Wellington Evening Post,* July 4, 1979; "Fight for and against 2,4,5-T Flares in USA and in NZ," *Wellington Evening Post,* October 3, 1979; clippings included in ANZ, ABQU 632, no. 49667: 1978–79.
101. "Correspondence: B. Faithful to MoH," April 2, 1979; "Correspondence: S. McKee to MoH," December 20, 1979; "Correspondence: Gair Reply to Faithful"; "Correspondence: Gair Reply to McKee," in Poisons: Substances—2,4,5-T, ANZ, ABQU 632, no. 49667: 1978–79. For Gair's evolving views on the media and the lens of

Vietnam, see "Correspondence, MoH," Poisons: Substances—2,4,5-T, ANZ, ABQU 632, no. 55162: 1979–80.
102. D. M. O. Bancroft, report to Health Ministry, March 28, 1979, Poisons: Substances—2,4,5-T, ANZ, ABQU 632, no. 49667: 1978–79.
103. "Department of Public Health Memorandum, January 18, 1979," Poisons: Substances—2,4,5-T, ANZ, ABQU 632, no. 49667: 1978–79.
104. "Correspondence: J. S. Roxbugh to Bates," March 8, 1979, Poisons: Substances—2,4,5-T, ANZ, ABQU 632, no. 49667: 1978–79.
105. "Cable: Embassy to MoH, Wellington," March 30, 1979; "Australians Give the Herbicide the All-Clear," *New Zealand Herald* (Auckland), March 27, 1979; Report of the Australian Weed Control Authority, May 1979; Poisons: Substances—2,4,5-T, ANZ, ABQU 632, no. 49667: 1978–79.
106. "Man Drinks 2,4,5-T," originally appeared in *The Australian,* July 9, 1980, reprinted in Christchurch (NZ) *Herald,* July 10, 1980, Poisons: Substances—2,4,5-T, ANZ, ABQU 632, no. 50989: 1980.
107. "Press Release: Study Finds No Health Problems among Workers at IWD Plant," October 6, 1980, Poisons: Substances—2,4,5-T, ANZ, ABQU 632, no. 52147: 1980.
108. "Assessment of the Toxic Hazards of the Herbicide 2,4,5-T in New Zealand," presentations made to the Royal Society of New Zealand on the Occasion of the Fellows' Annual Meeting, May 21, 1980, Poisons: Substances—2,4,5-T, ANZ, ABQU 632, no. 52147: 1980.
109. "Draft: The Use of 2,4,5-T in New Zealand," (October 1985), ANZ, "Air Pollution Control Records," ABQU 4452, no. 59301: 1985, p. 1; "Dioxin Burn Clean, But Opposition Continues," *New Zealand Herald,* April 5, 1985.
110. "Cregg clippings," ANZ, ABQU W4452, no. 60786: 1986; *Auckland Star,* June 26, 1986.
111. "Possible Health Effects of Manufacture of 2,4,5-T in New Plymouth: Report of Ministerial Committee of Inquiry to the Minister of Health," October 1986, ANZ, ABQU 4452, no. 62510.
112. "New Zealand Playcentre Federation to MoH," March 23, 1988, ANZ, ABQU W4452, no. 77949: 1983–91.
113. "Dr. Alan Parsons," Submission to Committee of Inquiry, July 15, 1986, ANZ, ABQU W4452, no. 60786: 1986.
114. "Correspondence: Food and Chemical Workers of New Zealand," July 25, 1985; "Federated Farmers of New Zealand," August 8, 1985; "2,4,5-T Clippings," all located in "The Use of 2,4,5-T in New Zealand," ANZ, AAUM W044, file 1: 1985–86.
115. "Correspondence: Peter Whitehouse to MK Moore," June 25, 1986, ANZ, ABQU W4452, no. 60786: 1986.
116. "Correspondence: MacKenzie-Komp to MoH Michael Bassett," June 21, 1986, ANZ, ABQU W4452, no. 60786: 1986.
117. "Possible Health Effects of Manufacture of 2,4,5-T in New Plymouth," 1.
118. Ibid., 3; 6; 7.
119. Ibid., 14.
120. Ibid., 15; 14.
121. Wildblood-Crawford, "Environmental (In)Justice and 'Expert Knowledge,'" 188.
122. "Parliamentary Petition Notes," Poisons—Substances, 2-4-5, ANZ, ABQU W4452, no. 77949: 1983–91.

CHAPTER FOUR. THE POLITICS OF UNCERTAINTY

1. "Correspondence, Rowe to 2,4,5-T Manufacturers," March 19, 1965, TTU, Burch collection, box 1, folder 1, p. 1.
2. "Correspondence, Rowe to 2,4,5-T Manufacturers," 1–2.
3. "Memo: Tricholorophenol Summary," TTU, Burch collection, box 1, folder 1, p. 2.
4. "Report on the Chloracne Problem Meeting on 3/24/65," TTU, Burch collection, box 1, folder 1, pp. 3–5.
5. Peter Schuck's *Agent Orange on Trial* (Cambridge: Harvard University Press, 1986) deals with this issue extensively. He notes that the original judge in the veterans' class action, Judge George Pratt, originally ruled that the government held "rather extensive knowledge" about the risks of Agent Orange, but the evidence he reviewed in making that decision was far from clear on the matter. The Court was never able to determine to what degree government officials knew about the risks to human health caused by dioxin exposure. In fashioning the eventual settlement in the case, Judge Jack Weinstein had been far less persuaded about the extent of government knowledge about dioxin. Schuck, *Agent Orange on Trial*, esp. 93–115.
6. "Memo: Rowe to Bioproducts Manager, Dow Canada (Sarnia)," June 24, 1965, TTU, Burch collection, box 1, folder 1, exhibit 24-37, pp. 1–2.
7. Testimony from Johnson in "Effects of 2,4,5-T on Man and the Environment," Hearings before the Subcommittee on Energy, Natural Resources, and the Environment, 91st Congress, 2nd session, April 7 and 15, 1970, (Washington: Government Printing Office, 1979), 1. In testimony given at several hearings in the late 1970s and early 1980s, several expert witnesses and members of Congress described provisions in the Federal Insecticide, Fungicide, and Rodenticide Act of 1947 that would have required Dow to inform the USDA of any health hazards associated with products it was selling commercially or producing as a contractor for any federal agency. While it is clear that Dow did not at any time inform the USDA of the presence of dioxin in 2,4,5-T, it is fairly clear from these hearings that few members of Congress were even aware of the Federal Insecticide, Fungicide, and Rodenticide Act of 1947.
8. Adriana Petryna, *Life Exposed: Biological Citizens after Chernobyl* (Princeton: Princeton University Press, 2002), 10.
9. Gregg Mitman, Michelle Murphy, and Christopher Sellers, "Introduction: A Cloud over History," in "Landscapes of Exposure: Knowledge and Illness in Modern Environments," ed. Mitman et al., special issue of *Osiris* 19 (2004): 14.
10. Ibid., 13.
11. This chapter also seeks to add the Agent Orange battles to a growing body of literature on chemical exposure and the so-called risk society. I have been particularly informed by Ulrich Beck's foundational work in this area, which suggests that we might think of efforts by veterans and their advocates as a way to counter modern science's denial of the subjective: "Science has risen to its present dominating image of technical power and objectivity by virtue of its repression of experience, so to speak. . . . Experience, which was once the main authority and judge of truth, has become the quintessence of the subjective, a relic, a source of illusions that attack the understanding and make a fool of it. In this view, it is not science but rather the subject and subjectivity that are wrong." Ulrich Beck, *Ecological Enlightenment: Essays on the Politics of the Risk Society* (New York: Humanity Books, 1995), 15; quoted in Stephen Couch and Steve Kroll-Smith, "Environmental Movements and Expert

Knowledge: Evidence for a New Populism," in *Illness and the Environment: A Reader in Contested Medicine,* ed. Kroll-Smith, Phil Brown, and Valerie Gunter (New York: New York University Press, 2000), 401. Beck's *Risk Society: Toward a New Modernity* (1982; reprint Thousand Oaks, CA: SAGE Publications, 1992) remains the foundational work in this area. A useful companion to Beck's volume is Jane Franklin, ed., *The Politics of Risk Society,* with contributions by Beck, Anthony Giddens, and others. The shifting epistemologies and ontologies of human bodies, chemicals, and nature can be found in a number of works as well. Among the most important figures in this literature are the science studies scholar Bruno Latour, particularly his *We Have Never Been Modern* (Cambridge: Harvard University Press, 1993) and *Pandora's Hope: Essays on the Reality of Science Studies* (Cambridge: Harvard University Press, 1999). Together with Latour, Donna Haraway, *Simians, Cyborgs, and Women: The Reinvention of Nature* (New York: Routledge, 1991), especially its famous essay "A Cyborg Manifesto," helped challenge the notion of individual bodies as being somehow separate from nature. More recently, these theories have been updated and applied by, among others, Christopher Sellers, *Hazards of the Job: From Industrial Disease to Environmental Health Science* (Chapel Hill: University of North Carolina Press, 1997); Joe Thornton, *Pandora's Poison: Chlorine, Health and a New Environmental Strategy* (Cambridge: MIT Press, 2000); and Mittman, Murphy, and Sellers, eds., "Landscapes of Exposure." A very useful overview can be found in Steve Kroll-Smith's and Worth Lancaster's review article "Bodies, Environments, and a New Style of Reasoning," in *Annals of the American Academy of Political and Social Science* 584 (2002): 203–12. For excellent applications of all of these ideas in specific case studies of particular environments and chemicals, see Michelle Murphy, *Sick Building Syndrome and the Problem of Uncertainty: Environmental Politics, Technoscience and Women Workers* (Durham: Duke University Press, 2006); Linda Nash, *Inescapable Ecologies: A History of Environment, Disease, and Knowledge* (Berkeley: University of California Press, 2007); and Nancy Langston, *Toxic Bodies: Hormone Disruptors and the Legacy of DES* (New Haven: Yale University Press, 2009).
12. "Effects of 2,4,5-T on Man and the Environment," Hearings before the Subcommittee on Energy, Natural Resources, and the Environment, 91st Congress, 2nd session, April 7 and 15, 1970, (Washington: Government Printing Office, 1979), 1.
13. Ibid., 23–24.
14. Ibid., 70.
15. *Unnatural Causes,* directed by Lamont Johnson (Los Angeles: ITC Entertainment Group, 1986).
16. "Maude DeVictor," in John Langston Gwaltney, *The Dissenters* (New York: Random House, 1986), 110.
17. "Interview of Dr. Alvin Young," April 7, 2001, ALY, box 222, folder 6554, p. 30. Young confirmed his account in several subsequent congressional hearings and again in an interview with the author on March 3, 2008.
18. DeVictor was eventually fired by the VA in 1984 for "conduct unbecoming a federal employee." She maintains that her termination was a result of her pressing the Agent Orange claims, but the Chicago VA has publicly stated that it was a result of her efforts to unionize the VA workers. For more on DeVictor's story, see Gwaltney, *The Dissenters,* 106–24; and "Maude DeVictor, Continuing Her Crusade to Expose Agent Orange," *WP,* November 10, 1986.

19. *Vietnam's Deadly Fog* (Chicago: WBBM TV, 1978). I would like to thank Bill Kurtis and the WBBM staff for providing me with a DVD copy of the program in 2010.
20. Bill Kurtis, *Bill Kurtis on Assignment* (Chicago: Rand MacNally, 1983), 45.
21. For a sampling of Young's studies on beach mice at Eglin Air Force Base, see Lorris Cockerham and Alvin L. Young, "The Absence of Hepatic Cellular Anomalies in TCDD-exposed Beach Mice: A Field Study," *Environmental Toxicology and Chemistry* 1, no. 4 (November 1982): 299–308; Lorris Cockerham and Alvin Young, "Ultrastructural Comparison of Liver Tissues from Field and Laboratory TCDD-Exposed Beach Mice," in *Human and Environmental Risk of Chlorinated Dioxins and Related Dioxins*, ed. Richard E. Tucker, Alvin L. Young, and Allan P. Gray (New York: Plenum Press, 1983); and A. L. Young, L. G. Cockerham, and C. E. Thalken, "A Long-term Study of Ecosystem Contamination with 2,3,7,8-tetrachlorodibenzo-p-dioxin," *Chemosphere* 16, nos. 8–9 (1987): 1791–1815. In an interview in 2008 Young responded to criticism that much of his early work on the toxicology and environmental fate of 2,3,7,8-TCDD was published in DOD technical reports rather than scholarly journals by claiming that toxicology journals refused to publish the reports because they did not trust military scientists (author's interview with Young [2008], 5).
22. "Interview of Alvin Young," 34.
23. Alvin Young, "Agent Orange: At the Crossroads of Science and Social Concern," 38 (emphasis added).
24. Ibid., 40.
25. Cited in ibid., 44.
26. Young confirmed these feelings in both his official oral history with the National Agricultural Library ("Interview of Alvin Young," 39) and in his interview with the author in 2008.
27. *Vietnam's Deadly Fog* (1978).
28. Kurtis, *On Assignment*, 80; Wilbur Scott, "Competing Paradigms in the Assessment of Latent Disorders: The Case of Agent Orange," in *Illness and the Environment: A Reader in Contested Medicine*, ed. Steve Kroll-Smith, Phil Brown, and Valerie Gunter (New York: New York University Press, 2000), 409–29, 414. Scott's book *Vietnam Veterans since the War* (Norman: University of Oklahoma Press, 2004) contains a much more detailed analysis of a number of the events described here.
29. "Herbicide 'Agent Orange,'" Hearings before the Subcommittee on Medical Facilities and Benefits, 95th Congress, 2nd session, October 11, 1978, (Washington: Government Printing Office, 1979), 27; "Agent Orange and Health," *WP*, March 25, 1978; "500 Vets Claims Herbicide Effects,'" *WP*, October 12, 1978.
30. Scott, "Competing Paradigms in the Assessment of Latent Disorders," 415.
31. The two best starting points for discussions about the case are Schuck, *Agent Orange on Trial*, and the discussion of the case in Scott, *Vietnam Veterans since the War*.
32. Schuck, *Agent Orange on Trial*, 85.
33. Ibid., 80–81.
34. Ibid., 82.
35. Ibid., 17n33. Schuck also refers vaguely to "testing at Edgewood Arsenal during the 1960s" and "knowledge of the dioxin contamination problem among government scientists in the mid to late 1960s," as evidence that surfaced at the trial, but the documentation for these claims is extremely thin: according to Schuck's notes, they came out only during the fairness hearings, and even the infamous Edgewood Task

Force Report referred to often in Agent Orange–related litigation that is ongoing in the twenty-first century, does not actually appear to be part of the record of the case. In 2007 I contacted Schuck; Constantine "Dean" Kokkoris, the lead litigator in the more recent Vietnamese Victims of Agent Orange suit; and the clerk of the Eastern District Court of New York, where the veterans' case was heard. None of them could locate the report, and neither Schuck nor Kokkoris had actually seen it. Kokkoris referred me to the Agent Orange collection at NARA in College Park. In more than five years of research in all the Agent Orange files at NARA and elsewhere, I have located a number of documents from personnel at Edgewood Arsenal, but I have been unable to locate the document in question or any other evidence that refers to such a document.

36. Ibid., 98.
37. Ibid., chap. 8, "Fashioning a Settlement," provides a fascinating, blow-by-blow account of the judge's machinations in the final hours before the case was set to go to trial.
38. Ibid., 165.
39. "Viet Vets' Herbicide Suit Settled," *WP*, May 8, 1984.
40. "Agent Orange Settlement Divides Vietnam Veterans," *WP*, August 5, 1984. Schuck, *Agent Orange on Trial*, 175.
41. "Agent Orange Settlement Divides Vietnam Veterans."
42. "The Agent Orange Settlement Fund," Department of Veterans' Affairs Website, www.vba.va.gov/bln/21/benefits/herbicide/AOn02.htm.
43. The reasons for these developments also involve shifting ideas of the relationship between scientific expertise and the law. See discussion in greater detail in chapter 5 below.
44. "Herbicide 'Agent Orange,'" Hearings before the Subcommittee on Medical Facilities and Benefits, 95th Congress, 2nd session, October 11, 1978 (Washington: Government Printing Office, 1979), 27.
45. "Oversight Hearing to Receive Testimony on Agent Orange," Hearings before the Subcommittee on Medical Facilities and Benefits, 96th Congress, 2nd session, February 25, 1980 (Washington: Government Printing Office, 1980), 69–70, 100.
46. Ibid., 95.
47. Ibid., 101.
48. Ibid., 112.
49. Ibid., 114.
50. "Oversight Hearing to Receive Testimony on Agent Orange," Hearings before the House Subcommittee on Medical Facilities and Benefits, 96th Congress, 2nd session, July 22, 1980 (Washington: Government Printing Office, 1981); on the various lawsuits and state resolutions, see "Agent Orange and 2,4,5-T: The States Respond," VVA press release (July 1980) and supplemental information, reprinted in "Oversight Hearings" (July 1980), 429–47.
51. Ibid., 55.
52. Ibid., 4.
53. Ibid., 232, 236–37.
54. Ibid., 263–71. Chief among the witnesses was Gabriel Brinsky of AMVETS, a veterans' group that has had, to put it mildly, a tense relationship with Vietnam veterans in general, rejecting the claims that they were different from veterans of other wars.

55. Letter from Daschle and Bonior to Max Cleland, March 18, 1980, reprinted in Ibid., 144.
56. Letter from Cleland to Daschle and Bonior, March 24, 1980, reprinted in Ibid., 146–47.
57. "U.S. Government Memorandum ['the unsigned memorandum']," October 12, 1977, reprinted in Ibid., 153–57. All misspellings are in the original.
58. Letter from Daschle and Bonior to Max Cleland, 145.
59. "Scientific Community Report on Agent Orange," Hearings before the Subcommittee on Medical Facilities and Benefits, 96th Congress, 2nd session, September 16, 1980 (Washington: Government Printing Office, 1981), esp. pp. 73–85; accessed through ALY, series VIII, subseries 1, no 05340.
60. Letter, Bonior and Daschle to General Hans Mark, Secretary of the Air Force, March 18, 1980, reprinted in *Oversight Hearings* (July 1980), 158–59.
61. Letter, Secretary Zengerle to Bonior and Daschle, reprinted in Ibid., 160–61.
62. "Problems with the Veterans Administration Agent Orange Registry," GAO report, (Washington: Government Printing Office, 1982); also see Kurtis, *On Assignment*, 79).
63. Patrick Hagopian, *The Vietnam War in American Memory: Veterans, Memorials, and the Politics of Healing* (Amherst: University of Massachusetts Press, 2009), 66–67. Hagopian's second chapter, "Something Dark and Bloody," offers a thorough overview of this period.
64. For more on the transformation, see ibid., chap. 8, "No Shame or Stigma." I also discuss the transformation of this image in *Invisible Enemies*, chap. 4. For an innovative and updated look at the place of the Vietnam War and the Vietnam veteran in American culture and memory, see David Kieran, *"Sundered by a Memory": Foreign Policy, Militarism, and the Vietnamization of American Memory, 1970–Present* (Amherst: University of Massachusetts Press, under advance contract).
65. "Memo, Bart Kull, Undersecretary for Intergovernmental Affairs, Department of Health and Human Service, to AOWG," August 18, 1983, ALY, series VIII, subseries 2, no. 5601.
66. Michael Gough, *Dioxin, Agent Orange: The Facts* (New York: Plenum Press, 1986), 95–96.
67. Ibid., 76.
68. For the AOWG critique of the initial Ranch Hand study, see "Review of VVA Critique of Ranch Hand Mortality Results," Science Panel, AOWG, April 16, 1984, ALY, series VIII, subseries 2, no. 5625.
69. Scott, *Vietnam Veterans since the War*, 192.
70. AOWG minutes, November 16, 1984. ALY, series VIII, subseries 2, box 199, no. 05630.
71. Alvin Young, Typescript: "Development of an Exposure Index for an Epidemiological Study of Ground Troops Exposed to Agent Orange during the Vietnam Conflict," ALY, series III, subseries 3, no. 01711; "Outline Report of the Agent Orange Working Group Science Panel Subcommittee on Exposure (November 1982), ALY, series VIII, subseries 2, no. 05591.
72. Scott, *Vietnam Veterans since the War*, 193.
73. "Agent Orange Projects Interim Report," CDC (February 1985), ALY, series III, subseries 3, no. 01735, p. 4.

74. Ibid., 18.
75. "Testimony before the Subcommittee on Hospitals and Health Care Committee on Veterans' Affairs of the House of Representatives, (July 31, 1986)," ALY, series VIII, subseries 2, no. 05526, p. 6.
76. Typescripts: "Comments on I. Introduction" and "II. Obtaining Unit Location Information," ALY, series VIII, subseries 2, no. 05734 (all emphasis in original).
77. "Exposure Assessment for the Agent Orange Study," Interim Report Number 2: Supplemental Information (December 1985), ALY, series VIII, subseries 2, no. 05653, 15.
78. Scott, *Vietnam Veterans since the War*, 195. Christian's comments came from an interview with Scott.
79. "Testimony by Mr. Richard Christian, Director, U.S. Army and Joint Services Environmental Support Group, before the Subcommittee on Hospitals and Veterans, July 31, 1986," ALY, series VIII, subseries 2, no. 05525, p. 6.
80. Gen. John Murray, "Report to the White House Agent Orange Working Group Science Subpanel on Exposure Assessment," May 27, 1986, ALY, series VIII, subseries 2, no. 5692 (hereafter Murray Report).
81. Ibid., 1.
82. Ibid., 44.
83. Ibid., 45; 52.
84. "Testimony by Mr. Richard Christian," 7.
85. Edgar, "Opening Statement" in "Testimony before the Subcommittee on Hospitals and Health Care Committee on Veterans' Affairs of the House of Representatives," 8; Scott, *Vietnam Veterans since the War*, 194.
86. Mittman, Murphy, and Sellers, "Landscapes of Exposure," 13.
87. Thornton, *Pandora's Poison*, 10.
88. Langston, *Toxic Bodies*, 148. Langston here is building on the work of, among others, Bruno Latour, Donna Haraway, Steve Kroll-Smith, and Worth Lancaster. See Langston, *Toxic Bodies*, 148nn39–40; and note 11 in this chapter.
89. Much of this increase is related to provisions contained in the Agent Orange Act of 1991, discussed below; the current benefits regime is discussed in chapter 5.
90. Jock McCulloch, *The Politics of Agent Orange* (Victoria, Australia: Heinemann Books, 1984), 146–49.
91. "Pesticides Used in Vietnam Hostilities and Their Use in Australian Agriculture: A Comparative Study," Australian Department of Health, Commonwealth Government Report, (Canberra, Australia: Australian Government Publishing Service, 1982).
92. "Report on the Use of Herbicides, Insecticides, and Other Chemicals by the Australian Army in South Vietnam," Australian Department of Health, Commonwealth Government Report (Canberra, Australia: Australian Government Publishing Service, 1982); available in ALY, series III, subseries 4, no. 01933.
93. For examples of these approaches, see VVAA, "Submission to the Royal Commission on the Use and Effects of Chemical Agents on Australian Personnel in Vietnam," November 1983, ALY, series III, subseries 4, no. 01950; and McCulloch, *The Politics of Agent Orange*, chap. 7. For more on "popular epidemiology," see Phil Brown, "Popular Epidemiology and Toxic Waste Contamination: Lay and Professional Ways of Knowing," in *Illness and the Environment*, 364–83, and Murphy, *Sick Building Syndrome*, chap. 4, "Indoor Pollution at the Encounter of Toxicology and Popular Epidemiology," 81–110.

94. VVAA, "Submission to the Royal Commission on the Use and Effects of Chemical Agents on Australian Personnel in Vietnam," 53.
95. Australian Broadcasting Company radio transcript, May 27, 1982, quoted in McCulloch, *The Politics of Agent Orange*, 167n62.
96. Senate Standing Committee on Science and the Environment, "First Report on Pesticides and the Health of Australian Vietnam Veterans," November 1982 (Canberra: Australian Government Publishing Service, 1982); available in ALY, series III, subseries 4, no. 01937; McCulloch, *The Politics of Agent Orange*, 215.
97. "Agent Orange: Department of Health Notes (July 1983)," Poisons-Substances-Agent Orange," ANZ, ABQU632 W4452, file 3: 1982–83.
98. "VR Johnson to Thomson," July 27, 1982, Poisons-Substances-Agent Orange, ANZ, ABQU632 W4452, file 3: 1982–83; "Agent Orange: Department of Health Notes."
99. VVAA, "Submission to the Royal Commission on Use and Effects of Chemical Agent on Australia Personnel in Vietnam" (November 1983), available at ALY, series III, subseries 4, box 75, no. 01950, p. 52.
100. "Final Report of the Royal Commission," vol. 1, chap. 1, ALY, series III, box 80, pp. 58–62.
101. Young, "Letter to Hon. Mr. Justice Phillip Evatt," December 22, 1983, ALY, series III, subseries IV, no. 01959.
102. Ibid.; also see Young, "Handwritten Notes Regarding Claims Filed/Study Costs in Australia Related to Agent Orange Exposure," ALY, series IV, subseries 3, no. 02105; and Young, "Letter to J. S. Coombs," January 1984, ALY, series IV, subseries 3, no. 01968.
103. "Final Report of the Royal Commission, vol. 1, I-53.
104. "Final Report of the Evatt Royal Commission," vol. 1, I-3.
105. "Final Report of the Royal Commission: vol. 1," chap. 1, 46. The report describes a number of these headlines ("Soldiers' Babies Death at Birth"; "Agents of Deformity at Birth"; "Shock Report on Agent Orange Babies"). For more samples of Australian press coverage, the ALY collection has a number of folders containing press clippings from Australia before, during, and after the Royal Commission, located in a number of folders in ALY, series IV, subseries 3.
106. "Final Report of the Royal Commission," vol. 1, chap. 3, pp. 4–5.
107. Ibid., chap. 4, pp. 152–53.
108. For more examples, see ibid., Introduction and Exposure Hearings, 144–80. The ALY collection also contains the entire transcripts of the informal public hearings, totaling over seven thousand pages, where veterans often testified about exposure. For a sampling of the testimony and veterans' descriptions of their exposure scenarios, see "Transcript of Proceedings: Royal Commission on the Use and Effects of Chemical Agents on Australian Personnel in Vietnam," June 1984, ALY, series IV, subseries 3, nos. 02023, 02024, 02025.
109. "Final Report of the Royal Commission," vol. 8, 39.
110. Ibid., vol. 1, xvii–xviii.
111. On the former charge, see Jock McCulloch, "The Ethics of the Royal Commission," and Brian Martin, "Pesticides, the Vietnam War, and the Evatt Commission," in *Evatt Revisited: Proceedings of a Conference which Re-examined the Findings of the Royal Commission on the Use and Effects of Chemical Agents on Australian Personnel in Vietnam*, ed. A. J. D. Bellett et al. (Sydney: Centre for Human Aspects of Science and Technology, University of Sydney, 1989).

112. "VR Johnson to Queen Elizabeth," Poisons-Substances-Agent Orange, ANZ, ABQU632 W4452, file 5: 1980–89.
113. Bellett et al., "Introduction," in *Evatt Revisited*, 1.
114. Particularly insightful essays include McCulloch, "The Ethics of the Royal Commission," and G. F. Humphrey, "The Royal Commission on Agent Orange as an Example of the Interface between Legal and Scientific Processes for Examination of Data," in *Evatt Revisited*.
115. Young, "Letter to Justice Phillip Evatt Regarding the Final Report of the Royal Commission," ALY, series IV, subseries 3, no 02014.
116. "Postservice Mortality among Vietnam Veterans: The Centers for Disease Control Vietnam Experience Study," *Journal of American Medicine* 257, no. 6 (February 13, 1987): 790–95; the full study, with additional statistical support, is available at ALY, series III, subseries 3, no. 01744; see also "Memorandum: Review of VA Mortality Study," Vernon N. Houk to Ronald W. Hart (September 11, 1987), ALY, series III, subseries 3, no. 01784.
117. "Vietnam Experience Study," ALY, series III, subseries 3, no. 01744, p. 29.
118. "Talking Paper on Agent Orange: Summary of Recent Studies," A. L. Young to Beverly Berger (April 30, 1987), ALY, series VIII, subseries 2, no. 05712. For a fuller discussion of the Kansas study and its implications, see Scott, *Vietnam Veterans since the War*, 197–99.
119. Scott, *Vietnam Veterans since the War*, 202–6. The hearing in question, of which Scott provides a thorough summary, is "Oversight Review of CDC's Agent Orange Study," Hearings before the House Subcommittee on Human Resources and Government Regulations, 101st Congress, 1st Session, July 11, 1989 (Washington: Government Printing Office, 1989). On the subcommittee report, see Scott, *Vietnam Veterans since the War*, 223–25.
120. "Agent Orange Down on the Farm," VVA *Veteran* 7, no. 12 (1987): 15.
121. "Agent Orange Study Called Botched or Rigged," *WP*, July 12, 1989, A6.
122. "The Agent Orange Coverup: A Case of Flawed Science and Political Manipulation," Report of the House Committee on Government Operations, 101st Congress, 2nd Session, August 9, 1990 (Washington: Government Printing Office, 1991), 3.
123. "Agent Orange Coverup," 2.
124. "The Story of Agent Orange," *U.S. Veteran News and Report*, Special Agent Orange Edition (November 1990), 2.
125. The Zumwalt story here is taken from the book the Zumwalts wrote together with John Pekkanene, *My Father, My Son* (New York: Macmillan, 1986); and "Agent Orange and the Anguish of an American Family," *New York Times Magazine*, August 24, 1986. *Big Z: The Life and Times and Admiral Elmo Russell Zumwalt, Jr.* (New York: Harper Collins, 2013) by the historian Larry Berman promises to be the definitive work on the Zumwalts.
126. Adm. E. R. Zumwalt, Jr., "Report to the Secretary of the Department of Veterans Affairs on the Association between Adverse Health Effects and Exposure to Agent Orange," May 5, 1990 (hereafter Zumwalt Report), available at TTU, virtual archive, no. 6150208001.
127. Clary Letter to Daschle, quoted in Zumwalt Report, 5n5. Despite my repeated attempts, the Daschle archives at South Dakota State University did not respond to my requests for or about the Clary letter. To my knowledge, the only existence of

the letter appears in Zumwalt's report and in Daschle's statement about the letter in the *Congressional Record,* November 21, 1989.
128. *Congressional Record,* November 21, 1989, S16541, available in ALY, series VIII, subseries 1, no. 05537.
129. Friendship Village Website, www.vietnamfriendship.org/wordpress/agent-orange/agent-orange-overview. A Google search for the Clary letter will turn up hundreds of Agent Orange–related websites and blogs using it as evidence of various conspiracies and coverups.
130. Jason Grotto and Tim Jones, "Agent Orange's Lethal Legacy: Defoliants More Dangerous than They Had to Be," *Chicago Tribune,* December 17, 2009. For other examples, see, for instance, Chris Arsenault, *Blowback: A Canadian History of Agent Orange at Home* (Black Point, Nova Scotia: Fernwood Publishers, 2009), 7–8; and Ben Quick, "Agent Orange: A Chapter from History That Just Won't End," *Orion Magazine,* March/April 2008, available at www.orionmagazine.org/index.php/articles/article/2862/.
131. For a more recent take on this approach, see "Preliminary Reflections on Administration of Complex Litigations," *DeNovo: Cardoza Law Review* 1, no. 1 (June 2009), by Jack Weinstein, the presiding judge in the class action suit and all subsequent Agent Orange litigation in U.S. federal courts.
132. *Agent Orange Act of 1991,* Public Law 102-4; also see www7.nationalacademies.org/ocga/laws/PL102-4.asp.

CHAPTER FIVE. "ALL THOSE OTHERS SO UNFORTUNATE"

1. Jeanne and Steven Stellman, Richard Christian, Tracy Weber, and Carrie Tomasallo, "The Extent and Patterns of Usage of Agent Orange and Other Herbicides in Vietnam," *Nature* 422 (April 2003): 681–87.
2. Ibid., 682.
3. Ibid., 684.
4. Ibid. The Stellmans elaborated on their methodology in several other pieces. Among the most useful are Steven D. Stellman and Jeanne Stellman, "Exposure Opportunity Models for Agent Orange, Dioxin, and Other Military Herbicides Used in Vietnam, 1961–1971," *Journal of Exposure Analysis and Environmental Epidemiology* 14 (2004): 354–62; and Stellman et al., "A Geographic Information System for Characterizing Exposure to Agent Orange and Other Herbicides in Vietnam," *Environmental Health Perspectives* 111, no. 3 (March 2003): 321–28.
5. Alvin Young, Paul Cecil, and John Guilmartin Jr., "Assessing Possible Exposures of Ground Troops to Agent Orange during the Vietnam War: The Use of Contemporary Military Records," *Environmental Science and Pollution Resolution* 11, no. 6 (2004): 349–59.
6. Young et al., "Environmental Fate and Bioavailability of Agent Orange and Its Associated Dioxin during the Vietnam War," *Environmental Science and Pollution Resolution* 11, no. 6 (2004): 359–70.
7. Ibid., 368–69.
8. In both pieces, Cecil's *Herbicidal Warfare* is regularly cited in support of such assertions. Other articles authored or coauthored by Al Young are also cited often in both pieces.

9. Steven and Jeanne Stellman, "Exposure Opportunity Models for Agent Orange, Dioxin, and Other Military Herbicides Used in Vietnam, 1961–1971."
10. Young et al., "Assessing Possible Exposures of Ground Troops to Agent Orange during the Vietnam War," 353.
11. Schecter, interview with author, February 15, 2008; Alvin Young, "TCDD Biomonitoring and Exposure to Agent Orange: Still the Gold Standard," *Environmental Science and Pollution Resolution* 11, no. 3 (2004): 143–46; Schecter et al., "A Comparison and Discussion of Two Differing Methods of Measuring Dioxin-Like Compounds: Gas Chromatography-Mass Spectrometry and the Calux Bioassay–Implications for Health Studies," *Organohalogen Compounds* 40 (1999); and Schecter and Thomas Gasiewicz, eds., *Dioxins and Health*, 2nd. ed. (Hoboken, NJ: Wiley and Sons, 2003).
12. Michael Ginevan, John Ross, and Deborah Watkins, "Assessing Exposure to Allied Ground Troops in the Vietnam War: A Comparison of AgDRIFT and Exposure Opportunity Index Models," *Journal of Exposure Science and Environmental Epidemiology* 19 (2009): 187–200.
13. Ibid., 199.
14. "The Toxicology, Environmental Fate, and Human Risk of Herbicide Orange and Its Associated Dioxin," USAF OEHL Technical Report, October 1978, ALY, series III, subseries 1, no. 01165, pp. III-1–III-6. For more on the testing, design, and development of the spray mechanisms, see ALY, series I and II, which contain numerous studies and documents by the USAF and by the DTIC, such as, "Supplement II to Technical Report 46: Basic Data from H-34/Hidal Calibration Trials, 1963," ALY, series II, no. 00062, and "Technical Report and Specifications: A/A45Y-1 Internal Defoliant Dispenser System," ALY, series I, no. 00039.
15. "Toxicology, Environmental Fate, and Human Risk of Herbicide Orange and Its Associated Dioxin," III-7. On the issue of working toward a measurement of 1 ppt, see Robert Baughman and Matthew Meselson, "An Analytical Method for Detecting TCDD (Dioxin): Levels of TCDD in Samples from Vietnam," *Environmental Health Perspectives* 5 (September 1973): 27–35.
16. Baughman and Meselson, "An Analytical Method for Detecting TCDD."
17. For Young's Eglin studies, see ALY, series VI, subseries 2 and 3, and Alvin Young, *The History, Use, Disposition, and Environmental Fate of Agent Orange* (New York: Springer Books, 2009), 238–43.
18. "Toxicology, Environmental Fate, and Human Risk of Herbicide Orange and Its Associated Dioxin," III-7.
19. Schecter et al., "Recent Dioxin Contamination from Agent Orange in Residents of a Southern Vietnam City," *Journal of Occupational and Environmental Medicine* 43, no. 5 (May 2001): 435–43.
20. Schecter et al., "Food as a Source of Dioxin Exposure in the Residents of Bien Hoa City, Vietnam," *Journal of Occupational and Environmental Medicine* 45, no. 8 (August 2003): 781–88.
21. In 2009 I witnessed this group firsthand when a number of Vietnamese-Americans, several of whom had fled Saigon in and after 1975, attended a conference on Agent Orange at the University of California-Riverside. In addition to a number of heated exchanges during the sessions, where the dangers of Agent Orange were regularly dismissed by these activists as communist propaganda, one Vietnamese-American scheduled to present was forced to drop out of the conference because of threats of

violence against him. The conference was held under increased security because of similar threats.
22. "Letters to the Editor," *Journal of Occupational and Environmental Medicine* 46, no. 5 (May 2004): 415–16.
23. Wayne Dwernychuk et al., "The Agent Orange Dioxin Issue in Vietnam: A Manageable Problem," paper presented at Dioxin 2006 Conference (Oslo, Norway), available at www.warlegacies.org/OsloPaper2006.pdf, 1–2 (emphasis added).
24. Wayne Dwernychuk et al., "Dioxin Reservoirs in Southern Vietnam: A Legacy of Agent Orange," *Chemosphere* 47, no. 2 (November/December 2002): 117–37.
25. Dwernychuk et al., "A Manageable Problem," 1–2.
26. Ibid., 2.
27. Quoted in Michael Martin, "Vietnamese Victims of Agent Orange and U.S.-Vietnam Relations," Congressional Research Service Report (Washington: Congressional Research Service, 2009), 7.
28. Martin, "Vietnamese Victims of Agent Orange and U.S.-Vietnam Relations," 2.
29. The United States initially imposed economic sanctions against "Communist controlled areas of Vietnam" in 1964. In April 1975 the administration of Gerald Ford extended and strengthened the sanctions against the entire country. I discuss the embargo and the normalization process at length in *Invisible Enemies: The American War on Vietnam, 1975–2000* (Amherst: University of Massachusetts Press, 2007).
30. Martin, "Vietnamese Victims of Agent Orange and U.S.-Vietnam Relations," 9.
31. Memo, "AMEMBASSY HANOI to SECSTATE WASHDC," February 13, 2003, available at the Fund for Reconciliation and Development, www.ffrd.org.
32. "Joint Statement between the Socialist Republic of Vietnam and the United States of America," White House press release, November 17, 2006 (available at www.warlegacies.org/Bush.pdf).
33. "Comprehensive Assessment of Dioxin Contamination in Da Nang Airport, Vietnam: Environmental Levels, Human Exposure and Options for Mitigating Impacts—Summary of Findings," report prepared by Hatfield Consultants, November 2009, available at www.hatfieldgroup.com.
34. "Comprehensive Assessment of Dioxin Contamination in Da Nang Airport, Vietnam: Environmental Levels, Human Exposure and Options for Mitigating Impacts—Final Report," 4–6.
35. Martin, "Vietnamese Victims of Agent Orange and U.S.-Vietnam Relations," 20.
36. "VN-US Cooperate in Dioxin Detoxification in Da Nang," Vietnam News Agency, December 31, 2010, available at http://english.vovne ws.vn/Home/VNUS-cooperate-in-dioxin-detoxification-in-Da-Nang/201012/122787.vov; "Vietnam Starts Joint Agent Orange Cleanup with U.S.," *USA Today,* June 20, 2011.
37. Dwernychuk et al., "Dioxin Reservoirs in Southern Vietnam: A Legacy of Agent Orange."
38. For more on the Pueblo incident, see Ed Brandt, *The Last Voyage of the USS Pueblo* (New York: Norton, 1969), and Frederick Schumacher, *Bridge of No Return: The Ordeal of the U.S.S. Pueblo* (New York: Harcourt Brace, 1971).
39. *Federal Register,* January 25, 2011, available at www.access.gpo.gov/su_docs/aces/fr-cont.html. For more on the decisions, see John Davis, "Agent Orange: Defoliated Korea's DMZ," *VFW Magazine* (February 2000), 20–23; "Tracking Agent Orange," *NYT,* November 3, 2000, A17; "New Row over Agent Orange," BBC News

Online, November 17, 1999, available at http://news.bbc.co.uk/1/hi/world/asia-pacific/524836; and "VA Publishes Final Regulation to Aid Korean War Veterans Exposed to Agent Orange," VA News Release, January 25, 2011, available at www.kwva.org/links/dva_agent_orange_final_110125.pdf.
40. Davis, "Agent Orange: Defoliated Korea's DMZ," 21.
41. Ibid.
42. "New Row over Agent Orange."
43. The two best sources on No Gun Ri, which differ significantly in their conclusions about the event, are Charles J. Hanley, Sang-Hun Choe, and Martha Mendoza, *The Bridge at No Gun Ri: A Hidden Nightmare from the Korean War* (New York: Henry Holt, 2001); Robert Bateman, *No Gun Ri: A Military History of the Korean War Incident* (Mechanicsburg, PA: Stackpole Books, 2002).
44. "South Korea's Vietnam Veterans Begin to be Heard," *NYT*, May 10, 1992. For more on the health of Korean Veterans of the Vietnam War, see Joung-Soon Kim et al., "Impact of Agent Orange Exposure among Korean Vietnam Veterans," *Industrial Health* 41, no. 3 (2003): 149–57, which found evidence for suggestive links between exposure and several health conditions.
45. Kim Jung Wook, "Koreans' Damage Related to Agent Orange," Proceedings of the International Conference of Victims of Agent Orange/Dioxin," Hanoi, March 28–29, 2006 (Hanoi: Culture and Information Publishing House, 2006), 42–48; "South Korea's Vietnam Veterans Begin to be Heard."
46. On the case against the Unites States, see "Korean Veterans Demand Compensation," BBC News online, http://news.bbc.co.uk/2/hi/asia-pacific/513811.stm. For additional background and the text of the ruling in the second case, see the Agent Orange Lawsuit Site, www.vn-agentorange.org/kaova_20060327.html.
47. "Korean Veterans Demand Compensation."
48. Lisa Brady, "Life in the DMZ: Turning a Diplomatic Failure into an Environmental Success," *Diplomatic History* 32, no. 4 (September 2008): 585–610.
49. "U.S. Taking Seriously USFK's Alleged Agent Orange Dumping," Yon Hap News Agency (South Korea), http://english.yonhapnews.co.kr/national/2011/05/25/13/0301000000AEN20110525000100315F.HTML, May 25, 2011.
50. "Dioxin Traces Found Near U.S. Base in South Korea," *NYT*, June 16, 2011.
51. Ibid. Major demonstrations against the United States took place in 2002 after two U.S. soldiers were acquitted in the death of two Korean girls and again in 2008 after the government lifted its ban on U.S. beef imports.
52. "Pollutants Found in Drainage Water from Yongsan Garrison," *Korea Herald* (online edition), June 16, 2011, available at www.koreaherald.com/national/Detail.jsp?newsMLId=20110607000741; "U.S. military: No Agent Orange at South Korea Base," *Stars and Stripes*, June 23, 2011, available at www.stripes.com/news/u-s-military-no-agent-orange-at-south-korea-base-1.147286; "No Trace of Agent Orange at U.S. Base in South Korea," *National Public Radio Online*, www.npr.org/2011/06/23/137361263/no-trace-of-agent-orange-at-u-s-base-in-south-korea.
53. Indeed, as this book went to press, there was a new row about the storage of Agent Orange at the U.S. military base on Okinawa, Japan. The revelations were once again made by U.S. veterans, and initial investigations were once again inconclusive. "Agent Orange Buried on Beach Strip," *Japan Times* (online edition), www.japantimes.co.jp/text/nn20111130a5.html.

54. "Government Probes Claims NZ Exported Agent Orange," *New Zealand Herald*, January 10, 2005; "New Zealand Confirms Supplying Agent Orange in Vietnam War," *Agence France Presse*, January 9, 2005; the inquiry in 1990 is discussed here in chapter 1.
55. "MPs to Reopen Issues over Agent Orange," New Zealand *Evening Post* (Wellington), May 29, 1989, located in Foreign Affairs and Defence Committee Inquiries—Manufacturing of Agent Orange in New Zealand (1987–1990), ANZ, ABGX 4731 16127 2777, batch A: Correspondence.
56. "Agent Orange Papers Released," *New Zealand Herald*, January 26, 2005.
57. Ian Wishart and Simon Jones, "Agent Orange: 'We Buried It under New Plymouth,'" *Investigate* (January/February 2001): 32–33.
58. "The Health Needs of the Children of Operation Grapple and Vietnam Veterans: A Critical Appraisal Undertaken for the Office of Veterans' Affairs, Ministry of Defence," (August 2001); hereafter McLeod Report, available at media.apn.co.nz/webcontent/document/pdf/mcleodreport.pdf, iii. For the Reeves report, see Advisory Committee on the Health of Veterans' Children, "Inquiry into the Health Status of Children of Vietnam and Operation Grapple Veterans," Department of Prime Minister and Cabinet (Wellington, NZ, 1999).
59. McLeod Report, iii.
60. "Army Brass Contradict Agent Orange Report," *New Zealand Herald*, April 12, 2003.
61. McLeod Report, ii.
62. "Agent Orange Researcher Backs Disputed Findings," *New Zealand Herald*, March 4, 2004.
63. "Inquiry into the Exposure of New Zealand Defence Personnel to Agent Orange and Other Defoliant Chemicals during the Vietnam War and Any Health Effects of That Exposure, and Transcripts of Evidence," Report of the Health Committee, 47th Parliament, October 2004, available at www.vvanz.com/pdfs/rpt-2004-chadwick/2004chadwick.pdf; hereafter Chadwick Report.
64. Ibid., 20.
65. Ibid., 40.
66. R. E. Rowlan, L. A. Edwards, and J. V. Podd, "Elevated Sister Chromatid Exchange Frequencies in New Zealand Vietnam War Veterans," *Cytogenetic Genome Research* 116, no. 4 (2007): 248–51; "Agent Orange Study to Show Significant Damage," *New Zealand Herald*, July 28, 2006. While news of this study created quite a stir in New Zealand, Canada, and the United States, its extremely small sample size (twenty-four subjects) limits its application severely.
67. "Agent Orange Campaigners Angry as Many Miss Out," *New Zealand Herald*, December 8, 2006; "4600 Vietnam Veterans Sign Up for Compensation Register," *New Zealand Herald*, December 20, 2007.
68. "Full Text of the Crown's Apology to Vietnam Veterans," *New Zealand Herald*, May 28, 2008.
69. "Sorry, Says Government to Soldiers," *Dominion Post* (Wellington, NZ), May 28, 2008; "Full Text of the Crown's Apology."
70. "Sorry, Says Government to Soldiers."
71. Louise Elliot, "N.B. Army Base Sprayed with Toxic Chemicals," June 13, 2005, available at www.cbc.ca/canada/story/2005/06/13/agent-orange050613.html; "Agent Orange and Agent Purple: In Depth," CBC News, August 21, 2007, available at www.

cbc.ca/news/background/agentorange/; Chris Arsenault, *Blowback: A Canadian History of Agent Orange at Home* (Black Point, Nova Scotia: Fernwood Publishers, 2009).

72. "Harper Woos with Agent Orange, Highway Promises," CBC News Online, January 12, 2006, available at www.cbc.ca/news/canada/newbrunswick/story/2006/01/12/nb_harpernb20060112.html.

73. As discussed here in chapter 1, both Louise Martin's CBC reports and Chris Arsenault's *Blowback* misstate a number of facts about Agents Orange and Purple.

74. "Agent Orange and Agent Purple: In Depth."

75. "Report: No Evidence to Link Cancer Rates to Agent Orange," CBC Online, at www.cbc.ca/news/canada/new-brunswick/story/2007/08/21/nb-agentorangereport.html.

76. "People Harmed by Agent Orange at Gagetown Offered $20K," CBC News Online, September 12, 2007, available at www.cbc.ca/news/canada/new-brunswick/story/2007/09/12/agent-orange.html.

77. "People Angry with Agent Orange Package Turn to Class-Action Lawsuit," Canadian Press Online, September 14, 2007, www.thecanadianpress.com/agent_orange.

78. "Judge Halts Agent Orange Class-Action Suit," *The Telegram* (St. John's, New Brunswick, Canada), April 1, 2010. Text of ruling from Claims Canada, *Ring v. Canada*, 2010 NLCA 20, www.claimscanada.ca/issues/article.aspx?aid=1000404982.

79. "Final Report: Toxicological Risk Assessment Pertaining to Potential Occupational and Related Exposures Associated with Herbicide Spraying Operations at CFB Gagetown," March 31, 2007, available at www.forces.gc.ca/site/reports-rapports/defoliant/index-eng.asp, v.

80. See, for instance, "Your Assignment: Agent Orange," CBC News Online, March 4, 2011, www.cbc.ca/news/yourcommunity/2011/03/your-assignment-agent-orange.html; and "Agent Orange 'Widely Used' in Ontario," CBC News Online, February 28, 2011, www.cbc.ca/news/canada/toronto/story/2011/02/28/agent-orange-ontario.html.

81. "Probe Sought on Agent Orange Use in Ontario," CBC News Online, March 2, 2011, www.cbc.ca/news/canada/toronto/story/2011/03/02/Ontario-agent-orange.html; "Agent Orange Calls Flood Hotline," CBC News Online, March 7, 2011, www.cbc.ca/news/canada/toronto/story/2011/03/07/agent-orange-hotline.html,.

82. "Original Complaint," available at www.ffrd.org/Lawsuit/Court.htm. Additional legal documents on the case are available through this site as well; also see the Agent Orange Record website's legal section, www.agentorangerecord.com/information/the_quest_for_additional_relief/.

83. "To the American People: Open to the American People from the Vietnam Association for Victims of Agent Orange/Dioxin," August 2004, included in packet provided by VAVA leadership during interview with author, Hanoi, May 5, 2008.

84. Quoted in Weinstein decision, 138–39, available at www.ffrd.org/AO/10_03_05_agentorange.pdf.

85. Agent Orange Record Website, www.agentorangerecord.com/information/the_quest_for_additional_relief/P2/.

86. Memorandum, Order, and Judgment, *In re "AGENT ORANGE,"* decision posted at www.ffrd.org/AO/10_03_05_agentorange.pdf.

87. Peter Schuck, *Agent Orange on Trial* (Cambridge: Harvard University Press, 1986), 311.

88. "Statement of the Vietnam Association for Agent Orange/Dioxin Victims on the Reject of Claim of Vietnamese Agent Orange Victims by Judge J. B. Weinstein," March 2005, included in packet provided by VAVA leadership during interview with author, Hanoi, May 5, 2008.
89. VAVA interview with the author, May 5, 2008.
90. Ibid.
91. Interview with the author, April 21, 2008.
92. Ibid.
93. Ibid., April 28, 2008.
94. Ibid.
95. Ibid.; VAVA interview with author, May 5, 2008.
96. Stellman et al., "Extent and Patterns of Usage of Agent Orange," 685.
97. "Vietnamese Victims of Agent Orange and U.S.-Vietnam Relations."
98. Agent Orange activists to review future plans, Vietnam News Agency Online, April 1, 2011, available at http://vietnamnews.vnagency.com.vn/Social-Isssues/209930/Agent-Orange-activists-to-review-future-plans.html (emphasis added).
99. Schecter interview with author.
100. "Table: Vietnam Red Cross List of Diseases Caused by Agent Orange/Dioxin," in "Vietnamese Victims of Agent Orange and U.S.-Vietnam Relations," 17.
101. Some early reports are discussed in Barry Weisberg, ed., *Ecocide in Indochina: The Ecology of War* (San Francisco: Canfield Press, 1970), 19–21. For a thoughtful exploration of this issue, see Diane Fox, "Agent Orange: Coming to Terms with a Transnational Legacy," in *Postwar Interventions: Transnational Legacies of the Second Indochina War*, ed. Scott Laderman and Edwin Martini (Durham: Duke University Press, forthcoming).
102. "Agent Orange's Bitter Harvest," *Science* 315, no. 5809 (January 12, 2007): 178.
103. Anh Ngo et al., "Association between Agent Orange and Birth Defects: Systematic Review and Meta-Analysis," *International Journal of Epidemiology* 35 (2006): 1220–30; 1223–24; 1228.
104. Arnold Schecter and John Constable, "Commentary: Agent Orange and Birth Defects in Vietnam," *International Journal of Epidemiology* 35 (2006): 1230–32.
105. Among the most famous recent examples of such photos is Phillip Jones Griffiths, *Agent Orange: Collateral Damage in Vietnam* (London: Trolley Books, 2003). The Vietnamese government has also produced *For the Victims of Agent Orange* (Hanoi: Vietnamese News Agency Publishing House, 2008), which contains hundreds of photographs depicting an array of birth defects and claims. In the introduction it is stated that "more than 4.8 million Vietnamese were exposed to dioxin, of whom about three million were seriously affected" (3).
106. "Our Forgotten Responsibility: What Can We Do to Help Victims of Agent Orange?" House Subcommittee on Asia, the Pacific, and the Global Environment, Committee on Foreign Affairs, 110th Congress, 2nd session, May 15, 2008 (Washington: Government Printing Office, 2008), 17–18.
107. "United States Supreme Court Hears Challenge to Twenty-Year-Old Agent Orange Class Action Settlement," FindLaw Website, http://library.findlaw.com/2003/Mar/7/132621.html.
108. *Veterans and Agent Orange: Health Effects of Herbicides Used in Vietnam,* Report of the Committee to Review the Health Effects in Vietnam Veterans of Exposure to

Herbicides (Washington: National Academy Press, 1994), 224 (emphasis added).
109. Ibid., 634.
110. Copies of *Agent Orange Review* are archived on the VA website, www.publichealth. va.gov/exposures/agentorange/newsletter_archive.asp.
111. *Agent Orange Review* 17, no. 1 (December 2000): 1–2.
112. *Veterans and Agent Orange: Herbicide/Dioxin Exposure and Type 2 Diabetes*, Institute of Medicine Report (2000), available at http://books.nap.edu/catalog/9982.html.
113. *Veterans and Agent Orange: Herbicide/Dioxin Exposure and Type 2 Diabetes*, 3–4.
114. Dr. Joel Michalek, who worked extensively on the Ranch Hand study, however, has consistently maintained that the diabetes issue among Vietnam veterans is actually *understated*. See his article "Diabetes and Cancer in Veterans of Operation Ranch Hand after Adjustment for Calendar Period, Days of Spraying, and Time Spent in Southeast Asia," *Journal of Occupational and Environmental Medicine* 50, no. 3 (March 2008): 331–40.
115. "Persisting Problems with Communication of Ranch Hand Study Data and Results," GAO Testimony before House Subcommittee on Government Reform, March 15, 2000.
116. "The History of the US Department of Defense Programs for the Testing, Evaluation, and Storage of Tactical Herbicides," available at www.dod.mil/pubs/foi/reading_room/TacticalHerbicides.pdf.
117. "Veterans Affairs: Health Care and Benefits for Veterans Exposed to Agent Orange," Congressional Research Service Report, February 11, 2008, available at www.fas.org/sgp/crs/misc/RL34370.pdf.
118. "Diabetes Now Tops Vietnam Vets' Claims," Associated Press Report, August 31, 2010.
119. "HR 2254: Agent Orange Equity Act of 2009," 111th Congress, 2009–10, available at www.govtrack.us/congress/bill.xpd?bill=h111-2254.
120. John Paul Rossie and Wallace Ward, "The Da Nang Harbor Report: Contamination of Da Nang Harbor: Blue Water Navy and Exposure to Herbicides in Vietnam," available at www.bluewaternavy.org/danangcombo2.pdf.
121. Ibid., 20–22.
122. "Inconclusive Agent Orange Study Is Conclusive Enough for Vet Groups," *NYT* "At War" Blog, June 14, 2011, available at http://atwar.blogs.nytimes.com/2011/06/14/inconclusive-agent-orange-study-is-conclusive-enough-for-vet-groups/.
123. Ibid. At the time of this writing, Sen. Kirsten Gillibrand (D-NY) had introduced a bill to amend the Agent Orange Act of 1991 to include Blue Water veterans.
124. "Oversight Hearing: VA Disability Compensation: Presumptive Disability Decision-Making United States Senate Committee on Veterans' Affairs," September 23, 2010, available at http://veterans.senate.gov/. Shinseki's testimony is also posted at www.va.gov/OCA/testimony/svac/09232010SecVA.asp.
125. Ibid.; "Diabetes Now Tops Vietnam Vets' Claims."
126. "Diabetes Now Tops Vets' Claims"; "Aging Vets' Costs Concern Obama's Deficit Co-Chair," Associated Press Report, August 31, 2010.
127. "Paul Reutershan and all those so unfortunate . . ." ALY, series VIII, subseries 1, no. 05551 (emphasis added).

CONCLUSION

1. Phung Tuu Boi, "Agent Orange and the Environment: From Research to Remediation," presentation at Agent Orange: Landscape, Body, Image Conference, Riverside, CA, May 2009, in possession of the author. My retelling of Boi's story also draws on an interview with the author, May 8, 2009, and on Christie Aschwanden, "Through the Forest, a Clearer View of the Needs of a People," *NYT,* September 18, 2007.
2. For more on such asymmetries, see Edwin Martini, *Invisible Enemies: The American War on Vietnam, 1975–2000* (Amherst: University of Massachusetts Press, 2007), epilogue. For more on the catfish episode, see Scott Laderman, "A Fishy Affair: Vietnamese Seafood and the Confrontation with U.S. Neoliberalism," in *Postwar Interventions: Transnational Legacies of the Second Indochina War,* ed. Scott Laderman and Edwin Martini (Durham: Duke University Press, forthcoming).
3. This estimate comes from Amy Belasco, "The Costs of Iraq, Afghanistan, and Other Global War on Terror Operations since 9/11," Congressional Research Report, September 2, 2010; and Center for Arms Control and Non-Proliferation, http://armscontrolcenter.org/policy/securityspending/articles/gwot_spending_burn_rate.

Index

Advanced Research Project Agency (ARPA), 22, 25, 41, 46–49, 57, 91–92
Agena, Robin, 156
Agent Blue: chemical makeup of, *23*; defoliation in Vietnam and, 33, 47–48, 54, 243; disposal of, 110; domestic herbicide use and, 35; Hurricane Camille and, 105, 116; propaganda and, 75; testing of, 32; transportation of, 27, *28*; veterans' health claims and, 167–69
Agent Green, *23*, 57
Agent Orange: ARVN handling of, 29–31; banning of, 34–35, 101–3; barrels and, 27–31, 103–4, *104*, 105–11, *121–22*, 123, *124*, 126; birth defects and, 3–4, 11, 112, 131, 137–39, 151, 153, 167, 213–14, 226–28, 231, 238–39, 246; carcinogenic capacities of, 11, 15, 215–19; chemical makeup of, 2, *23*, 32, 97, 143, 147–49; civilian complaints and, 29–30, 50, 100; congressional hearings on, 150–62, 190, 192, 229; crop destruction missions and, 18, 243; defoliating effects of, 17; diabetes and, 1, 232–34, 288n114; disposal of, 6, 28–30, 97–123, 131; domestic herbicide use and, 35, 108–9, 148; environmentalism and, 3–4, 98, 100–111, 121, 126–37; epidemiological explorations of, 11, 180–88; exposure-proving and, 14–15, 149–50, 155–56, 162–79, 196; forest fire operations and, 48; global or transnational nature of, 3–5, 8, 26–40, 137–45, 179–238; governmental knowledge about, 4, 97–98, 100–101, 107–8, 128, 146–47, 160–61, 187, 191–96, 211, 215, 217–18, 239–42; Hurricane Camille and, 105, 116; indemnification process and, 77–80, 88, 96; international law and, 114, 222–37, 240, 242–44; journalistic accounts of, 34, 37; Korea and, 3, 8, 145, 183, 197, 210–14; legal battles over, 9, 100, 129, 222–37, 273n5; maps of, *16*; media and, 154–62, 170, 179–80; Operation Ranch Hand and, 6, 17–18; producers of, 2–3, 18–19, 23, 26–27, 31–37, 100, 126, 146–47; propaganda and, 18, 75, 204–5; protective equipment and, 63; redrumming and, 104; remediation for, 2, 14, 207–14, 246; scholarship on, 6, 8–9, 62–63, 66; soil and water contamination and, 198–238; spills and, 29, *124*; spraying of, *118*, 203; storage of, 6, 28, *28*, 28–30, 101, *104*; symbolic resonances of, 4–5, 12–15, 41, 143, 239; testing of, 21, 23–24, 31–32, 34, 219–22; transportation of, 27, 99; trauma and, 180; veterans' lawsuits and, 98, 129, 131–32, 152–88, 210–14, 244–47. *See also* cancer; dioxins; Geneva Accords (1954); international law; Operation Pacer HO; Operation Pacer IVY; Operation Ranch Hand; veterans; *specific air bases, missions, scholars, and scientists*
"Agent Orange: At the Crossroads of Science and Social Concern" (Young), 157
Agent Orange: the Bitter Harvest (Dux and Young), 37
Agent Orange: The Human Harvest, 159
Agent Orange: The View from Vietnam, 159
Agent Orange: Vietnam's Deadly Fog (documentary), 154–62, 179, 185
Agent Orange Act of 1991, 195–96, 198, 230–32, 236

[291]

INDEX

Agent Orange Association of Canada, 220
"The Agent Orange Cover Up" (report), 190
Agent Orange Equity Act of 2009, 234
Agent Orange on Trial (Schuck), 224, 273n5
Agent Orange Registry, 170
Agent Orange Review (VA), 231
Agent Orange Study, 171
Agent Orange Update (VA), 232
Agent Orange Working Group (AOWG), 153, 158, 171–75, 178, 183, 186, 189–90
Agent Pink, 23, 24, 57
Agent Purple: chemical makeup of, 23, 33; deployments of, 57, 243; dioxin content of, 34–35; testing of, 24, 32, 34, 219–22
Agent White: chemical makeup of, 23; defoliation in Vietnam, 35–36; defoliation in Vietnam and, 33, 48; disposal of, 110; domestic herbicide use and, 35; testing of, 34, 219; transportation of, 27
Air War College, 56
Alien Tort Claims Act, 222, 224
Alsea, Oregon, 137–40, 183
A Luoi, 204, 206, 208, 210, 238
American Association for the Advancement of Science (AAAS), 100, 241
American Medical Association, 131
AnChem (company), 38
An Xuyen province, 77, 79–80
Ap Bac (battle), 71
Arkansas, 32
Armed Forces Pest Control Board, 110
Arsenault, Chris, 34–35
ARVN (South Vietnamese Army): Agent Orange handling and, 29, 101, 104; crop destruction and, 57; military effectiveness of, 66; propaganda efforts and, 75; strategic hamlet program and, 71–73; USFS and, 50
"Assessing Exposure to Allied Ground Troops in the Vietnam War" (Ginevan, Ross, and Watkins), 202

Australia: Agent Orange's production and, 4–5, 9, 36–37, 40, 140–41; Vietnam veterans and, 98, 129, 179–96

Baker, Mike, 249n1
Ball, George, 57
Baltimore, 27
Ba Ria province, 64, 69
barrels (of herbicides): deterioration of, 103, 124, *205*; disposal of, 30, 122–24; photos of, *104, 121–22, 124*; redrumming and, 104–11, 126; repurposing of, 28, 30–31; residual dioxin content of, 27, 123. *See also* Operation Pacer HO; Operation Pacer IVY
Bates, Michael, 140
Bayley, Ned, 152
Beck, Ulrich, 273n11
Bernstein, Joan, 163
Betts, Russell, 54, 85, 87
Bien Hoa: as Agent Orange staging point, 27, *28*, 29, 31, 103–4; Laotian missions and, 32; Operation Flyswatter and, 44; redrumming operations on, 105; soil and water contamination and, 30, 125, 204, 209–10, 226, 245–47
Binh Dinh province, 17, 63, 69, 89
Binh Duong province, 90
Bionetics Research Laboratory, 100, 102, 228, 241
birth defects: Agent Orange and, 3–4, 131, 151, 153, 155–56, 167, 213–14, 227–28, 231, 238–39, 246; Environmental Impact Statements, 112; herbicide production locations and, 137–39; studies on, 100, 138–40; Vietnamese exposure models and, 207–8, 226. *See also* Agent Orange; dioxins
Bliss, Russell, 97–98, 126–37, 270n80
Blowback (Arsenault), 34
Blue Spruce Chemical Company, 109
Blue Water navy servicemen, 234–37
Boi Loi Woods, 46
Bonior, David, 165–70
Brady, Lisa, 213
Brazil, 109
Brinkman, G. L., 143–44
"Brother Nam" cartoons, *75, 76,* 77, 263n81

Brown, James, 21, 25, 45
Bryant, Ferris, 36
Buckingham, William, 25, 31, 36, 51, 54, 57, 60–61, 80, 102
Bundy, McGeorge, 53
Bundy, William, 61
Bunker, Ellsworth, 94
Bush, George H. W., 207
Bush, George W., 208

Ca Mau province, 64
Cambodia, 18, 32–33, 45, 48, 255n55
Camp Carroll, 214
Camp Drum, 23–24
Canada, 3, 8–9, 32, 34, 140–41, 183, 197, 219–22
Canadian Broadcasting Corporation, 34–35, 219, 221–22
cancer: epidemiological studies of, 11, 171–79, 189–91, 232; suggestive links to Agent Orange and, 15, 131, 220–22, 232, 236–37; veterans' Agent Orange exposure and, 153, 189
Carmichael, Richard, 104–5
Carson, Rachel, 10, 53, 103
Carter, Jimmy, 163, 171
Cat Son (spraying of), 17
Cecil, Paul, 29, 54, 81, 200–201, 250n5
Centers for Disease Control (CDC), 98, 126–28, 134, 166, 171–79, 183, 189–91
Chadwick, Steve, 217–18
Cha La (hamlet), 79–80
Chemical Corps, 25, 30–32, 71, 109, 201
chemical war: definition of, 13; domestic environmental problems and, 126–37; forest fire and, 19–20, 44–46, 48–49; global or transnational infrastructure of, 19, 26–40; insecticides and, 53; international law and, 55–57, 61–62, 101–2; military tactics and, 2, 4; origins of, 20–26; scholarship on, 54; uncertainty's role in and after, 150–62; United States legislation and, 114; Vietnamese civilians and, 84–96. *See also* Agent Orange; Operation Ranch Hand; science (discourse); United States; Vietnam War; *specific companies*
Chemical Warfare Service, 22

Chernobyl, 149
Chieu Hoi program, 49, 62
chloracne, 138, 147, 177
Christian, Richard, 173–78
Chul, Hwang Myung, 212
Chu Pong Mountain, 47–48
Cisco, James, 134
Civil Operations and Revolutionary Development Support (CORDS), 29, 78–79, 199–200
Clark, Helen, 218
Clary, James, 191–92, 241, 256n64, 280n127
Clean Air and Clean Water Acts, 103, 113
Cleland, Max, 163–64, 167–69
Clinton, Bill, 207
Cobb, Michael, 141
Coca-Cola Company, 39–40
Cold War: assumptions and mindsets of, 3–6, 18, 20–26, 240; decolonization and, 55; propaganda and, 18–19, 56. *See also* Kennedy, John F.; propaganda; Vietnam War
Colombia, 109
Colston, Ted, 190
Combined Intelligence Center, Vietnam (CICV), 81–83
Coming Home (film), 170
Commoner, Barry, 159
"Concern over Unclear Threat of Dioxin" (article), 131
conspiracy theories, 166–71, 188–96, 215–18, 239–42
Constable, John, 204, 229
Coolidge, Calvin, 112
coverups, 166–71, 188–96, 215–18, 239–42
Cregg, Sheila, 142
crop destruction missions: Agent Orange's use in, 33, 243; assessments of, 54, 57, 63–64; ending of, 95; indemnification process and, 77–80, 88, 96; military strategy and, 18, 21–22, 58–59, 84–90, 261n32; politics of, 53, 55–59, 61–70, 79–96; propaganda and, 70–77, 241, 263n81; scholarship on, 54; Vietnamese civilians and, 14, 50, 57–59, 62–96. *See also* Agent Orange; National Liberation Front

crop destruction missions (*continued*) (NLF); Operation Ranch Hand; Vietnam War
Cuba, 18

Da Nang: Agent Orange's shipping and storage and, 27–31, 103, 146; herbicide missions and, 32, 90; Operation Flyswatter and, 44; Pacer IVY and, 104, 118; soil and water contamination and, 125, 204, 206–10, 226, 234–35, 245–47
Da Nang Harbor Report, 234–35
Daniel, Raymond, 186
Daschle, Tom, 164, 166–70, 192
Davis, California, 32
The Deer Hunter (film), 170
Defense Intelligence Agency (DIA), 90, 92
Defense Threat Reduction Agency (DTRA), 113
defoliation: chemical effects and, 17, 47–48; disposal of, 100–111; effectiveness of, 84–96; indemnification processes and, 77–80, 88, 96; military use of, 2, 13–14, 17–19, 24–26, 31–33, 41–46, 60–61; National Liberation Front's goals and, 2, 41–46, 53–54; politics of, 53, 56–57, 60–61, 66, 70–77, 84–95. *See also* Agent Orange; Operation Ranch Hand; Vietnam War
"Defoliation-Incidents Correlation Study" (report), 91
Delmore, Fred, 25
demilitarized zones (DMZ), 33. *See also* Korea
Denton, Frank, 54, 85, 87, 92–94
Department of Health and Veterans Affairs (New Zealand), 182, 218
Department of Veterans Affairs, Australian (DVA), 180
DeVictor, Maude, 152–55, 157, 159–60, 168–69, 185
diabetes (Type 2), 1, 232–34, 236, 245, 288n114
Diamond Alkali, 147, 265n2
Dinh Tuong province, 71–77
"Dioxin: Quandary for the '80s" (series), 145

dioxins: barrels of Agent Orange and, 27, 123; chemical and biological weapon definitions and, 114–15; congressional hearings on, 150–62, 190–92, 229; emotive responses to, 140, 157–62, 166–67, 169–71; environmental contamination and, 6–7, 9, 98, 108, 115, 128–37, 270n80; health consequences associated with, 1–2, 26, 55–56, 97, 100–101, 106, 108, 115, 126, 128, 133, 137–38, 140, 144, 147–48, 155, 159, 161, 171–79, 197–98, 220–22, 226–37; hot spots of, 202–10, 237–38; incineration and, 106–23, 131, 134–36, 208–10; safe and harmful levels of, 110, 125, 134, 147, 165; soil and water contamination and, 12, 27, 30, 33–34, 100–101, 116, 124–37, 187, 198–238, 245–47; uncertainty and conflicting information regarding, 127–45, 149, 154. *See also* Agent Orange; barrels (of herbicides); Dow Chemical; Operation Ranch Hand; Times Beach (Missouri)
"Dioxin's Peril to Humans" (article), 131
Dow Chemical: Agent Orange's production and, 26–27, 110, 146–47, 265n2; congressional hearings and, 152; exposure modeling and, 202; herbicides' testing and, 32; military contracts of, 35–36, 110–11, 160, 223–25; public relations and, 131; veterans' lawsuits and, 159–62, 183, 202, 212
drums (of Agent Orange). *See* barrels (of herbicides)
Duval, Merlin, 123
Dux, John, 37
Dwernychuk, Wayne, 205

Economist, 132
Edgewood Task Force Report, 275n35
"Effects of 2,4,5-T on Man and Environment" (hearings), 150–51
Eglin Air Force Base: Agent Orange testing and, 34, 108–9, 157; cleanup and, 124; pilot training and, 33, 192, 203
Eisenhower, Dwight, 19–21, 51–52
Elliot, David, 66
Elliot, Louise, 34–35, 219

Elliot, Mai, 84, 88, 94
Ellsberg, Daniel, 94
England, 140–41
Environmental Impact Statements (EIS), 99, 107, 111–19, 123–26, 199, 268n40
environmentalism: Agent Orange's storage and disposal and, 6, 98–99, 115, 202–10; bodily divide and, 61–62, 99–100, 150, 178–79; ecological disasters and, 103, 106–7, 126–38, 140, 142, 152, 270n80; "emotive" responses and, 140, 157–62, 166–71; historical scholarship and, 3–4; nature's control and, 41–52; political rise of, 10, 13. *See also* Agent Orange; Nixon, Richard; *specific legislation*
Environmental Protection Agency (EPA), 99; Agent Orange's domestic use and, 107–10; creation of, 10, 103; dioxin's danger and, 131–32, 157; disposal of Agent Orange and, 111–23, 125–37; PCB contamination and, 107, 129; scientific uncertainty and, 152; Times Beach and, 129, 134–36
Environmental Science and Pollution Resolution (journal), 200
"An Evaluation of Chemical Crop Destruction in Vietnam" (Betts and Denton), 54, 85–87
Evans, Charles, 153, 155
Evatt, Phillip, 183
Evatt Revisited (essay collection), 188
Evatt Royal Commission on the Use and Effects of Chemical Agents on Australian Personnel in Vietnam, 179, 181–88, 194
experience (as challenge to scientific objectivity), 10–12, 171–79, 181–96, 198
exposure (to Agent Orange): Australian models of, 180; causality and, 1, 152–62; experiential evidence for, 171–79; health effects and, 144; modeling of, 171, 196, 200–210, 227–37; proving of, 14, 130, 136–45, 149, 162–71, 264n2; symptoms of, 138, 177. *See also* veterans; Veterans Administration (VA); *specific studies*

Faithful, Barbara, 139
Faleomavaega, Eni, 229
First International Conference on Agent Orange and Dioxin, 229
First World War, 13, 137
Firth, Jerry, 121
Food and Chemical Workers Union, 143
Ford Foundation, 208
Foreign Claims Act, 77–80
forest fire (as military tool), 19–20, 44–49
Fort Detrick, 22–23, 30–32
Fox, Diane, 250n6
Friendship Village, 192
Furlong, Ray, 50

Gagetown Base (Canada), 34–35, 219–22
Gair, George, 139–40
Galston, Arthur, 152
Geneva Accords (1954), 55, 61–62
Geneva Protocols on Chemical and Biological Weapons (1925), 101–2, 114–15, 223, 240, 242–44
Geographic Information Systems (GIS), 199, 201
Georgia, 32
Germany, 143
Gibson, James, 41
Ginevan, Michael, 202
Gough, Michael, 171, 188
Government Accountability Office, 159, 170–71, 233
Greenpeace, 142–43
Greenville, Mississippi, 32
Griggs, David, 88
Grozon (pesticide), 144
Gulfport, Mississippi, 27, 101, 105, 110, 115–17, 119, 122, 124–26. *See also* Naval Construction Battalion Center (NCBC)

Haas, Jonathan, 234–37
Haber, Paul, 163
Hagopian, Patrick, 170
Hamilton, Neil, 121
Hamlet Evaluation System, 92, 199–200. *See also* strategic hamlet program
Hammer, Sid, 130
Hance, Ernest, Jr., 130, 132

Harriman, Averell, 58
Hart, Phillip, 151–52
Hatfield Consultants, 204, 206–8, 210
Hau Nghia province, 68, 89
Hawaii, 32
Herbicidal Warfare (Cecil), 54, 81
Herbicide Policy Review, 78–79, 94–95
herbicides: dioxin content of, 34–35, 86, 137; disposal of, 100–111; domestic uses of, 22–23, 31, 36–37, 47, 107–8; environmental legislation and, 100–123; handling of, 105–11; nature's unruliness and, 41–52; New Zealand's use of, 136–45; production of, 33–38; spraying devices and, 203; testing of, 21, 23, 26, 31, 40; U.S. governmental knowledge about, 13, 15, 41, 55–57; weaponization of, 1–2, 4–6, 17, 19, 21–26, 61, 240. *See also* Agent Orange; defoliation; dioxins; Operation Pacer HO; Operation Pacer IVY; Operation Ranch Hand; United States
HERBS tapes, 29, 33, 180, 189, 198–201, 255n39
Hercules (company), 147
Hill, Bonnie, 138–39
Hilsman, Roger, 58, 241
Ho Chi Minh, 222–23
Ho Chi Minh City, 225, 229. *See also* Saigon
Ho Chi Minh Trail, 32–33, 41, 50
Hoffman-Taff (company), 98, 161
Hooker Chemical, 147
House Committee on Government Operations, 190
House Veterans' Affairs Committee, 121
Hull, Warren, 105
Hunt, David, 66
Hurricane Camille, 105, 116

Illinois, 107, 135
incineration (of Agent Orange/dioxin), 106–23, 131, 134–36, 208–10
"Inconclusive Agent Orange Study Is Conclusive Enough for Vet Groups" (article), 235
indemnification (for crop destruction), 77–80, 88, 96

Independent Petrochemical Corporation, 97–98
Inouye, Daniel, 116
insecticides, 53, 180–81. *See also* dioxins; malathion
International Control Commission, 57
International Journal of Epidemiology, 228–29
international law, 55–57, 61–62, 114–15, 222–37, 240, 242–44
International Society of Toxicology, 147
The Invention of Ecocide (Zierler), 3–4
Iran, 42
Ivon Watkins Dow (IWD), 37–38, 40, 137–45, 215–16

Japan, 112, 143
Johnson, Adam, 134
Johnson, Alexis, 60
Johnson, Julius, 148
Johnson, Lyndon B., 18, 21, 52–53, 60, 243
Johnson, V. R., 39, 182, 188
Johnston Atoll, 105–6, 111–13, 115, 117, 119, 121–22, 124–25, 135, 233–34, 268n40
Joint Chiefs of Staff (JCS): Agent Orange's banning and, 101; crop destruction programs and, 58–60, 88–90, 92–93; forest fire operations and, 46; herbicide war's approval, 25, 87
Kastenmeier, Robert, 61

Keffler, Bill, 133
Kennedy, John F.: assassination of, 20; "best and brightest" mindset and, 5–6, 18–19, 25, 41, 51–52, 91; "flexible response" approach of, 21, 25, 55; herbicide weaponization and, 5, 56–57, 60; international perception and, 53, 59–61, 70, 102–3, 240–43; Vietnam War's escalation and, 21, 254n26
Kokkoris, Constantine, 275n35
Korea, 56, 145, 183, 197, 210–14, 233; Agent Orange uses in, 3, 8
Korean Vietnam Veterans Association, 212
Korean War, 212
Kraus, E. J., 41
Kurtis, Bill, 154–59, 185

Lake Sen, 208–9
Langston, Nancy, 179
Laos, 18, 21, 32–33, 49, 255n55
Lavelle, Rita, 132–33
Le Cao Dai, 204
Lemnitzer, Lyman, 60
"Lessons Learned: Vietnam" (MACV), 28–29
Lewis, James, 50
"Life in the DMZ" (Brady), 213
Long An province, 68
Long Dinh province (and resettlement hamlet), 68, 71, 74
Love Canal disaster, 99–100, 103, 106, 133–34
Luu Van Dat, 224–25

MacDonald, G. C., 109
Mackenzie-Komp, Kayla, 143
Mai, Truyet, 204
malaria, 42, 44
malathion, 42, 44, 137, 180–81
Mansker, Donna, 130, 132
Marine, Michael, 207
Marine Protection, Research and Sanctuaries Act (MPRSA), 114
Maryland, 32
McCulloch, Jock, 182
McKee, Stephanie, 139
McLeod, Deborah, 216–17
McMillan, W. G., 49
McNamara, Robert S., 19–21, 25, 36, 56–59, 87–88, 91–94, 241
McNamara Line, 41
McNew, George, 41
McVeigh, Timothy, 216
Meramec River, 126, 135
Messelson, Matthew, 102, 159
Messner, A. J., 181
Michalek, Joel, 288n114
Midland, Michigan, 18, 27, 146–48, 193
Military Assistance Command, Vietnam (MACV): Agent Orange's banning and, 101; crop destruction programs and, 58–60, 82, 89, 91–94; defoliation efforts and, 73; drum disposal and, 28; forest fire operations and, 46, 49; herbicide war and, 25, 33, 87;

indemnification process and, 77–79. *See also* Vietnam War
Miller Products (company), 38
Minarik, Charles, 109–10
Mississippi Air and Water Pollution Control Commission, 116
Missouri, 126–37, 145; Agent Orange storage and disposal in, 3
Mobile, Alabama, 27
"Modification and/or Destruction of Defoliants" (presentation), 110–11
Moller, John, 216
Monsanto (chemical company): Agent Orange's production and, 26–27, 147–48; disposal of Agent Orange and, 107, 135; exposure modeling and, 202; government contracts of, 32, 223–25; herbicide disposal and, 107; lawsuits and, 160, 183, 187, 212
Montagnards (ethnic minority), 58, 60
Moore, Michael, 215
Mui Lon hamlet, 89
Muldoon, Robert, 215
Muller, Bobby, 164–66, 190
Murray, John, 176–78
Murrow, Edward R., 53
mustard gas, 13
My Lai massacre, 18
My Tho province, 66, 71

napalm, 44–48, 267n24
National Academy of Sciences, 155, 162, 166, 195, 198, 203
National Cancer Institute (NCI), 189–90
National Council of the Churches of Christ, 165
National Institute of Medicine (IOM), 190–91, 228, 231
National Interrogation Center (NIC), 89
National Liberation Front (NLF): crop destruction missions and, 14, 18, 21–22, 54–56, 58–60, 62–70, 75, 80, 82, 85, 96; defoliation campaigns and, 2, 41–42, 44–46; food shortages and, 64–65, 67; forest fire attacks on, 44–47; growth of, 20–21; interrogations of, 88–91, 261n32; morale assessments and, 81, 84; People's Chemical Teams

National Liberation Front (*continued*)
  and, 68; strategic hamlet program and, 41, 52, 71–77; Vietnamese civilians and, 62–70. *See also* Agent Orange; crop destruction missions; Vietnam War
National Security Memorandum 115, 25
Nature (journal), 27, 198–200, 227
Naval Construction Battalion Center (NCBC), 101, 105, 110, 116–17, 124–25
Naval Defensive Sea Areas, 112
Naval Ordnance Laboratory, 51
Neal, Leo and Suzanne, 139
Nelson, Otis "Barney," 135
New Orleans, 27
New Plymouth (New Zealand), 7, 37, 99, 136–45, 215–19
Newton, Michael, 108
New Yorker, 53
New York Times, 45, 61, 130–31, 134, 191, 212, 235
New Zealand: herbicide production in, 3–5, 9, 37–40; Vietnam veterans and, 98, 129, 136–45, 182, 214–19, 245. *See also specific politicians and sites*
New Zealand Environmental Council, 142
Ngo, Anh, 228–29
Ngo Dinh Diem, 20–21, 52, 55, 57, 60
Nguyen Dinh Thuan, 60
Nguyen Minh Triet, 208
Nguyen Trong Nhan, 224
Nha Trang, 103
NIMBY (Not In My Backyard), 100, 106–7, 115, 122–23, 135–36
9/11 Commission, 179
Nixon, Richard, 6, 95, 97, 101–3, 115
No Gun Ri, 212
Northeastern Pharmaceutical and Chemical Company (NEPACCO), 128, 135, 270n80
North Korea. *See* Korea
Nuclear Regulatory Commission, 113

Obama, Barack, 232, 235–36
Office of Emergency Planning, 36
Office of Technology Assessment (OTA), 171, 173
Olenchuk, Peter, 71–73
Operation Dominic, 112

Operation Flyswatter, 42–44
Operation Grapple, 216
Operation Hot Tip, 45, 47–48
Operation Pacer HO, 99, 113–23, *120–22*, 123, 125–26, 128, 130, 135–36, 145, 148–49, 155, 199, 203, 209, 234, 268n40
Operation Pacer IVY, 99–111, 115, 118, 122, 126, 128, 148–49, 208, 234
Operation Pink Rose, 45, 48
Operation Popeye, 49
Operation Ranch Hand: accidents and mishaps in, 79–80; Agent Orange's spraying missions and, 6; airbases of, 27; approval of, 56–57; assessments of, 54, 57, 81, 83–96; cancellation of, 95, 97, 100–101, 149, 162, 241; environmentalist thinking and, 10; escalation of, 29–30, 32, 39, 60–61, 89, 128, 211; exposure questions and, 171–79, 200–201, 227; extent of, 50; insecticides and, 42, 53; Kennedy's technocratic mindset and, 19–20; missions of, 17, 243; as part of larger chemical war, 49–52; photos of, *43*; pilot training and, 33; policymakers' "evil" and, 13, 101; politics of, 53–54; propaganda and, 18, 42–43, 66, 71–77, 79–81; theory and practice discrepancy and, 24–26, 29, 33, 71, 198–210
Operation Ranch Hand (Buckingham), 54
Operation Rolling Thunder, 69, 146
Operation Sherwood Forest, 45–46, 48

Parkinson's disease, 232, 236
Parsons, Alan, 142
Patterson, Richard, 110–11
Pennsalt (company), 32
Pentagon Papers, 94
People's Chemical Teams, 68
Perlstein, Rick, 103
Petryna, Adriana, 149
Phu Cat, 27, 29, 103, 204, 209
Phu My Orphanage for Agent Orange Victims, 1
Phung Tuu Boi, 238
Phuoc Long province, 48
Phuoc Toy province, 217
Phu Yen province, 65

Piatt, Judy, 98, 128, 156
politics. *See* Agent Orange; environmentalism; propaganda; science (discourse); uncertainty; United States; veterans
The Politics of Agent Orange (McCulloch), 182
"Post-service Mortality Study among Vietnam Veterans" (study), 188–89
Pratt, George, 160–61, 273n5
Program of Action for South Vietnam, 21
The Progressive, 157
Project Agile, 91
Project EMOTE, 46
propaganda: crop destruction programs and, 54, 56, 58–60, 63–77, 76, 77, 82, 84–96; importance of, 18, 240; Psychological Operations and, 42, 44, 47, 60, 243, 263n81; Vietnamese-American Science and Technology Society and, 204–5, 282n21. *See also* Kennedy, John F.; Vietnam War
Proxmire, William, 121
Psychological Operations (psyops), 42, 44, 46–47, 60, 75, 89, 94, 96, 243
Puerto Rico, 32, 233
Pure New Zealand, 143

Quang Ngai province, 89
Quang Tri province, 226–27

rainbow herbicides, 23. *See also specific herbicides*
Rampton, Calvin, 106
RAND corporation, 54, 62–69, 81–92, 94, 96, 261n32
"Re: Defoliation Operations from 1965–1969 during Vietnam War" (memo), 167
Reagan, Ronald, 171
Red Cross, 208, 227
redrumming, 104–11, 126. *See also* barrels (of herbicides)
Reeves Report, 216–17
Remote Area Conflict Information Center (RACIC), 91–92
"Report on the Use of Herbicides and Insecticides and Other Chemicals by the Australian Army in South Vietnam," 36, 180
"Report to the White House Agent Orange Working Group Science Subpanel on Exposure Assessment" (Murray), 176
Republic of Vietnam (RVN). *See* ARVN (South Vietnamese Army); Ngo Dinh Diem; Vietnam (RVN and Socialist Republic of); *specific airfields, cities, and provinces*
Residents Against Dioxins (RAD), 141–42
Reutershan, Paul, 152–53, 159–62, 237
"Review and Evaluation of the Defoliation Program" (ARPA), 57
Rogers, William, 102
Rolling Stone, 185
Roosevelt, Franklin, 61, 112
Ross, John, 202
Ross, Milton, 155–56
Ross, Richard, 155–56
Rostow, Walt, 21, 252n4
Round-Up, 157
Rowe, Vernon T., 147–48
Rowen, Faron, 130
Rusk, Dean, 56, 59, 241
Russo, Anthony, 54, 86–88

Saigon, 1, 24, 27–28
St. Louis, 107
St. Louis Post-Dispatch, 131, 145
San Antonio Air Materiel Area, 109, 111
San Antonio *Light*, 108
Sauget, Illinois, 107, 135
Schecter, Arnold, 202, 204, 227, 229–30
Schuck, Peter, 160, 162, 223–24, 273n5, 275n35
science (discourse): Agent Orange's toxicity and, 149–50; experiential
science (discourse) (*continued*) evidence and, 10–12, 171–79, 181–98; in the legal system, 157–71; limitations of, 10–12, 100, 230–37, 273n11; policy and, 162–71; uncertainty and, 7, 128, 132, 136–45, 150–62. *See also* Kennedy, John F.; Vietnam War; *specific scientists and studies*

Scott, Wilbur, 159, 173, 175, 189–90
Seamans, Robert, 107, 116
Second World War, 22, 41–42, 61, 88, 112, 136, 146, 191
Sellers, Christopher, 99–100
Senate Energy Committee, 148
Seveso (Italy), 9, 137–38, 140, 142, 156, 183
Shell Oil, 107
Shinseki, Eric, 235–36
Sikes, Robert, 108
Silent Spring (Carson), 10, 53, 62, 103
Simmons, James, 154–55
Socialist Republic of Vietnam, 125, 152, 197–210, 223–37, 245–47. *See also* Vietnam; *specific airfields, cities, and provinces*
Sokal, Alan, 11
South Korea. *See* Korea
Spey, Jack, 236
spina bifida, 217, 228, 233
Spivey, Gary, 129
Staley, Eugene, 21
Stanton, Shelby, 175
"A Statistical Analysis of the U.S. Crop Spraying Program in South Vietnam" (Russo), 54, 86–87
Stellman, Jeanne and Steven: dioxin estimates of, 27; exposure models of, 166–67, 189–90, 201–2, 227, 281n4; Ranch Hand flight path calculations of, 33, 198–99
Stephenson, Daniel, 230
Stevens, John Paul, 230
strategic hamlet program, 41, 52, 70–77, 89, 92, 199–200, 257n84
Stump, Bob, 121
Sydney *Sun Herald*, 185
Syntex Corporation, 135

Taxi Driver (film), 170
Taylor, Maxwell, 21
Tay Ninh province, 48, 78
TCDD (2,3,7,8-tetrachlorodibenzo-p-dioxin): Agent Orange's composition and, 2, 33–34, 148, 164–65; chemical properties and degradation of, 200–204; environmental accidents and, 142; soil and water contamination and, 124–26, 136; teratogenic effects of, 100, 236; testing of, 40, 101. *See also* Agent Orange; dioxins
Tennessee, 32
Tet Offensive, 91, 95
Texas, 32, 106, 108–11, 267n24
Thailand, 32, 45, 233
Thalken, C. E., 124
Thi Phuong-Lan Bui, 70, 250n6
Thomas, B. R., 40
Thornberry, Douglas, 107
Thornton, Joe, 178–79
Three Mile Island disaster, 103, 106, 133
Times Beach (Missouri), 7, 97–99, 126–37, 145, 152–56, 183, 209, 270n80
Ton That Tung, 157
Tordon-101, 32, 68. *See also* Agent White
"Toxic Hazards of 2,4,5-T," 141
Turley, William, 20–21, 253n6
Turner, C. A., 39
Tuy Hoa, 104–5
2,4-dichlorophenoxyacetic acid (herbicide): Agent Orange's composition and, 2, 24; carcinogenic qualities and, 189; congressional hearings on, 151–62; degradation of, 203; domestic use of, 22–23, 31, 109, 163, 180, 189, 221; New Zealand agriculture and, 136–45; NLF's identification of, 68; soil and water contamination and, 116, 198–238; testing of, 33–35, 167. *See also* Agent Orange; dioxins
2,4,5-trichlorophenoxyacetic acid (herbicide): Agent Orange's composition and, 2, 24, 97; banning of, 97, 103, 108–9, 115, 138–43, 152, 211, 214–15, 219; chemical companies' knowledge about, 146–47, 160–61, 187; congressional hearings on, 148, 150–62, 190, 192, 229; degradation of, 203; domestic use of, 22–23, 31, 108, 148, 156, 163, 180, 221; manufacture of, 37–38, 40; New Zealand agriculture and, 136–45, 214–19; NLF's identification of, 68; soil and water contamination and, 116, 198–238; testing of, 33–35, 100–101, 167. *See also* Agent Orange; dioxins

U Minh forest, 45
uncertainty: causality and, 14–15, 161,

INDEX

[301]

220–21; experiential evidence and, 10–12, 149, 171–79, 181–98; exposure models and, 171–79; legal ramifications of, 160–62, 230–37; political uses and misuses of, 7, 128–30, 132, 136–45, 150–71, 237, 241–42; questions of exposure and, 130; scientific discourse and, 7, 10–12, 197–98, 273n11; suggestive links and, 1–2, 14–15; veterans' Agent Orange claims and, 154–62, 180–88, 244–47; Vietnamese exposure claims and, 207–10, 244–47. *See also* science (discourse); veterans

United Nations, 208–9

United States: Agent Orange lawsuits and, 98, 129, 131–32, 135, 149, 157–71, 183, 210–19; chemical warfare and, 13, 26–32, 34–35; controlling impulses of, 6, 19–20, 41–52; counterinsurgency strategies and, 18, 21, 25–26, 32, 35, 41; coverup insinuations and, 166–71, 188–96, 215, 217–18, 239–42; DOD (Department of Defense), 27, 33, 35, 59, 103, 110, 211, 233; domestic politics of, 100–111, 132–37; DOS of (Department of State), 5–6, 55–60, 70–71, 242; herbicides' production and, 23, 35–38; herbicides' weaponization and, 1–2; international law and, 55–57, 61, 100–102, 114–15, 222–23; military's knowledge of Agent Orange and, 4, 13, 15, 21, 34–35; propaganda and, 53; responsibilities of, regarding Agent Orange, 15; technocratic mindset of, 20–26, 41–52; USDA (Department of Agriculture), 23, 32, 97, 148, 152; USFS (United States Forest Service), 46–50, 108; Vietnamese damage claims and, 77–80, 207–10, 222–37, 246. *See also* Agent Orange; environmentalism; veterans; Vietnam War; *specific departments, legislation, and politicians*

United States Agency for International Development (USAID), 30, 50, 109, 209

United States Air Force (USAF): Agent Orange's storage and disposal and, 6, 99, 101, 104, 111–23, 125, 134, 162; Laotian theater and, 32, 50, 255n55; Pacer IVY and, 100–111; records of, 198; training and safety procedures of, 29, 203; veterans' Agent Orange claims and, 162–71

United States Army Chemical Warfare Center, 21

United States Information Agency, 53, 56, 73

United States Navy, 112, 234–37

Unnatural Causes (film), 153

U.S.-Vietnam Dialogue Group, 227

"The U.S. Army Biological Laboratories Presents: Vegetation Control Testing, Vietnam" (newsreel), 24

"The Use of 2,4,5-T in New Zealand" (report), 142

USS Pueblo, 211

Utah, 106

Venezuela, 109

Verona, Missouri, 97, 126

Veteran (journal), 190–91

veterans: birth defects and, 155–56; congressional hearings and, 164–71, 177, 192, 229; diseases associated with Agent Orange and, 1; environmentalism's growth and, 115, 117, 145; experiential evidence and, 149–50, 171–79, 189–96; exposure and causality claims of, 7, 129, 152–55, 157–71, 181–88, 200–201, 230–37; interviews with, 5; lobbying of, 232–37, 244–45; memory and, 157–62, 170–88, 277n64; psychosomatization and, 180–88; U.S. government relations with, 2

Veterans Administration (VA): Agent Orange claims and, 1, 15, 152–54, 162–63, 165–71, 178, 195, 231–37; Australian veterans and, 181, 183, 185; scientific uncertainty and, 165–71, 190, 198; studies conducted by, 129

Veterans and Agent Orange (report), 231

Viet Cong (VC). *See* National Liberation Front (NLF)

Vietnam Association of Victims of Agent Orange/Dioxin (VAVA), 5, 202, 222–37, 243–47

Vietnam Experience Study (VES), 171, 188–96

Vietnam (RVN and Socialist Republic of): Agent Orange exposure claims and, 1–2, 152, 197, 199–200, 207–10, 222–37; civilian perceptions of crop destruction and, 54–55, 84–95, 127–28; corruption and, 78–79, 207–8; creation of, 51–52; government of, 60; health concerns of citizenry and, 117; herbicides' disposal in, 102; indemnification process and, 77–80, 88, 96; maps of, 16; National Interrogation Center of, 89; pacification processes and, 71–77; propaganda and, 66–67; soil and water contamination in, 30, 125, 198–210, 226, 234–35, 245–47; United States' political relations to, 94, 101, 246. *See also* Socialist Republic of Vietnam; Vietnam War; *specific airfields, cities, and provinces*

Vietnam Veterans Association of Australia (VVAA), 180–88, 194

Vietnam Veterans Association of New Zealand (VVANZ), 37, 39, 188, 216

Vietnam Veterans of America (VVA), 164, 166, 190–91

Vietnam Veterans since the War (Scott), 173

Vietnam War: American veterans of, 1, 98, 115, 131; antiwar movement and, 100; Australian veterans and, 179–96; as chemical and military laboratory, 23–24, 100, 176, 240; civilians and, 17–18; crop destruction missions, 14; data-driven conducting of, 84–96; defoliation campaigns and, 2, 13–14, 17–19, 24–26, 31–33; early expansion of, 20–26, 146; environmental consciousness and, 19; experiential evidence of Agent Orange exposure and, 10–12, 171–79, 181–98; forest fires and, 19–20, 44–49; international law and, 222–37; Laotian and Cambodian aspects of, 32–33; memory and, 157–62, 180–88, 193–94, 277n64; New Zealand veterans of, 37, 98, 136–45, 214–19; Pentagon Papers and, 94; strategic hamlet program and, 41, 70–77, 89, 92, 199–200, 257n84; symbolic import of, 128–29, 143, 239; as technological and military problem, 6, 19–20, 41–52; Tet Offensive and, 91, 95. *See also* Agent Orange; crop destruction missions; defoliation; Military Assistance Command, Vietnam (MACV); National Liberation Front (NLF); Operation Ranch Hand

The Vietnam War in American Memory (Hagopian), 170

Vinh Long province, 225

Vulcanus (incineration ship), 113, 117, 119, 122, 123

Wake Island, 112
"War on the Farm" (film), 137
Warren Commission, 179
Washington Post, 80, 129, 133, 162, 190
Watkins, Deborah, 202
Webb, Jim, 236
Weed-B-Gone, 157
Weed Control Authority, 141
Weinstein, Jack, 161, 223–24
Wellford, Harrison, 151–52
Wellington *Dominion Post*, 219
Wells, John, 235
Wells-Dang, Andrew, 255n55
Westad, Odd Arne, 252n3
Westmoreland, William, 49, 66, 87–88, 90–91
Whitehouse, Peter, 143
Wolfe, W. H., 158
Wook, Kim Jung, 213
World Trade Organization, 207

Yannacone, Victor, 161
Yarram, Australia, 180
Young, Alvin: Agent Orange studies and, 29, 157–60, 167–68, 173, 175–76, 183–84, 187–88, 200–201, 203, 251n8; AOWG and, 153–54, 199–200; disposal of Agent Orange and, 104–5, 117
Young, P. J., 37

Zengerle, Joseph, 169
Zierler, David, 3, 22, 25, 41, 100, 251n7
Zumwalt, Elmo, Jr., 191–92, 194, 241, 256n64